Lecture Notes in Computer Science 12644

More information about this subseries at http://www.springer.com/series/7412

Andrea Torsello · Luca Rossi ·
Marcello Pelillo · Battista Biggio ·
Antonio Robles-Kelly (Eds.)

Structural, Syntactic, and Statistical Pattern Recognition

Joint IAPR International Workshops, S+SSPR 2020
Padua, Italy, January 21–22, 2021
Proceedings

 Springer

Editors
Andrea Torsello
Ca' Foscari University of Venice
Venezia, Italy

Luca Rossi
Queen Mary University of London
London, UK

Marcello Pelillo
Ca' Foscari University of Venice
Venezia, Italy

Battista Biggio
University of Cagliari
Cagliari, Italy

Antonio Robles-Kelly
Deakin University
Burwood, VIC, Australia

ISSN 0302-9743   ISSN 1611-3349   (electronic)
Lecture Notes in Computer Science
ISBN 978-3-030-73972-0   ISBN 978-3-030-73973-7   (eBook)
https://doi.org/10.1007/978-3-030-73973-7

LNCS Sublibrary: SL6 – Image Processing, Computer Vision, Pattern Recognition, and Graphics

This Springer imprint is published by the registered company Springer Nature Switzerland AG
The registered company address is: Gewerbestrasse 11, 6330 Cham, Switzerland

Preface

This volume contains the papers presented at the joint IAPR International Workshops on Structural and Syntactic Pattern Recognition (SSPR 2020) and Statistical Techniques in Pattern Recognition (SPR 2020). S+SSPR 2020 was jointly organized by Technical Committee 1 (Statistical Techniques in Pattern Recognition, chaired by Battista Biggio) and Technical Committee 2 (Structural and Syntactical Pattern Recognition, chaired by Antonio Robles-Kelly) of the International Association of Pattern Recognition (IAPR).

Originally set to be held in Padua, Italy, in September 2020, due to the COVID-19 pandemic, the conference was first postponed to January 2021 and then moved to an online format in a rich two-day event spanning January 21–22, 2021.

We received 81 submissions from 29 countries across 5 continents. Each submission was reviewed by at least two and usually three Program Committee members. In total, 35 papers were accepted for presentation at the conference. The accepted papers cover the major topics of current interest in pattern recognition, including classification and clustering, deep learning, structural matching and graph-theoretic methods, and multimedia analysis and understanding.

We were delighted to have four prominent keynote speakers: Nicholas Carlini, from Google, USA; Prof. Michael Bronstein, from Twitter/Imperial College London, UK/University of Lugano, Switzerland; Prof. Max Welling, from Qualcomm/University of Amsterdam, Netherlands; and Fabio Roli, from the University of Cagliari, Italy, the IAPR TC1 Pierre Devijver Award winner for 2020. All the presentations have been made available on the conference YouTube channel (S+SSPR 2020).

We would like to thank all the Program Committee members for their help in the review process. We also wish to thank all the local organizers. They where ready to adapt to the shifting global landscape caused by the pandemic and without their contributions S+SSPR 2020 could not have happened. Finally, we express our appreciation to Springer for publishing this volume. More information about the workshops and organization can be found on the website: http://iapr.org/ssspr2020.

February 2021

Luca Rossi
Andrea Torsello

Organization

General Chairs

Marcello Pelillo   Ca' Foscari University of Venice, Italy
Luca Rossi   Queen Mary University of London, UK
Andrea Torsello   Ca' Foscari University of Venice, Italy

SPR and SSPR Chairs

Battista Biggio   University of Cagliari, Italy
Antonio Robles-Kelly   Deakin University, Australia

Honorary Chair

Edwin R. Hancock   University of York, UK

Local Chairs

Filippo Bergamasco   Ca' Foscari University of Venice, Italy
Sebastiano Vascon   Ca' Foscari University of Venice, Italy

Web Chair

Filippo Bergamasco   Ca' Foscari University of Venice, Italy

Program Committee

Gady Agam   Illinois Institute of Technology, USA
Ethem Alpaydin   Ozyegin University, Turkey
Silvia Biasotti   CNR-IMATI, Italy
Manuele Bicego   University of Verona, Italy
Luc Brun   National Graduate School of Engineering, France
Ambra Demontis   University of Cagliari, Italy
Francisco Escolano   University of Alicante, Spain
Francesc J. Ferri   University of Valencia, Spain
Pasi Fränti   University of Eastern Finland, Finland
Giorgio Fumera   University of Cagliari, Italy
Michal Haindl   Czech Academy of Sciences, Czech Republic
Laurent Heutte   University of Rouen, France
Jose M. Iñiesta   University of Alicante, Spain
Atsushi Imiya   Chiba University, Japan
Francois Jacquenet   University of Saint-Etienne, France

Xiaoyi Jiang	University of Münster, Germany
Nils Kriege	TU Dortmund, Germany
Jesse Krijthe	Leiden University, Netherlands
Adam Krzyzak	Concordia University, Canada
Marco Loog	Delft University of Technology, Netherlands
Mauricio Orozco-Alzate	National University of Colombia, Colombia
Nikunj Oza	NASA, USA
Marcos Quiles	Federal University of Sao Paulo, Brazil
Eraldo Ribeiro	Florida Institute of Technology, USA
Jairo Rocha	University of the Balearic Islands, Spain
Samuel Rota Bulò	Facebook, USA
Punam Kumar Saha	University of Iowa, USA
Carlo Sansone	University of Naples Federico II, Italy
Simone Scardapane	Sapienza University, Italy
Frank-Michael Schleif	University of Applied Sciences Würzburg-Schweinfurt, Germany
Francesc Serratosa	University Rovira i Virgili, Spain
J. Humberto Sossa A.	National Polytechnic Institute, Mexico
Salvatore Tabbone	University of Lorraine, France
Kar-Ann Toh	Yonsei University, South Korea
Ventzeslav Valev	Bulgarian Academy of Sciences, Bulgaria
Mario Vento	University of Salerno, Italy
Richard C. Wilson	University of York, UK
Terry Windeatt	University of Surrey, UK
Jing-Hao Xue	University College London, UK
De-Chuan Zhan	Nanjing University, PRC
Lichi Zhang	Shanghai Jiao Tong University, PRC
Zhihong Zhang	Xiamen University, PRC
Jun Zhou	Griffith University, Australia

Contents

Graph-Theoretic Methods

Multimedia Analysis and Understanding

Classification and Data Processing

Target Robust Discriminant Analysis

Wouter M. Kouw[1]([✉]) and Marco Loog[2,3]

[1] TU Eindhoven, Groene Loper 19, Eindhoven, The Netherlands
`w.m.kouw@tue.nl`
[2] TU Delft, Van Mourik Broekmanweg 6, Delft, The Netherlands
[3] University of Copenhagen, Universitetsparken 1, Copenhagen, Denmark

Abstract. In practice, the data distribution at test time often differs, to a smaller or larger extent, from that of the original training data. Consequentially, the so-called source classifier, trained on the available labelled data, deteriorates on the test, or *target*, data. Domain adaptive classifiers aim to combat this problem, but typically assume some particular form of domain shift. Most are not robust to violations of domain shift assumptions and may even perform worse than their non-adaptive counterparts. We construct robust parameter estimators for discriminant analysis that *guarantee* performance improvements of the adaptive classifier over the non-adaptive source classifier.

Keywords: Domain adaptation · Robustness · Discriminant analysis

1 Introduction

Domain adaptation is a supervised learning setting where labelled training data is drawn from one distribution (*source domain*) and unlabelled test data is drawn from another distribution (*target domain*) [2,8]. Often, adapting a source domain classifier, i.e., changing predictions to suit the target domain, is the only means by which one can potentially obtain satisfactory performance. Unfortunately, many domain adaptive classifiers assume some relationship between the domains, such as that only the covariates have shifted between domains and not the posterior distributions. They are not robust to violations of such assumptions and can subsequently perform *worse* than non-adaptive classifiers.

We formulate a conservative adaptive classifier that always performs at least as well as the non-adaptive one. More specifically, a core contribution of this paper is that we construct estimators that produce estimates with an empirical target risk that is *always* smaller or equal to the target risk of the source classifier. Since only the performance of the given target samples is considered, our result is transductive in nature [16]. Importantly, our guarantees are obtained without making *any* domain shift assumptions such as covariate shift or the existence of a domain-invariant subspace [8]. Furthermore, we show that in the case of classical likelihood-based discriminant analyses [12], the estimator will produce *strictly* smaller risks (i.e. larger log-likelihoods) *almost surely*, i.e., with probability 1. To the best of our knowledge, this is the first demonstration of a performance guarantee for a target classifier compared to the source classifier.

© Springer Nature Switzerland AG 2021
A. Torsello et al. (Eds.): S+SSPR 2020, LNCS 12644, pp. 3–13, 2021.
https://doi.org/10.1007/978-3-030-73973-7_1

2 Robust Target Domain Estimator

Consider a feature space $\mathcal{X} \subseteq \mathbb{R}^D$ and class labels $\mathcal{Y} = \{1, \ldots, K\}$. Let $\mathcal{S}$ denote the source domain, with n samples drawn from the source domain's joint distribution, $p_\mathcal{S}(x, y)$, collected as the set $\{(x_i, y_i)\}_{i=1}^n$. Similarly, let $\mathcal{T}$ denote the target domain, with m samples drawn from the target domain's joint distribution, $p_\mathcal{T}(x, y)$, collected as $\{(z_j, u_j)\}_{j=1}^m$. The goal is to predict the unknown target labels u (transductive setting), using only the unlabelled target samples $\{z_j\}_{j=1}^m$ and the labelled source samples $\{(x_i, y_i)\}_{i=1}^n$.

The empirical risk in the source domain is defined as the average loss ℓ of a classification function h over the source samples: $\hat{R}(h \mid x, y) \triangleq \frac{1}{n} \sum_{i=1}^n \ell(h \mid x_i, y_i)$. The *source classifier* is the classifier that minimizes the empirical source risk:

$$\hat{h}^\mathcal{S} \triangleq \arg\min_{h \in \mathcal{H}} \hat{R}(h \mid x, y), \tag{1}$$

where $\mathcal{H}$ refers to the hypothesis space. Evaluating the source classifier is typically done through the classification error. Arguably, a more appropriate evaluation is to consider the risk itself, given that it is the surrogate loss that is being optimized [11]. We evaluate $\hat{h}^\mathcal{S}$ based on the empirical target risk:

$$\hat{R}(\hat{h}^\mathcal{S} \mid z, u) = \frac{1}{m} \sum_{j=1}^m \ell(\hat{h}^\mathcal{S} \mid z_j, u_j). \tag{2}$$

In our objective function, the source classifier's target risk in (2) is subtracted from the target risk of a prospective target classifier h:

$$\hat{R}(h \mid z, u) - \hat{R}(\hat{h}^\mathcal{S} \mid z, u). \tag{3}$$

The target risk of the source classifier will act as a bound on the hypothesis space during minimization:

Lemma 1. *For fixed samples z and labels u, the difference in empirical target risks between a classifier $h \in \mathcal{H}$ and $h^\mathcal{S} \in \mathcal{H}$ is less than or equal to 0:*

$$\min_{h \in \mathcal{H}} \hat{R}(h \mid z, u) - \hat{R}(\hat{h}^\mathcal{S} \mid z, u) \le 0. \tag{4}$$

Proof. Let $\tilde{h} \triangleq \min_{h \in \mathcal{H}} \hat{R}(h \mid z, u) - \hat{R}(\hat{h}^\mathcal{S} \mid z, u)$. Since $\tilde{h}$ and $\hat{h}^\mathcal{S}$ are elements of the same hypothesis space, $\tilde{h} = h^\mathcal{S}$ is a potential solution to the minimization problem. In that case, the difference in target risks would be 0. The estimate $\hat{h}^\mathcal{S}$ can always be recovered, which implies all $h \in \mathcal{H}$ that would lead to risk differences greater than 0 are not valid minimizers. □

Equation 3 still contains the unknown target labels u. To be able to guarantee a better or equal performance to that of the source classifier, we use a worst-case labelling, achieved by introducing a hypothetical labelling q and *maximizing* the difference in risks: $\hat{R}(h \mid z, u) \le \max_q \hat{R}(h \mid z, q)$.

Note that for any classifier h, the risk with respect to this worst-case labelling will always be larger than the risk with respect to the true target labelling. Combining the difference in risks from Eq. 3 with the hypothetical labelling q results in the following risk function:

$$\hat{R}^T\left(h \mid \hat{h}^S, z, q\right) \triangleq \hat{R}(h \mid z, q) - \hat{R}(\hat{h}^S \mid z, q).\tag{5}$$

We refer to the risk in Eq. 5 as the Target Robust (TR) risk. Minimizing it with respect to a classifier h and maximizing it with respect to the hypothetical labelling q, leads to a TR classifier:

$$\hat{h}^T \triangleq \arg\min_{h \in \mathcal{H}} \max_{q \in \mathcal{Y}^m} \hat{R}^T\left(h \mid \hat{h}^S, z, q\right).\tag{6}$$

The TR risk only considers the given target samples $\{z_j\}_{j=1}^m$ and is, therefore, a transductive approach [16].

The maximization over a set of discrete labels is a difficult combinatorial problem. Therefore, we apply a relaxation and represent the hypothetical labelling probabilistically, $q_{jk} := p(u_j = k \mid z_j)$. That is, q_j is a non-negative vector of K elements that sum to 1. As such, it represents an element of the standard $K-1$ simplex Δ_{K-1}. For m samples, an m-dimensional $K-1$ simplex Δ_{K-1}^m is taken. This means that the loss for every element z_j becomes a weighted sum,

$$\ell\left(h \mid z_j, q_j\right) = \sum_{k \in \mathcal{Y}} q_{jk}\, \ell\left(h \mid z_j, k\right),\tag{7}$$

and the maximization in Eq. 6 will be over $q \in \Delta_{K-1}^m$ instead of $q \in \mathcal{Y}^m$. Known, deterministic labels can of course also be represented probabilistically, for example $y_i = 1 \Leftrightarrow p(y_i = 1 \mid x_i) = 1$ and $p(y_i \neq 1 \mid x_i) = 0$. Hence, in practice, both y_i and u_j can be represented as $1 \times K$-vectors with the k-th element marking the probability that sample i or j belongs to class k (a.k.a. a one-hot encoding).

With the relaxation from "hard" to "soft" labels, we can say the following:

Lemma 2. *For fixed samples z and source classifier $\hat{h}^S$, the Target Robust risk will be lower or equal to 0 at its saddle point with respect to both h and q:*

$$\min_{h \in \mathcal{H}} \max_{q \in \Delta_{K-1}^m} \hat{R}^T\left(h \mid \hat{h}^S, z, q\right) \leq 0.\tag{8}$$

Proof. In the given minimax problem, we first go over all $h \in \mathcal{H}$ and find, for every h, the $q_h \in \Delta_{K-1}^m$ that maximizes the Target Robust risk. In a second step, we take the minimum of $\hat{R}^T\left(h \mid \hat{h}^S, z, q_h\right)$ over all h. Given that q_h is fixed, Lemma 1 applies, which means the resulting Target Robust risk will be less than or equal to 0. ◻

3   Discriminant Analyses

Lemma 2 tells us that our target classifier performs at least as well as the source classifier on the given target samples. For classical linear and quadratic discriminant analysis, we are able to show that strict improvements are obtained *almost surely*, i.e., with probability 1.

In discriminant analysis, the data from each class is modelled with a Gaussian distribution, proportional to the class prior [12]. We maintain a parameter vector θ_k for each class, consisting of the prior, mean and covariance matrix; $\theta_k = (\pi_k, \mu_k, \Sigma_k)$. One obtains an empirical risk minimization formulation by taking the negative log-likelihoods as the loss function: $\ell(\theta \mid x, y) = \sum_{k=1}^{K} -y_k \log \left[\pi_k \, \mathcal{N}(x \mid \mu_k, \Sigma_k) \right]$. Note the resemblance to Eq. 7.

If each class is modelled with a separate covariance matrix, the resulting classifier is known as *quadratic discriminant analysis* (QDA) [12]. For target data z and probabilistic labels q, the risk is formulated as:

$$\hat{R}_{\text{QDA}}(\theta \mid z, q) \triangleq \frac{1}{m} \sum_{j=1}^{m} \sum_{k=1}^{K} -q_{jk} \log \left[\pi_k \, \mathcal{N}(z_j \mid \mu_k, \Sigma_k) \right] . \tag{9}$$

Note that the risk is now expressed in terms of classifier parameters θ, as opposed to the classifier h. Plugging the risk from (9) into (5), the full TR-QDA risk becomes:

$$\hat{R}_{\text{QDA}}^{\mathcal{T}}(\theta \mid \hat{\theta}^{\mathcal{S}}, z, q) \triangleq \hat{R}_{\text{QDA}}(\theta \mid z, q) - \hat{R}_{\text{QDA}}(\hat{\theta}^{\mathcal{S}} \mid z, q) , \tag{10}$$

where the estimate itself is:

$$\hat{\theta}^{\mathcal{T}} \triangleq \arg \min_{\theta \in \Theta} \max_{q \in \Delta_{K-1}^{m}} \hat{R}_{\text{QDA}}^{\mathcal{T}}(\theta \mid \hat{\theta}^{\mathcal{S}}, z, q). \tag{11}$$

If the model is constrained to share a single covariance matrix for each class, the resulting classifier is a linear function of the feature values and hence is termed *linear discriminant analysis* (LDA). The optimal overall class-covariance matrix Σ can be determined with $\Sigma = \sum_{k=1}^{K} \pi_k \Sigma_k$.

3.1   Performance Improvement Guarantee

Discriminant analysis has a special property: it obtains a *strictly* smaller risk. In other words, this parameter estimator is *guaranteed to improve its performance* - on the given target samples, and in terms of risk - over the source classifier.

Theorem 1. *Let the number of target samples from a continuous target distribution be greater than its number of features. The empirical discriminant analysis risk $\hat{R}_{DA}$, i.e., the negative log-likelihood over the target samples, of the TR estimated parameters $\hat{\theta}^{\mathcal{T}}$ is almost surely strictly smaller than for the source parameters $\hat{\theta}^{\mathcal{S}}$. In other words, with probability one we have the strict inequality*

$$\hat{R}_{DA}(\hat{\theta}^{\mathcal{T}} \mid z, u) \; < \; \hat{R}_{DA}(\hat{\theta}^{\mathcal{S}} \mid z, u) .$$

Proof. Let $\{(x_i, y_i)\}_{i=1}^n$ be a data set of size n drawn *i.i.d.* from a continuous source distribution defined over feature space $\mathcal{X} \subseteq \mathbb{R}^D$ and label space $\mathcal{Y} = \{y = \{0,1\}^K : \sum_{k=1}^K y_k = 1\}$. Similarly, let $\{(z_j, u_j)\}_{j=1}^m$ be a data set of size $m > D$, drawn *i.i.d.* from a continuous target distribution defined over $\mathcal{X} \times \mathcal{Y}$. Consider a discriminant analysis model parametrized by $\theta = (\pi_1, .., \pi_K, \mu_1, .., \mu_K, \Sigma_1, ..\Sigma_K)$ with empirical risk defined as

$$\hat{R}_{\text{QDA}}(\theta \mid x, y) = \frac{1}{m} \sum_{j=1}^m \sum_{k=1}^K -y_{ik} \log[\pi_k \, \mathcal{N}(x_i \mid \mu_k, \Sigma_k)] \, . \tag{12}$$

Let $\hat{\theta}^\mathcal{S}$ be the parameters estimated on labelled source data

$$\hat{\theta}^\mathcal{S} = \underset{\theta \in \Theta}{\arg\min} \; \hat{R}_{\text{QDA}}(\theta \mid x, y) \tag{13}$$

and let $(\hat{\theta}^\mathcal{T}, q^*)$ be the parameters and worst-case labelling estimated by mini-maximizing the Target Robust risk:

$$\hat{\theta}^\mathcal{T}, q^* = \underset{\theta \in \Theta}{\arg\min} \; \underset{q \in \Delta_{K-1}^m}{\arg\max} \; \hat{R}_{\text{QDA}}(\theta \mid z, q) - \hat{R}_{\text{QDA}}(\hat{\theta}^\mathcal{S} \mid z, q) \, . \tag{14}$$

Lemma 2 tells us that

$$\hat{R}_{\text{QDA}}(\hat{\theta}^\mathcal{T} \mid z, q^*) - \hat{R}_{\text{QDA}}(\hat{\theta}^\mathcal{S} \mid z, q^*) \le 0 \, . \tag{15}$$

Since this holds for the worst-case labelling q^*, it must also hold for the true labelling u:

$$\hat{R}_{\text{QDA}}(\hat{\theta}^\mathcal{T} \mid z, u) \; \le \; \hat{R}_{\text{QDA}}(\hat{\theta}^\mathcal{S} \mid z, u) \, . \tag{16}$$

The Equality in (16) occurs with probability 0, which can be shown as follows. Firstly, note that the total mean for the source classifier consists of the weighted combination of the class means, resulting in the overall source sample average

$$\hat{\mu}^\mathcal{S} = \sum_{k=1}^K \hat{\pi}_k^\mathcal{S} \, \hat{\mu}_k^\mathcal{S} = \sum_{k=1}^K \frac{\sum_{i=1}^n y_{ik}}{n} \left[\frac{1}{\sum_{i=1}^n y_{ik}} \sum_{i=1}^n y_{ik} x_i \right] = \frac{1}{n} \sum_{i=1}^n x_i \, . \tag{17}$$

The total mean for the TP-QDA estimator is similarly defined, resulting in the overall target sample average:

$$\hat{\mu}^\mathcal{T} = \sum_{k=1}^K \hat{\pi}_k^\mathcal{T} \, \hat{\mu}_k^\mathcal{T} = \sum_{k=1}^K \frac{\sum_{j=1}^m q_{jk}^*}{m} \left[\frac{1}{\sum_{j=1}^m q_{jk}^*} \sum_{j=1}^m q_{jk}^* z_j \right] = \sum_{k=1}^K \frac{1}{m} \sum_{j=1}^m q_{jk}^* z_j = \frac{1}{m} \sum_{j=1}^m z_j \, . \tag{18}$$

Because q^* consists of probabilities, the sum over classes $\sum_{k=1}^K q_{jk}^*$ in Eq. 18 is 1, for every sample j.

Secondly, the TR objective function is quasi-convex-concave. In fact, it is linear in terms of q. Since its domain Δ_{K-1}^m is compact, Sion's theorem holds which allows for interchanging the order of the minimization and the maximization [15]. This implies that the minimax solution is a saddle point and that the optimal parameter estimates $\hat{\theta}^{\mathcal{T}}$ for the discriminant analysis are unique, because the objective function is strictly (quasi-)convex in terms of these parameters when $m > D$ [13].

Now, equal risks for the source and target parameter sets on the worst-case labelling q^*, i.e., $\hat{R}_{\mathrm{QDA}}(\hat{\theta}^{\mathcal{T}} \mid z, q^*) = \hat{R}_{\mathrm{QDA}}(\hat{\theta}^{\mathcal{S}} \mid z, q^*)$, implies equality of the total means, $\hat{\mu}^{\mathcal{T}} = \hat{\mu}^{\mathcal{S}}$, because $\hat{\theta}^{\mathcal{T}}$ is the unique minimizer of a strictly convex risk. By Eqs. 17 and 18, equal total means implies equal sample averages: $\frac{1}{m} \sum_{j=1}^{m} z_j = \frac{1}{n} \sum_{i=1}^{n} x_i$. Given a set of source samples, drawing a set of target samples such that its average is *exactly equal* to the average of the source samples, is an event that has probability 0 under continuous distributions. Therefore, a strictly smaller risk occurs almost surely. In other words, with probability 1, we have that

$$\hat{R}_{\mathrm{QDA}}(\hat{\theta}^{\mathcal{T}} \mid z, u) < \hat{R}_{\mathrm{QDA}}(\hat{\theta}^{\mathcal{S}} \mid z, u). \tag{19}$$

This concludes the proof for the case of QDA. The proof for LDA follows from plugging in Σ for Σ_k. Since this does not alter the mean estimators, the Equality in (16) still occurs with probability 0. □

3.2   Optimization

As pointed out in the proof of Theorem 1, we seek a saddle point to a quasi-convex-linear problem. That can be found by first performing a gradient descent step with respect to h (or, equivalently, θ), followed by a gradient ascent step with respect to q. For discriminant analyses models, the minimization with respect to θ has a closed-form solution:

$$\pi_k = \frac{1}{m} \sum_{j=1}^{m} q_{jk}, \qquad \mu_k = \left(\sum_{j=1}^{m} q_{jk} \right)^{-1} \sum_{j=1}^{m} q_{jk} z_j,$$

$$\Sigma_k = \left(\sum_{j=1}^{m} q_{jk} \right)^{-1} \sum_{j=1}^{m} q_{jk} (z_j - \mu_k)(z_j - \mu_k)^{\top}. \tag{20}$$

One encounters the same solutions in the M step of EM-based Gaussian mixture modelling, where data points also have probabilistic class assignments [13]. To ensure the updated q remains on the simplex, it is projected back after each gradient step. The projection $\mathcal{P}$ maps a point outside the simplex a to the point b on the simplex that is closest in terms of Euclidean distance: $\mathcal{P}(a) = \arg\min_{b \in \Delta} \|a - b\|_2$ [4]. The projection complicates the computation of the step size, which we replace by a learning rate α^t decreasing over iterations t. This results in the overall update: $q^{t+1} \leftarrow \mathcal{P}(q^t + \alpha^t \nabla q^t)$.

A gradient descent-gradient ascent procedure for globally convex-linear objectives is guaranteed to converge to a saddle point (c.f. Proposition 4.4 and Corollary 4.5 in [3]).

4 Experiments

Our contribution is first and foremost theoretical. Nevertheless, we perform an experiment on a natural data set comparing the empirical target risks of our TR classifiers with source classifiers (S-LDA and S-QDA) as well as classifiers trained on labelled target data (T-LDA and T-QDA), which represent the best possible performance of the models. Furthermore, we perform an experiment comparing our estimator to other domain-adaptive classifiers. Since these do not incorporate the same loss as the DA models, we measure performance in area under the ROC-curve (AUC).

The data set we used is split geographically into domains. The goal is to predict heart disease in patients from 4 different hospitals [6]. These are located in Hungary, Switzerland, California and Ohio. Each hospital can be considered a domain because patients are measured on the same biometrics but the local patient populations differ. For example, the age distributions are shifted between countries. The data set was pre-processed using z-scoring.

We compared to Kernel Mean Matching (KMM) [7], Robust Covariate Shift Adjustment (RCSA) [17], the Robust Bias-Aware (RBA) classifier [9] and Transfer Component Analysis (TCA) [14]. KMM represents a standard importance-weighted classifier, which assumes covariate shift between domains. RBA and RCSA still assume covariate shift, but incorporate robust importance weight estimators. TCA represents an alternative domain shift assumption, namely the existence of a feature subspace common to both domains. These methods are discussed further in the Related Work section (Sect. 5.1). All methods were trained with both a logistic and quadratic loss, and the better performing loss was chosen. For RCSA, we used the authors' implementation, which incorporates a support vector machine with Gaussian kernel. All methods use L^2-regularization. Since no labelled target data is available for validation, the regularization parameter was set to 0.01 for logistic and $0.01n$ for quadratic losses.

4.1 Results

Table 1 lists target risks for source, Target Robust and target classifiers for each possible pair of domains in the data set. In most cases, the TR classifier is quite close to the optimal target risk. Note that the source classifier performs terribly in some settings (with target risks in the positive hundreds), while it is does not differ much from the target classifier in others.

Table 2 lists AUCs of different classifiers in the heart disease data set. Perhaps the most striking observation is that AUC's are sometimes below 0.5; these scenarios represent domain shifts so large that source classifiers perform worse than chance in the target domain. Adaptive classifiers will not perform much better if their shift assumptions are violated. Our own TR classifiers are merely built to improve over their source counterpart: when the source classifier is poor to begin with, a "better" performance may still be below chance. A few more things to note: firstly, TR-LDA generally outperforms TR-QDA, indicating that the additional flexibility of QDA does not outweigh the increase the complexity.

Table 1. Target risks (average negative log-likelihoods) for all pairwise combinations of domains in heart disease data set (O = 'Ohio', C = 'California', H = 'Hungary' and S = 'Switzerland'). Smaller values are better.

S	T	S-LDA	TR-LDA	T-LDA	S-QDA	TR-QDA	T-QDA
O	H	−53.55	−57.18	−57.35	−53.55	−57.20	−57.62
O	S	−8.293	−16.76	−17.54	−8.293	−16.76	−17.54
O	C	−37.84	−53.88	−54.69	−37.83	−53.73	−54.89
H	S	−12.50	−16.08	−17.54	−12.80	−16.44	−17.54
H	C	−41.70	−53.91	−54.69	−40.08	−54.45	−54.89
S	C	494.9	−54.49	−54.69	498.9	−54.44	−54.89
H	O	−48.91	−55.08	−55.23	−49.20	−54.84	−55.53
S	O	709.9	−54.07	−55.23	709.9	−54.10	−55.53
C	O	−49.21	−55.00	−55.23	−49.17	−55.05	−55.53
S	H	649.9	−56.09	−57.35	650.3	−56.19	−57.62
C	H	−53.05	−57.19	−57.35	−53.15	−57.17	−57.62
C	S	−15.45	−17.43	−17.54	−15.47	−17.44	−17.54

Table 2. AUC for all pairwise combinations of domains in heart disease data set (O = 'Ohio', H = 'Hungary', S = 'Switzerland' and C = 'California').

S	T	S-LDA	S-QDA	TCA	KMM	RCSA	RBA	TR-LDA	TR-QDA
O	H	0.866	0.829	0.674	0.709	0.646	0.502	0.864	0.822
O	S	0.674	0.674	0.597	0.591	0.667	0.670	0.675	0.675
O	C	0.658	0.503	0.500	0.460	0.572	0.430	0.653	0.500
H	S	0.671	0.660	0.453	0.503	0.641	0.636	0.673	0.661
H	C	0.726	0.668	0.466	0.568	0.483	0.423	0.725	0.660
S	C	0.527	0.484	0.530	0.552	0.459	0.582	0.555	0.432
H	O	0.866	0.840	0.544	0.742	0.749	0.556	0.867	0.841
S	O	0.500	0.500	0.439	0.302	0.626	0.366	0.424	0.422
C	O	0.830	0.811	0.693	0.294	0.651	0.523	0.831	0.813
S	H	0.559	0.502	0.408	0.345	0.685	0.396	0.717	0.565
C	H	0.883	0.834	0.661	0.290	0.647	0.597	0.882	0.847
C	S	0.440	0.452	0.572	0.508	0.343	0.412	0.447	0.414
	Avg	0.683	0.647	0.545	0.489	0.597	0.508	0.693	0.638

Secondly, TR-LDA and TR-QDA are either performing similarly or better than S-LDA and S-QDA. Note that cases where the source classifiers perform well correspond to cases where the source classifier's target risk was small and close to that of the target classifier (compare to Table 1). Thirdly, RCSA and RBA do not always outperform KMM, indicating that robust weight estimation is not always beneficial. Fourthly, TCA's performance varies around chance level, which means that its assumption is likely violated.

5 Discussion

As could be seen in the experimental results, an improvement in terms of the classifier's intrinsic loss does *not* imply an improvement in AUC. This is due to the difference between optimizing a surrogate loss, here the negative log-likelihood, and evaluating the 0/1-loss [1,11]. They do not necessarily have the same minimizers. Note that the 0/1-loss is not differentiable, and cannot be optimized over directly. We therefore argue that guarantees in terms of intrinsic losses are the most one can expect.

One advantage of our estimator is that we do not explicitly require source samples at training time. Our approach is therefore more memory-efficient than other domain-adaptive classifiers and more suited to privacy-sensitive supervised learning settings, such as federated learning.

5.1 Related Work

Most methods for domain adaptation rely on an assumption of how the domains have shifted [8]. Examples of such assumptions include low joint-domain-error [2], the existence of a domain-invariant subspace [14] and the assumption that only the covariates have shifted but not the posterior distributions [5,7]. These assumptions may be implicit, for example domain-adversarial neural networks simultaneously minimize the divergence between the domains and train a source classifier which amounts to the low joint-domain error assumption [2]. Violations of assumptions mean adaptation could deteriorate performance. For example, Transfer Component Analysis assumes a domain-invariant latent representation where class separability is preserved [14]. When that assumption does not hold, mapping data onto transfer components will mix the class-conditional distributions and classification will become harder.

Research into robust domain adaptation tends to revolve around importance weight estimators for methods assuming covariate shift. Unfortunately, importance weight estimators may assign few samples large weights and many samples near-zero weights, greatly reducing effective sample size and producing pathological importance-weighted classifiers [5]. Robust Covariate Shift Adjustment builds an importance-weighted classifier that is robust to poor importance weight estimates by first maximizing risk with respect to the importance-weights and subsequently minimizing with respect to classifier parameters [17]. However, it can perform worse than standard importance-weighted classifiers when it *unnecessarily* considers worst-case weights. The Robust Bias-Aware classifier employs a similar mini-max strategy, but avoids accounting for worst-case importance weights. It attempts to match the statistics, specifically the moments, of the importance-weighted classifier's labelling of the target samples with the statistics of the source labels [9]. This favours more stable importance-weighted classifiers, but the RBA classifier loses predictive power in areas of feature space where the source distribution's support is limited.

Similar research concerning improvement guarantees has been carried out: Maximum Contrastive Pessimistic Likelihood estimation is a worst-case approach to semi-supervised learning that ensures complete robustness to the labelling of the unlabelled samples [10]. It also comes with performance guarantees in terms of the objective that the classifier actually optimizes, such as log-likelihood, hinge loss, logistic loss, etc. [11].

6   Conclusion

We have designed a risk minimization formulation for a domain-adaptive classifier whose performance, in terms of empirical target risk, is always at least as good as that of the non-adaptive source classifier. Furthermore, for the discriminant analysis case, its risk is always strictly smaller. An experiment on data gathered under a geographical bias supports the claim empirically and shows competitive performance compared to other robust domain-adaptive classifiers.

References

1. Bartlett, P.L., Jordan, M.I., McAuliffe, J.D.: Convexity, classification, and risk bounds. J. Am. Stat. Assoc. **101**(473), 138–156 (2006)
2. Ben-David, S., Blitzer, J., Crammer, K., Kulesza, A., Pereira, F., Vaughan, J.W.: A theory of learning from different domains. Mach. Learn. **79**(1–2), 151–175 (2009). https://doi.org/10.1007/s10994-009-5152-4
3. Cherukuri, A., Gharesifard, B., Cortes, J.: Saddle-point dynamics: conditions for asymptotic stability of saddle points. SIAM J. Control. Optim. **55**(1), 486–511 (2017)
4. Condat, L.: Fast projection onto the simplex and the l_1 ball. Math. Program. **158**(1–2), 575–585 (2016)
5. Cortes, C., Mohri, M.: Domain adaptation and sample bias correction theory and algorithm for regression. Theor. Comput. Sci. **519**, 103–126 (2014)
6. Dua, D., Graff, C.: UCI Machine Learning Repository (2017)
7. Huang, J., Smola, A.J., Gretton, A., Borgwardt, K.M., Schölkopf, B., et al.: Correcting sample selection bias by unlabeled data. In: Advances in Neural Information Processing Systems, p. 601 (2007)
8. Kouw, W.M., Loog, M.: A review of domain adaptation without target labels. IEEE Trans. Pattern Anal. Mach. Intell. **43**, 766–785 (2019)
9. Liu, A., Ziebart, B.: Robust classification under sample selection bias. In: Advances in Neural Information Processing Systems, pp. 37–45 (2014)

10. Loog, M.: Contrastive pessimistic likelihood estimation for semi-supervised classi-fication. IEEE Trans. Pattern Anal. Mach. Intell. **38**(3), 462–475 (2016)
11. Loog, M., Krijthe, J.H., Jensen, A.C.: On measuring and quantifying perfor-mance: error rates, surrogate loss, and an example in semi-supervised learning. In: Chen, C.H. (ed.) Handbook of Pattern Recognition and Computer Vision, 5th edn., pp. 53–68. World Scientific, Singapore (2016). doi: https://doi.org/10.1142/9789814656535_0003
12. McLachlan, G.J.: Discriminant Analysis and Statistical Pattern Recognition, 2nd edn. Wiley, Hoboken, New Jersey (2004)
13. McLachlan, G.J., Peel, D.: Finite Mixture Models. Wiley, Hoboken, New Jersey (2000)
14. Pan, S.J., Tsang, I.W., Kwok, J.T., Yang, Q.: Domain adaptation via transfer component analysis. IEEE Trans. Neural Netw. **22**(2), 199–210 (2011)
15. Sion, M.: On general minimax theorems. Pac. J. Math. **8**, 171–176 (1958)
16. Vapnik, V.: Statistical Learning Theory. Wiley, New York (1998)
17. Wen, J., Yu, C.N., Greiner, R.: Robust learning under uncertain test distributions: relating covariate shift to model misspecification. In: International Conference on Machine Learning, pp. 631–639 (2014)

Complex-Valued Embeddings of Generic Proximity Data

Maximilian Münch[1,2(✉)] (ID), Michiel Straat[2] (ID), Michael Biehl[2] (ID),
and Frank-Michael Schleif[1] (ID)

[1] Department of Computer Science and Business Information Systems,
University of Applied Sciences Würzburg-Schweinfurt, 97074 Würzburg, Germany
{maximilian.muench,frank-michael.schleif}@fhws.de
[2] Bernoulli Institute for Mathematics, Computer Science and Artificial Intelligence,
University of Groningen, P.O. Box 407, 9700 AK Groningen, The Netherlands
{m.j.c.straat,m.biehl}@rug.nl

Abstract. Proximities are at the heart of almost all machine learning methods. In a more generic view, objects are compared by a (symmetric) similarity or dissimilarity measure, which may not obey particular mathematical properties. This renders many machine learning methods invalid, leading to convergence problems and the loss of generalization behavior. In many cases, the preferred dissimilarity measure is not metric. If the input data are non-vectorial, like text sequences, proximity-based learning is used or embedding techniques can be applied. Standard embeddings lead to the desired fixed-length vector encoding, but are costly and are limited in preserving the full information. As an information preserving alternative, we propose a complex-valued vector embedding of proximity data, to be used in respective learning approaches. In particular, we address supervised learning and use extensions of prototype-based learning. The proposed approach is evaluated on a variety of standard benchmarks showing good performance compared to traditional techniques in processing non-metric or non-psd proximity data.

Keywords: Proximity learning · Embedding · Complex valued data · Learning vector quantizer · Krein space

1 Introduction

Machine learning has a growing impact in various fields and the considered input data become more and more generic [9,12]. In particular non-vectorial data like

MM is supported by the ESF (WiT-HuB 4/2014–2020), project KI-trifft-KMU, StMBW-W-IX.4-6-190065. M.B. and M.S. acknowledge support through the Northern Netherlands Region of Smart Factories (RoSF) consortium, lead by Noordelijke Ontwikkelings en Investerings Maatschappij (NOM), The Netherlands, see http://www.rosf.nl.

A. Torsello et al. (Eds.): S+SSPR 2020, LNCS 12644, pp. 14–23, 2021.
https://doi.org/10.1007/978-3-030-73973-7_2

text data, biological sequence data, graphs, and other input formats are used [14]. The vast majority of learning algorithms expect fixed-length real value vector data as inputs and can not directly be used on non-standard data [12].

Using embedding approaches is one strategy to obtain a vectorial embedding, but this is costly, needs large amounts of data to train the embedding and information is only partially preserved [12]. In a more generic scenario, proximity measures, like alignment functions, can be applied to compare non-vectorial objects to obtain a proximity score between two objects. If all N input objects are pairwise compared, we obtain a proximity matrix $P \in \mathbf{R}^{N \times N}$. If the measure is a metric dissimilarity measure, we have a distance matrix, which can be used for the nearest-mean classifier. In the case of inner products like the Euclidean inner product, a kernel matrix is obtained. If this kernel matrix is positive semidefinite (psd), multiple kernel methods can be used [16]. Also, so-called empirical feature space approaches have been considered, but with the drawback of high model complexity and inherent data transformations [7].

Here we consider non-vectorial input data given either by a non-metric dissimilarity measure or a non-standard inner product, leading to an indefinite kernel function. As detailed in [14], learning models can be calculated on these generic proximity data in very different ways. Most often, the proximities are transformed to fit into classical machine learning algorithms, with a number of limitations [14]. In this work, we propose the application of a complex-valued embedding on these data to overcome some of the limitations. Recently different classical learning algorithms have been extended to complex-valued inputs [18]. It is now possible to preserve the information provided in the generic proximity data while learning in a fixed-length vector space using a highly effective, well-understood learning algorithm. The respective procedures are detailed in the following and evaluated on classical benchmark data with strong results.

2   Background and Basic Notation

Consider a collection of N objects $\mathbf{x}_i$, $i = \{1, 2, ..., N\}$, in some input space $\mathcal{X}$. Given a similarity function or inner product on $\mathcal{X}$, corresponding to a metric, one can construct a Mercer kernel acting on pairs of points from $\mathcal{X}$. For example, if $\mathcal{X}$ is a finite-dimensional vector space, a classical similarity function in this space is the Euclidean inner product (corresponding to the Euclidean distance).

2.1   Positive Definite Kernels - Hilbert Space

The Euclidean inner product is also known as linear kernel with $k(\mathbf{x}, \mathbf{x}') = \langle \phi(\mathbf{x}), \phi(\mathbf{x}') \rangle$, where ϕ is the identity mapping. Another prominent kernel function is $k(\mathbf{x}, \mathbf{x}') = \exp\left(-\frac{\|\mathbf{x}-\mathbf{x}'\|^2}{2\sigma^2}\right)$, with $\sigma > 0$ as a free scale parameter. In any case, it is assumed that the kernel function $k(\mathbf{x}, \mathbf{x}')$ is psd.

The transformation ϕ is, in general, a *non-linear* mapping to a high-dimensional Hilbert space $\mathcal{H}$ and may not be given in an explicit form, but allowing *linear* techniques in $\mathcal{H}$. Instead of providing an explicit mapping, a kernel function $k : \mathcal{X} \times \mathcal{X} \mapsto \mathbb{R}$ is given, which encodes the inner product in $\mathcal{H}$. The kernel k is a positive (semi-) definite function such that $k(\mathbf{x}, \mathbf{x}') = \langle \phi(\mathbf{x}), \phi(\mathbf{x}') \rangle_{\mathcal{H}}$, for any $\mathbf{x}, \mathbf{x}' \in \mathcal{X}$. The matrix $K_{i,j} := k(\mathbf{x}_i, \mathbf{x}_j)$ is an $N \times N$ kernel (Gram) matrix derived from the data. For general similarity measures, we use $\mathbf{S}$ to describe the respective similarity matrix.

Kernelized methods process the embedded data points in a feature space utilizing only the inner products $\langle \cdot, \cdot \rangle_{\mathcal{H}}$ [16], without the need to explicitly calculate ϕ, known as *kernel trick*. Explicit mappings of psd kernel function are also frequently used to employ linear methods. However, the underlying similarity function may not be metric, but a domain-specific similarity measure, as mentioned before. Such similarity measures imply *indefinite* kernels, preventing standard "kernel-trick" methods developed for Mercer kernels to be applied.

2.2   Non-positive Definite Kernels - Krein Space

A Krein space is an *indefinite* inner product space endowed with a Hilbertian topology. Let $\mathcal{K}$ be a real vector space. An inner product space with an indefinite inner product $\langle \cdot, \cdot \rangle_{\mathcal{K}}$ on $\mathcal{K}$ is a bi-linear form where all $f, g, h \in \mathcal{K}$ and $\alpha \in \mathbb{R}$ obey the following conditions:

- Symmetry: $\langle f, g \rangle_{\mathcal{K}} = \langle g, f \rangle_{\mathcal{K}}$;
- linearity: $\langle \alpha f + g, h \rangle_{\mathcal{K}} = \alpha \langle f, h \rangle_{\mathcal{K}} + \langle g, h \rangle_{\mathcal{K}}$;
- $\langle f, g \rangle_{\mathcal{K}} = 0$ implies $f = 0$.

A vector space $\mathcal{K}$ with inner product $\langle \cdot, \cdot \rangle_{\mathcal{K}}$ is called an inner product space. An inner product space $(\mathcal{K}, \langle \cdot, \cdot \rangle_{\mathcal{K}})$ is a Krein space if we have two Hilbert spaces $\mathcal{H}_+$ and $\mathcal{H}_-$ spanning $\mathcal{K}$ such that $\forall f \in \mathcal{K}$ we have $f = f_+ + f_-$ with $f_+ \in \mathcal{H}_+$ and $f_- \in \mathcal{H}_-$ and $\forall f, g \in \mathcal{K}$, $\langle f, g \rangle_{\mathcal{K}} = \langle f_+, g_+ \rangle_{\mathcal{H}_+} - \langle f_-, g_- \rangle_{\mathcal{H}_-}$.

Indefinite kernels are typically observed by means of domain-specific non-metric similarity functions (such as alignment functions used in biology [17]), by specific kernel functions - e.g., the Manhattan kernel $k(x, x') = -||x - x'||_1$ or others. A finite-dimensional Krein-space is a so-called pseudo-Euclidean space.

3   Embedding for Non-psd Proximities

Embedding of a proximity matrix into a vector space is not a new consideration, see e.g. [5], but was shown to be valid so far only in case of psd kernel functions. Given a symmetric *dissimilarity* matrix with zero diagonal, an embedding of the data in a pseudo-Euclidean vector space, determined by the eigenvector decomposition of the associated similarity matrix $\mathbf{S}$, is always possible [3][1]. Given the

---

[1] The associated similarity matrix can be obtained by double centering [12] of the dissimilarity matrix. $\mathbf{S} = -\mathbf{JDJ}/2$ with $\mathbf{J} = (\mathbf{I} - \mathbf{1}\mathbf{1}^\top/N)$, identity matrix $\mathbf{I}$ and vector of ones $\mathbf{1}$.

eigendecomposition of $\mathbf{S} = \mathbf{U}\mathbf{\Lambda}\mathbf{U}^\top$, we can compute the corresponding vectorial representation $\mathbf{V}$ in the pseudo-Euclidean space by

$$\mathbf{V} = \mathbf{U}_{p+q+z} \left| \mathbf{\Lambda}_{p+q+z} \right|^{1/2} \tag{1}$$

where $\mathbf{\Lambda}_{p+q+z}$ consists of p positive, q negative non-zero eigenvalues, and z zero eigenvalues. $\mathbf{U}_{p+q+z}$ consists of the corresponding eigenvectors. The triplet (p, q, z) is also referred to as the signature of the pseudo-Euclidean space. The crucial point in Eq. (1) is the *absolute* operator used in the embedding, which is also called a flip operation in the field of indefinite learning [14]. This very costly operation makes the data metric and alters the underlying data structure [9].

The transformation of dissimilarities to obey metric properties, or of similarities to be psd is at least technically useful because it permits to employ many mathematical concepts [5], not available otherwise. We remove the absolute function from the embedding in Eq. (1) and obtain Eq. (2), and show later how the embedding can be made computational effective also for non-psd inputs. Apparently, the new embedding does not modify the data, in particular, an inner product of the embedded data reveals the input's information again.

$$\mathbf{V} = \mathbf{U}_{p+q+z}\mathbf{\Lambda}_{p+q+z}^{1/2} \tag{2}$$

The real-valued embedding in Eq. (1), leading to a psd formulation, and the complex-valued embedding in Eq. (2) is straight forward but extremely costly. Already in [5], the costs in Eq. (1) were addressed by using the Nyström approximation, applicable to the psd case only. This approach can not be used directly in our setting since the input is non-psd.

In our former work [2] (simplified in [11]), we have shown that the Nyström approximation remains valid for generic proximity data, in particular non-psd similarities. Hence the Nyström approximation becomes available to approximate a non-psd matrix. In [2], we have further shown how the Nyström approximation can also be used to have an approximated Double-Centering for dissimilarity data. Our work helps twofold to permit an effective embedding in Eq. (2):

1. the input needs not to be a kernel but can also be a dissimilarity matrix
2. the Nyström matrix approximation can also be applied on non-psd similarities which reduced the costs of the embedding

In the Nyström approximation, we have to specify the number m of landmarks with $m \ll N$. The landmarks can be selected for non-psd matrices randomly or by kmeans++ as shown in [11]. Our efficient approach to get an approximated complex-valued, vectorial embedding of a non-psd matrix is shown in Algorithm 1.

In the first step of Algorithm 1, the input matrix is approximated using the Nyström approximation (potentially with an integrated double centering). this can be done with linear costs and with guaranteed approximation bounds [2,11]. Subsequently, we calculate the essential part of the embedding function in Eq. (2) combined with the projection matrix of the Nyström approximation, by

---

Algorithm 1. Complex valued embedding of non-psd proximities

Embed_proximities(P, m)
if P is dissimilarity **then**
  $Knm, Kmm :=$ ApproximatedDoubleCentering(P, m) using [2] and kmeans++
else
  $Knm, Kmm :=$ Approximate(P, m) using [2] for similarities and kmeans++
end if
$[C, A] :=$ eig(Kmm); with eigenvectors C and eigenvalues in A (diagonal)
$W :=$ diag(sqrt(1./diag(A))) $\cdot$ C'   complex-valued Nyström projection matrix
$M := W \cdot$ Knm'   complex-valued embedding
$K^* := M' \cdot M$ reconstruction (optional)
return M

---

taking the square root of the (pseudo-) inverse of the eigenvalue decomposition of Kmm, done with linear costs. Details to costs and approximation procedure are shown in [2]. The vectorial embedding M is finally done by mapping the rectangular Nyström part Knm of the similarities to the projection matrix W^2 If the similarity matrix K is non-psd, A contains negative eigenvalues and the embedding becomes complex-valued.

We now have an approximated complex-valued fixed-length vectorial embedding of the proximity data P, whereby the respective reconstruction is exact if the rank of P equals to the number of non-vanishing eigenvalues in A. Algorithm 1 has a linear complexity (without K^*) as long as the number of landmarks $m \ll N$, which is, in general, the case. The embedding procedure has a straight forward out of sample extension. The mapping in Algorithm 1 can be done for new points by evaluating the proximity function for the landmark point and using the respective projection function.

For the complex-valued embedding (so far), only a limited number of machine learning algorithms is available, like the complex-valued support vector machine (cSVM) [22], the complex-valued generalized learning vector quantization (cGM-LVQ) [18], or a complex-valued neural network (cNNet) [19]. Further, a nearest neighbor (NN) classifier can be used by employing a standard norm operator. While cSVM, cGMLVQ, cNNet are parametric methods, the NN classifier is parameter-free and can be used directly. In particular, after applying the norm, the obtained dissimilarity values are metric. Due to its good performance and simplicity, we focus on cGMLVQ, briefly reviewed in the following.

4   Complex-Valued Generalized Learning Vector Quantization

In Learning Vector Quantization (LVQ), the classification scheme is parameterized by a set of labeled prototypes and a distance measure $d(\cdot, \cdot)$. New data is classified according to the nearest prototype's label with respect to the distance

---

[2] Some heuristic ideas on Landmark MDS, which is imprecise, are discussed in [2].

measure $d(\cdot, \cdot)$. In contrast to the NN classifier in which the full dataset is used, the classes in LVQ schemes are represented by only very few prototypes. Hence, in the algorithm's working phase, LVQ methods require less computational effort and storage. Moreover, LVQ is often praised for its white-box character, which is beneficial in many applications [20].

4.1 Training an LVQ Classifier

Given a training dataset of N labeled inputs $(\mathbf{x}_i, y_i)_{i=1}^{N}$, in which $\mathbf{x}_i \in \mathcal{R}^d$ is an input vector and $y_i \in \{1, 2, ..., K\}$ its class label. The aim of the training procedure is the adaptation of M labeled prototypes $\{(\mathbf{w}_k, y_k)\}_{k=1}^{M}$, such that the resulting classification scheme gives high classification accuracy with respect to unseen data. The distance measure $d(\cdot, \cdot)$ is of central importance in the training- and classification procedure. A common choice is squared Euclidean distance measure $(\mathbf{x} - \mathbf{w})^T(\mathbf{x} - \mathbf{w})$. In [13], a valid cost function for the LVQ heuristic was proposed, that can be minimized by, e.g., gradient descent:

$$E_{GLVQ} = \sum_{i=1}^{N} \Phi(\mu_i), \text{ with } \mu_i = \frac{d_+(\mathbf{x}_i) - d_-(\mathbf{x}_i)}{d_+(\mathbf{x}_i) + d_-(\mathbf{x}_i)}. \tag{3}$$

The argument μ_i is based on the difference between the distance $d_+(\mathbf{x}_i)$ from its position to the closest prototype with the same label and the distance $d_-(\mathbf{x}_i)$ to the closest prototype with a different label, normalized to the range $\mu_i \in [-1, 1]$. The function $\Phi(\cdot)$ is monotonically increasing and is usually chosen to be identity $\Phi(x) = x$ or the logistic function $\Phi(x) = 1/(1 + \exp(-x))$. The standard Euclidean distance does not account for differences in the classification importance of the dimensions. To improve classification accuracy, matrix *relevance learning* was introduced [15]. A full matrix of adaptive relevances $\mathbf{\Lambda} = \mathbf{\Omega}^T\mathbf{\Omega}$ is introduced in the distance measure:

$$d^{\Lambda}(\mathbf{w}, \mathbf{x}_i) = (\mathbf{x}_i - \mathbf{w})^T \mathbf{\Omega}^T \mathbf{\Omega}(\mathbf{x}_i - \mathbf{w}), \tag{4}$$

The linear projection defined by the matrix $\mathbf{\Omega}$ is adapted during training to reflect the importance of the features and to account for correlations between features.

The above cost function in Eq. (3) is minimized with respect to the prototypes $\{\mathbf{w}_k\}_{k=1}^{M}$ and the linear projection matrix $\mathbf{\Omega}$ by either batch- or stochastic gradient descent. To formulate the gradient descent update rules with respect to $\mathbf{w}_{\pm}$ and $\mathbf{\Omega}$ for an example $\mathbf{x}_i$, one applies the chain rule:

$$\mathbf{w}_{\pm} = \mathbf{w}_{\pm} - \alpha \Phi'(\mu_i) \frac{\partial \mu_i}{\partial d_{\pm}} \frac{\partial d_{\pm}}{\partial \mathbf{w}_{\pm}}, \quad \mathbf{\Omega}_{\pm} = \mathbf{\Omega}_{\pm} - \beta \Phi'(\mu_i) \frac{\partial \mu_i}{\partial d_{\pm}} \frac{\partial d_{\pm}}{\partial \mathbf{\Omega}_{\pm}} \tag{5}$$

with the learning rates α and β. For all results reported in the following, we have set $\alpha = 0.01$ and $\beta = 0.001$.

4.2 Learning Rules for Complex-Valued Data

When the data lives in the complex-valued space $\mathbb{C}^d$ and one uses the Hermitian transpose in Eq. (4), the distance is always real-valued, since it is a sum of squared magnitudes. Hence, only the innermost derivatives of the distance measure in Eq. (5) have to be considered with respect to the complex-valued variables. These can be done using the Wirtinger differential operators [21] as proposed in [18]:

$$\frac{\partial}{\partial z} = \frac{1}{2}\left(\frac{\partial}{\partial x} - i\frac{\partial}{\partial y}\right) , \quad \frac{\partial}{\partial z^*} = \frac{1}{2}\left(\frac{\partial}{\partial x} + i\frac{\partial}{\partial y}\right) , \tag{6}$$

in which $z = x + iy$ and $z^* = x - iy$, the complex conjugate. Using the differential operator with respect to z^*, the inner most derivatives in Eq. (5) are as follows:

$$\frac{\partial d}{\partial \mathbf{w}_{\pm}^*} = -\mathbf{\Omega}^H \mathbf{\Omega}(\mathbf{x}_i - \mathbf{w}_{\pm}), \quad \frac{\partial d}{\partial \mathbf{\Omega}^*} = \mathbf{\Omega}(\mathbf{x}_i - \mathbf{w}_{\pm})(\mathbf{x}_i - \mathbf{w}_{\pm})^H , \tag{7}$$

which are conceptually similar to the derivatives of the real-valued variables. The model update is implicitly done in a Krein space, while the predictions are guided by the metric dissimilarities from the employed norm operator. The cG(M)LVQ model can be trained with linear costs on the vectorial data.

5 Experiments

In this section, we show the effectiveness of the proposed embedding approach on a set of benchmark data typically used in the area of proximity-based supervised learning and by employing appropriate classified models. The following section contains a brief description of the datasets with details in the references. Subsequently, we evaluate the performance of our embedding approach on these datasets compared to some baseline classifier.

5.1 Datasets

All data sets used in this experimental setup are indefinite with different spectral properties. If the data are given as dissimilarities, a corresponding similarity matrix can be obtained by double centering, as mentioned before [12]. The datasets used for the experiments are described in the following, with details given in the references.

1. **Balls3d/50d** has 200/2000 samples in 2/4 classes. Dissimilarities are generated between balls with the shortest distance on the surfaces [12].
2. The Copenhagen **Chromosomes** data consist of 4,200 human chromosomes from 21 classes represented by grey-valued images. These are transferred to strings measuring their silhouettes and compared using an edit distance [10].
3. The **Delft gestures** (1500 points, 20 classes, signature: (963,536,1)), taken from [1], is a set of dissimilarities generated from a sign-language interpretation problem. The dissimilarities are computed by dynamic time-warping.

4. The **Flowcyto** dataset is based on 612 FL3-A DNA flowcytometer histograms from breast cancer tissues labeled in 3 classes. Dissimilarities are computed between normalized histograms using the L1 norm [1].
5. **Protein**: the Protein data set consists of sequence-alignment similarities for 213 proteins and is used for comparing and classifying protein sequences according to its four classes of globins. The signature is (170,40,3), where class one through four contains 72, 72, 39, and 30 points, respectively [4].
6. **Sonatas** dataset consists of 1068 sonatas from 5 composers (classes) encoded as MIDI data and transformed by normalized compression distance [8].
7. **Zongker** dataset is a digit dissimilarity dataset. The dissimilarity measure was computed between 2000 digits in 10 classes, 200 entries each [6].

5.2   Results

We evaluate the performance of the proposed embedding using cG(M)LVQ from Sect. 4, with the fixed length complex-valued embedded data as inputs. The cGLVQ was parametrized once with and once without relevance learning. To show that our approach performs at least as well as classical embedding, the methods were tested equally on data sets with classical embedding following the same Nyström procedure as the complex embedding. We used one prototype per class for the cG(M)LVQ. In the embedding step of Algorithm 1 we set the meta parameter m (# of landmarks) by a rule of thumb: if $N < 1000$, $m = 40$, if $1000 < N < 5000$, $m = 70$, otherwise $m = 100$. As a baseline, we used a nearest neighbor classifier (NN), which is valid for generic, uncorrected input data and embedded data, but very costly due to the storage of the full input matrix. This is particular unattractive if the dataset is large. Experiments were run in a ten-fold cross-validation. Mean prediction accuracy on the holdout test data and the respective standard deviation is reported in Table 1.

If the data were left uncorrected, we often obtained a rather poor result using the nearest neighbor classifier, sometimes even significantly worse compared to (c)GLVQ and (c)GMLVQ (see balls3d, protein, zongker). In some cases, NN had equal or slightly better performance than the two (c)G(M)LVQ variants (Chromosomes, Sonatas). This is due to the spectrum of eigenvalues: Chromosomes has many eigenvalues, which are almost negligible and close to zero. Sonatas has only a few negative eigenvalues and these eigenvalues are also close to zero. Relevance learning (cGMLVQ) significantly improves the results compared to cGLVQ without relevance learning. However, even the mere use of the cGLVQ without relevance learning leads to a significant increase in performance compared to the NN with uncorrected data. Therefore, we assume that our embedding approach, is indeed useful since the use of uncorrected non-psd data shows a clear drop in accuracy using NN and the vectorial embedding permits a more flexible weighting of input contributions. In summary, the presented approach, applying an embedding of the indefinite input data into a complex-valued vector space, shows promising results on a variety of data sets.

Table 1. Prediction accuracy (mean ± standard-deviation) for the GLVQ/cGLVQ variants and the nearest neighbor classifier.

Dataset	Without embedding NN	Classic (real) embedding NN	GLVQ	GMLVQ	Proposed (complex) embedding NN	cGLVQ	cGMLVQ
Balls3d	0.49 ±0.06	0.54 ±0.13	0.78 ±0.08	0.98 ±0.02	0.54 ±0.13	0.67 ±0.12	**1.0 ±0.0**
Balls50d	0.25 ±0.04	0.26 ±0.04	0.52 ±0.04	**0.78 ±0.18**	0.26 ±0.04	0.28 ±0.03	0.54 ±0.11
Chromosomes	**0.95 ±0.01**	0.92 ±0.02	0.91 ±0.01	0.94 ±0.02	0.92 ±0.02	0.92 ±0.01	0.94 ±0.01
DelftGestures	**0.96 ±0.02**	0.88 ±0.01	0.94 ±0.01	0.96 ±0.02	0.87 ±0.04	0.95 ±0.01	**0.96 ±0.02**
Flowcyto	0.62 ±0.08	0.63 ±0.06	0.62 ±0.06	0.67 ±0.07	0.59 ±0.04	0.66 ±0.07	**0.70 ±0.05**
Protein	0.23 ±0.1	**0.98 ±0.02**	0.93 ±0.05	0.97 ±0.05	**0.98 ±0.02**	0.92 ±0.07	**0.98 ±0.02**
Sonatas	0.89 ±0.02	0.87 ±0.02	0.82 ±0.03	0.88 ±0.03	0.87 ±0.02	0.80 ±0.03	**0.90 ±0.02**
Zongker	0.58 ±0.05	0.68 ±0.09	0.88 ±0.02	0.92 ±0.02	0.70 ±0.06	0.89 ±0.02	**0.93 ±0.02**

6 Conclusions

In this work, we proposed an efficient, complex-valued embedding and a processing pipeline to analyze non-metric or non-psd proximity data. The approach shows very promising performance on a variety of datasets and is easy to employ. A careful combination of approximation techniques, derived by the authors in former work, permits a valid and still effective calculation of the embedding matrix. By processing the embedding matrix, a straight forward, non-modifying out of sample extension is obtained, not available otherwise. The low-rank embedding is fast and has the benefit that the reconstructed matrix approximates the original indefinite kernel with low error; hence all major information in the original data is preserved. In particular we can omit additional modifiers or eigenvalue corrections which are costly and substantially alter the data. The model of the proposed complex embedding implicitly exists in the Krein space. Using learning algorithms for complex-valued data, predictive models can be obtained with low computational costs. In this initial work, we focused on complex-valued G(M)LVQ and Nearest Neighbor to calculate classification models, but this will be extended to other models in future work. Our initial findings show that the suggested complex-valued embedding of indefinite proximity data, combined with complex-valued classifier models, is a very effective and promising approach favorable over classical alternatives. Therefore, experimental comparisons to more classical eigenspectrum approaches (like clipping, flipping or shifting negative eigenvalues) or to models working in Krein space [11] are also interesting for further research.

References

1. Duin, R.P.: PRTools (2012). http://www.prtools.org
2. Gisbrecht, A., Schleif, F.: Metric and non-metric proximity transformations at linear costs. Neurocomputing **167**, 643–657 (2015)
3. Goldfarb, L.: A unified approach to pattern recognition. Pattern Recogn. **17**(5), 575–582 (1984)
4. Hofmann, T., Buhmann, J.M.: Pairwise data clustering by deterministic annealing. IEEE Trans. Pattern Anal. Mach. Intell. **19**(1), 1–14 (1997)
5. Iosifidis, A., Gabbouj, M.: Nyström-based approximate kernel subspace learning. Pattern Recogn. **57**, 190–197 (2016)
6. Jain, A., Zongker, D.: Representation and recognition of handwritten digits using deformable templates. IEEE Trans. Pattern Anal. Mach. Intell. **19**(12), 1386–1391 (1997)
7. Kar, P., Jain, P.: Similarity-based learning via data driven embeddings. In: Proceedings of Advances in Neural Information Processing Systems 24: 25th NIPS 2011, Granada, Spain, pp. 1998–2006 (2011)
8. Mokbel, B.: Dissimilarity-based learning for complex data. Ph.D. thesis, Bielefeld University (2016). https://nbn-resolving.de/urn:nbn:de:hbz:361-29004254
9. Münch, M., Raab, C., Biehl, M., Schleif, F.: Structure preserving encoding of non-euclidean similarity data. In: Proceedings of 9th ICPRAM 2020, pp. 43–51 (2020)
10. Neuhaus, M., Bunke, H.: Edit distance based kernel functions for structural pattern classification. Pattern Recogn. **39**(10), 1852–1863 (2006)
11. Oglic, D., Gärtner, T.: Scalable learning in reproducing Kernel Krein spaces. In: Proceedings of the 36th ICML 2019, USA, pp. 4912–4921 (2019)
12. Pekalska, E., Duin, R.: The Dissimilarity Representation for Pattern Recognition. World Scientific (2005)
13. Sato, A., Yamada, K.: Generalized Learning Vector Quantization. In: Proceedings of 8th NIPS 1995 (NIPS'95), pp. 423–429. MIT Press, Cambridge, MA, USA (1995)
14. Schleif, F., Tiño, P.: Indefinite proximity learning: a review. Neural Comput. **27**(10), 2039–2096 (2015)
15. Schneider, P., Biehl, M., Hammer, B.: Adaptive relevance matrices in learning vector quantization. Neural Comput. **21**(12), 3532–3561 (2009)
16. Shawe-Taylor, J., Cristianini, N.: Kernel Methods for Pattern Analysis and Discovery. Cambridge University Press, Cambridge (2004)
17. Smith, T.F., Waterman, M.S.: Identification of common molecular subsequences. J. Mol. Biol. **147**(1), 195–197 (1981)
18. Straat, M., et al.: Learning vector quantization and relevances in complex coefficient space. Neural Comput. Appl. **32**, 18085–18099 (2019)
19. Trabelsi, C., et al.: Deep complex networks. In: 6th ICLR 2018 (2018)
20. van Veen, R., et al.: An application of generalized matrix learning vector quantization in neuroimaging. Comp. Meth. Progr. Biomed. **197**, 105708 (2020)
21. Wirtinger, W.: Zur formalen Theorie der Funktionen von mehr komplexen Veränderlichen. Math. Ann. **97**, 357–376 (1927)
22. Zhang, L., Zhou, W., Jiao, L.: Complex-valued support vector classifiers. Digit. Signal Process. **20**(3), 944–955 (2010)

Metric Learning for Multi-label Classification

Marco Brighi$^{(\boxtimes)}$, Annalisa Franco, and Dario Maio

DISI, University of Bologna, Cesena, Italy
{marco.brighi6,annalisa.franco,dario.maio}@unibo.it

Abstract. This paper proposes an approach for multi-label classification based on metric learning. The approach has been designed to deal with general classification problems, without any assumption on the specific kind of data used (images, text, etc.) or semantic meaning assigned to labels (tags, categories, etc.). It is based on clustering and metric learning algorithm aimed at constructing a space capable of facilitating and improving the task of classifiers. The experimental results obtained on public benchmarks of different nature confirm the effectiveness of the proposal.

Keywords: Metric learning · Multi-label classification · Clustering · Supervised learning.

1 Introduction and Related Works

Multi-label classification plays an important role in the context of data analysis in many different applications, ranging from text classification to multimedia annotation or bioinformatics [9]. The general problem can be stated as follows. Let $Y = \{y_j, j = 1, .., m\}$ be a finite set of m class labels and let $D = \{(\mathbf{x}_i, Y_{\mathbf{x}_i}), i = 1, .., n\}$ be a generic set of labeled patterns, where $\mathbf{x}_i$ is a data pattern and $Y_{\mathbf{x}_i} \subseteq Y$ the set of related labels. In multi-label classification each pattern $\mathbf{x}_i$ is therefore generally assigned to more than one class. The objective of multi-label classification is to determine a function $f(\mathbf{x}_i)$ able to correctly identify all the labels $Y_{\mathbf{x}_i}$ that can be associated to each pattern in the domain. Traditional classifiers are not feasible in this case, and an extension to the multi-label case is needed. Most of the commonly used algorithms have their own multi-label variant; as an alternative, some works in the literature propose approaches able to transform multi-label problems into more canonical multi-class ones [4]. Interested readers can refer to [5] for a description of the existing approaches. Recently, some approaches for multi-label classification have been proposed, based on neural networks and deep learning, such as [18] and [13]. These techniques achieve in general good performance but require, for a proper training, a large amounts of data that are not always available. The solution proposed in this work is aimed at designing a method for improving performance of natively multi-label classifiers for general pattern classification problems, in the presence of datasets of limited dimensions.

A. Torsello et al. (Eds.): S+SSPR 2020, LNCS 12644, pp. 24–33, 2021.
https://doi.org/10.1007/978-3-030-73973-7_3

In an ideal situation, just like in a traditional multi-class problem, the patterns sharing the same labels should lay in the same sub-region of the feature space. Unfortunately, this desirable situation seldom occurs in real cases, since labels can generally represent high-level concepts used by humans for classification that do not have a direct and immediate counterpart in terms of feature similarity; of course, the multi-label scenario further complicates the problem. A common approach to derive a better representation space is based on the use of algorithms capable of learning a specific metric, which can properly group patterns belonging to the same classes. To this purpose, several techniques based on supervised learning have been proposed such as Local Fisher Discriminant Analysis [16] (LFDA) or Large Margin Nearest Neighbor Metric Learning [19] (LMNN). Unfortunately, such approaches cannot be easily extended to the multi-label case. An easy and intuitive solution to this problem is to consider super-sets of labels by transforming the multi-label problem into a multi-class one. This approach is commonly referred to as *label powerset* (see for instance [17]) and it maps each combination of labels to a unique class changing the nature of the problem; each new class created corresponds to a list of labels of the original problem. This approach is well-suited for problems where the number of labels and their possible combinations are limited, otherwise the risk is to obtain an extremely high number of super-sets, i.e. to many classes to deal with.

This paper proposes a novel approach aimed at improving the multi-label classifier accuracy by the adoption of a metric learning algorithm. The authors of [11] present a method in order to learn a new distance metric maximizing the margin present between different instances. To achieve the expected result, they project data and labels into the same embedding space imposing constraints on distances such that instances with very different multiple labels will be moved far away. Unlike the previous solutions, [8] proposes an innovative loss function based on the Jaccard index able to determine how different two instances are based on their labels set. Using the Adam optimisation algorithm, they learn a new metric able to improve the classification performance.

The approach proposed in this work is based on label clustering, meaning that the patterns in the dataset are clustered according to their labels rather than on the basis of the related feature vectors. The information produced by clustering is exploited to learn a metric aimed at improving classification accuracy. In [9] the authors use clustering algorithms for pruning infrequent multi labels. They assume that the elimination of infrequently multi label instances from the training set leads to the identification of better label power sets. Another interesting use of clustering can be found in [20]. The authors address the multi-label classification problem using the classifier chain approach, training a single classifier for each label. The use of a clustering algorithm, *k-means* in this case, allows to discover any correlations present between the labels. This information allows to identify the correct order in which to arrange the classifiers, maximizing the final performance.

The proposed approach has been designed as a general solution capable of dealing with multi-label classification problems. For this reason the experimental evaluation will consider heterogeneous datasets containing data of different nature. The paper is organized as follows: Sect. 2 presents the proposed approach, its characteristics and steps, Sect. 3 shows a complete overview of the results achieved and Sect. 4 draws some conclusions and presents possible future research directions.

2 The Proposed Approach

The main steps of the approach can be summarized as follows (detailed description in the following subsections):

1. *Clustering:* the patterns within the dataset are grouped on the basis of label sharing;
2. *Metric Learning:* the original space is replaced thanks to the learning of a metric capable of bringing together patterns belonging to the same cluster;
3. *Multi-label Classification:* a suitably trained classifier identifies the labels to be associated with each pattern inside the new space.

2.1 Clustering

The idea of exploiting label clustering in this work is motivated by the observation that, in real life scenarios, patterns sharing the same set of labels not always can be considered similar from the point of view of their respective feature descriptors. Labels represent very general high-level concepts assigned by humans, especially in case of multi-label classification; clustering techniques based on data descriptors may work well for specific applications (e.g. image retrieval in [14]), but could present limited generalization capabilities when dealing with classification problems of different nature. Label clustering, on the contrary, allows to group together data with similar labels, even when they are spread over the feature space.

The membership of a generic pattern $\mathbf{x}$ to the available labels can be described through a binary vector $\mathbf{l} = \{l_i, i = 1, .., m\}$ of size m equal to the number of possible labels; each element of the vector $l_i \in \{0, 1\}$, representing the membership of $\mathbf{x}$ to class y_i. This kind of encoding defines a real space of the labels, different from that of the features used to describe the single patterns.

The proposed approach applies a clustering algorithm to the label space, ignoring at all the feature vectors associated to patterns, thus ensuring that clusters are defined according to class membership regardless of the descriptors used to encode them. The metric used for clustering is the *hamming distance*, more suited than the Euclidean distance to deal with the specific label encoding adopted. Clustering is performed by the HDBSCAN [3] technique, a density-based clustering algorithm proposed by the same authors of DBSCAN, of which it represents an evolution. The algorithm identifies clusters as high-density areas in space. The traditional DBSCAN algorithm uses a single density value to identify all clusters, while HDBSCAN extends this concept allowing to identify clusters with different density levels.

The first step of the algorithm requires the estimation of the local density for each pattern $\mathbf{x}_i$. The easiest way to get a density estimate is to evaluate the neighborhood of a point. If a large number of neighbors are placed in a very narrow radius, that particular area can be considered dense. The distance $d_{core-k}(\mathbf{x}_i)$ is simply defined as the distance between $\mathbf{x}_i$ and its k-th nearest neighbor, using any metric. Based on this measure, the *mutual reachability distance* $d_{reach-k}(\mathbf{x}_i, \mathbf{x}_j)$ between two points $\mathbf{x}_i$ and $\mathbf{x}_j$ is defined as follows:

$$d_{reach-k}(\mathbf{x}_i, \mathbf{x}_j) = \max\{d_{core-k}(\mathbf{x}_i), d_{core-k}(\mathbf{x}_j), d(\mathbf{x}_i, \mathbf{x}_j)\} \qquad (1)$$

The practical effect of this metric is to preserve the distances of points laying in dense regions and increasing the reciprocal distances of the points in sparse ones. Clustering is then based on a weighted graph representation of the points in the dataset. The points assume the role of the vertices, while the arc connecting two points $\mathbf{x}_i$ and $\mathbf{x}_j$ is assigned a weight equal to $d_{reach-k}(\mathbf{x}_i, \mathbf{x}_j)$. The idea is to eliminate the weakest arches, thus identifying the high density areas. This can be accomplished extremely efficiently by calculating the minimum spanning tree using a proper algorithm. Given the minimum spanning tree, the next step is to convert that representation into a hierarchy of connected components by grouping neighboring points. DBSCAN cuts the hierarchy tree horizontally by identifying a single threshold used to locate clusters. This represents a strong limitation if there are areas with heterogeneous densities in the dataset. HDB-SCAN uses a different approach using the notion of *minimum cluster size*, which is taken as input parameter for the algorithm. Each split in the tree represents a subdivision of the data into multiple clusters. In fact, the root of the tree treats all data as a single cluster. The deeper you go down, the more clusters you divide your data with. Browsing through the tree, HDBSCAN makes an evaluation on the validity of each split present. Only splits into two clusters each at least as large as the minimum cluster size can be considered valid. Otherwise, that split is eliminated by condensing that part of the tree. After walking through the whole hierarchy, we end up with a much smaller condensed tree with a reduced number of nodes. Given the tree, it is only a matter of choosing those nodes/clusters to save.

2.2   Metric Learning

The output of the previous clustering stage is a set of clusters containing data sharing large part of the respective labels. These new labels, found out by a clustering algorithm, are used to learn a new metric in a supervised fashion from the feature space. The metric learning step is aimed at reducing intra-cluster distances while maximizing inter-cluster distances. The literature proposed several supervised metric learning algorithms. The approach adopted in this work is Neighbourhood components analysis (NCA) [7]. It is an algorithm that learns a linear transformation, in a supervised fashion, to improve the classification accuracy of a stochastic nearest neighbors rule in the transformed space. The goal of NCA is to learn an optimal linear transformation matrix A such that the average *leave-one-out (LOO)* classification performance is maximized. It identifies the optimal transformation matrix by maximizing the sum over all samples of the probability of being correctly classified according to LOO classification. This type of classification tries to predict the class label of a single data point by consensus of its k-nearest neighbours, using a given distance metric. Unfortunately, it is not so simple to identify the optimal matrix as any objective function based on neighborhood points would be not differentiable. In particular, the set of neighbors for a point may undergo discrete changes in response to regular changes in the elements of A. This difficulty is overcome by adopting an approach based on stochastic gradient descent. The entire transformed dataset is considered as stochastic nearest neighbours using a softmax function of the squared euclidean distance between a point and each other point in the space. Considering C_i as the set of points in the same class as sample i, the probability of sample i being correctly classified is:

$$p_i = \sum_{j \in C_i} p_{ij} \quad where \quad p_{ij} = \frac{\exp(-||A\mathbf{x}_i - A\mathbf{x}_j||^2)}{\sum_{z \neq i} \exp{-(||A\mathbf{x}_i - A\mathbf{x}_z||^2)}}, \quad p_{ii} = 0 \quad (2)$$

At each iteration the algorithm adequately modifies the parameters of matrix A in order to approach patterns belonging to the same class. It is possible to proceed until convergence or until a maximum number of iterations is reached. By limiting the number of iterations, the impact of the algorithm on the new metric can be modulated, preserving part of the old space. NCA also offers the possibility of reducing the dimensionality of the data if deemed excessive.

2.3   Multi-label Classification

The final step of the approach is represented by multi-label classification, which takes place in the new space created by the metric learning algorithm. From the results of this last phase it is possible to verify the effectiveness of the entire proposal, which should lead to an improvement in the classification performance. Among the different approaches in the literature, we decided to adopt in this work two classifiers, Multi-label K-Nearest Neighbors [21] and Random Forest [2], to evaluate the effectiveness of the proposed approach.

Multi-label K-Nearest Neighbors Classifier (ML-kNN). It is a multi-label classification algorithm based on an extension of the well known K-Nearest Neighbor. It assigns labels to a pattern by evaluating the labels of its neighborhood based on a simple approach: it identifies the k closest patterns present in the training set and then uses Bayesian inference to select the labels to assign. Considering the main characteristics of this algorithm, it is essential to have an adequate metric able to aggregate the patterns with the same labels in space.

Random Forest. It is a meta classifier algorithm that fits a number of decision tree classifiers on various sub-samples of the dataset. The final result of the classifier is the average of the results obtained by individual trees. This strategy allows to increase the classification accuracy and reduce the occurrence of over-fitting. The random forest algorithm uses the technique of *bagging*. Every tree is trained selecting a random sample with replacement of the training set. Feature bagging is also used to reduce correlation between trees: during the learning process only a random subset of the available features is used. The use of these techniques leads to better model performance because it decreases the variance of the model, without increasing the bias. While the predictions of a single tree are very sensitive to noise in its training set, the average of many trees is not, as long as the trees are not correlated. It is important to notice that this algorithm is natively multi-label. Given a generic pattern $\mathbf{x}$, each of the trees available will return the predicted class $y \in Y$. The set of these labels represents the output of the entire algorithm. Some more robust implementations only consider those labels voted by a minimal number of classifiers.

3 Experiments and Results

3.1 Evaluation Protocol

To evaluate the effectiveness of this approach, it was chosen to use the main and most known indicators in the field of multi-label classification. They represent a multi-label variant of the most commonly used indicators in classification problems [6]. Let D be a multi-label test set. Let $Y_\mathbf{x}$ be the set of labels associated with the generic pattern $\mathbf{x}$ and $Z_\mathbf{x}$ those predicted by the classification algorithm. Accuracy, Precision and Recall are three of the most important indicators for evaluating a classification approach. In their multi-label version they are defined as:

$$Accuracy = \frac{1}{|D|} \sum_{i=1}^{|D|} \frac{|Y_{\mathbf{x}_i} \cap Z_{\mathbf{x}_i}|}{|Y_{\mathbf{x}_i} \cup Z_{\mathbf{x}_i}|} \tag{3}$$

$$Precision = \frac{1}{|D|} \sum_{i=1}^{|D|} \frac{|Y_{\mathbf{x}_i} \cap Z_{\mathbf{x}_i}|}{|Z_{\mathbf{x}_i}|} \tag{4}$$

$$Recall = \frac{1}{|D|} \sum_{i=1}^{|D|} \frac{|Y_{\mathbf{x}_i} \cap Z_{\mathbf{x}_i}|}{|Y_{\mathbf{x}_i}|} \tag{5}$$

Accuracy measures the overall ability of the classifier of correctly classifying patterns. Precision measures the portion of predicted labels that are correct, while recall measures the portion of real labels that were correctly predicted. Another key indicator within the multi-label classification is the hamming loss. Its definition recalls the concept of hamming distance:

$$HammingLoss = \frac{1}{|D|} \sum_{i=1}^{|D|} \frac{|Y_{\mathbf{x}_i} \oplus Z_{\mathbf{x}_i}|}{|Y_{\mathbf{x}_i}|} \tag{6}$$

where $\oplus$ is the XOR operator. This indicator is extremely significant in the multi-label context as it summarizes the difference between the real label set and the one predicted by the classifier. Despite the presence of all these specific indicators for multi-label classification, it is still possible to evaluate the quality of the results by counting the percentage of *exact matches*, i.e. the percentage of test data for which all the required labels have been correctly predicted.

3.2 Datasets

Extensive experiments have been carried out to evaluate the effectiveness of our proposal as well as its generality. To this purpose, three heterogeneous multi-label datasets (see Table 1) have been selected:

- *enron* [15], a dataset of emails labelled with a set of categories; each email is encoded using the Bag Of Words encoding;
- *scene* [1], an image dataset; the six labels available identify the characteristics of the landscape depicted in the image itself; each image is described with visual numeric features;
- *bibtex* [10], containing bibliographic data from the BibSonomy social bookmark and publication sharing system, annotated with a subset of the tags assigned by BibSonomy users; it uses the Bag Of Words model.

The three datasets are quite different from different points of view such as content (text/image), features representation, size, number of labels, and how they are distributed. Each dataset assigns a different meaning to the labels (category, description, tag), so that they represent an interesting test bed. Table 1 also reports the training/testing partitioning suggested in the literature for the datasets. Of course clustering, metric learning and classifier training were done on the training set, while the test set was used to measure classification results.

Table 1. The main characteristics of the datasets.

Dataset	Domain	Instances (train/test)	Attributes	Labels	Avg/Max labels per instance
Enron	Text	1702 (1123/579)	1001	53	3.38 / 12
Scene	Image	2407 (1211/1196)	294	6	1.07 / 3
Bibtex	Text	7395 (4880/2515)	1836	159	2.4 / 28

3.3 Results

Before analysing the classification performance, we briefly report the result produced in the clustering phase by the HDBSCAN algorithm. The minimum cluster size required as an input parameter by the algorithm was arbitrarily fixed to 5 for our experiments. Table 2 shows some results related to clustering such as the number of clusters identified in each dataset and their average size in terms of patterns. The number of clusters obtained is in line with the size and the number of labels in the datasets. It is interesting to note the limited number of clusters identified in the *scene* dataset, quite close to the number of labels. Another important information reported by the table is the average number of clusters in which a single label ends. It should provide an indication of how the patterns presenting that label have been grouped. For example, if a label is present on the majority of the dataset is highly unlikely that all patterns indicating the label end up in a single cluster. Conversely, if a single or a group of labels characterize a certain portion of the dataset, those data will probably shape a single cluster. Of course, the occurrence of these situations is closely related to the meaning that each dataset assigns to its labels. On average, each label in the *enron* and *scene* datasets ends up in a very similar number of clusters (4/3) despite the huge difference in the number of labels available. This is due to the different distribution of the labels within the dataset.

Table 2. The results of clustering.

Dataset	No. of clusters	Cluster average size	Avg No. of Clusters per label
Enron	39	16.38	4.04
Scene	11	89.82	3.00
Bibtex	226	19.11	19.01

The second step of the method is focused on the NCA algorithm for learning a better metric. The maximum number of iterations to be performed during learning has been fixed for all datasets to 30, to avoid excessive flattening of data on clusters, which proved to be counterproductive in terms of results. No dimensionality reduction has been applied in our experiments. As regards the classification, the two algorithms illustrated above were used: ML-kNN and Random Forest. The results obtained by ML-kNN using the proposed approach are compared with the benchmarks indicated in [12], following the same protocol. In particular, the best results in the new space were obtained by assigning the value 10 to the hyper-parameter k. As for Random Forest, it was made a comparison between the original space of the datasets and the built one using a model with 100 decision trees. Thanks to these comparisons, it is possible to evaluate the effectiveness of the proposed method. The classification results, obtained using the indicators shown above, are reported in Table 3.

The results clearly show that the proposed method is able to improve classification performance. In particular, ML-kNN improves significantly, especially for the *bibtex* dataset. Given the characteristics of the classifier, it was reasonable to expect a positive effect from the construction of a space more suitable for classification. The results obtained show the ability of the previously identified clusters to group patterns with the same labels. In general, Random Forest also seems to get benefits, though in a limited form. While on the one hand the increase in performance on *scene* is remarkable, on the other hand the results

Table 3. Classification results on selected datasets.

		Accuracy	Precision	Recall	Hamming D.	Exact match
Enron	ML-kNN (Benchmark)	0.319	0.587	0.358	0.051	0.062
	ML-kNN (Our method)	0.413	0.66	0.465	0.048	0.123
	Random forest (Original Space)	0.401	0.694	0.449	0.047	0.119
	Random forest (Our method)	0.413	0.7	0.456	0.046	0.123
Scene	ML-kNN (Benchmark)	0.629	0.661	0.655	0.099	0.573
	ML-kNN (Our method)	0.698	0.732	0.715	0.090	0.646
	Random forest (Original space)	0.540	0.565	0.540	0.093	0.514
	Random forest (Our method)	0.657	0.688	0.659	0.087	0.624
Bibtex	ML-kNN (Benchmark)	0.129	0.254	0.132	0.014	0.056
	ML-kNN (Our method)	0.257	0.390	0.271	0.013	0.156
	Random forest (Original space)	0.217	0.362	0.219	0.013	0.135
	Random forest (Our method)	0.186	0.316	0.188	0.013	0.114

on the other datasets are not so good. This is not accidental if we consider the characteristics of the two datasets and those of the classifier. On average, *enron* and *bibtex* have a much higher number of labels for each instance than *scene*. The new space, by grouping different patterns, has evidently reduced the ability of the trees to discriminate with respect to a single label, causing not exactly exciting results. In summary, contrary to what happened with ML-kNN, the new learned metric hindered the task of the Random Forest classifier.

Overall, the results obtained are very good and prove the capability of the proposed method to build a more effective space for multi-label classification.

4   Conclusions and Future Works

In this work a metric-learning approach aimed at improving the performance of a multi-label classifier has been presented. In particular, the use of a cluster algorithm applied to pattern labels allowed to identify groups of data that share most of their labels, subsequently used to build a new, more effective, representation space. The results obtained on three different multi-label datasets show how the new space built by the proposed approach is able to significantly increase the classification results. Future researches will be devoted to investigate the scalability of this approach to larger datasets. Another future development involves the introduction, in the metric learning step, of a neural network capable of building a more suitable space for classifiers.

References

1. Boutell, M.R., Luo, J., Shen, X., Brown, C.M.: Learning multi-label scene classification. Pattern Recogn. **37**(9), 1757–1771 (2004)
2. Breiman, L.: Random forests. Mach. Learn. **45**(1), 5–32 (2001)

3. Campello, R.J.G.B., Moulavi, D., Sander, J.: Density-based clustering based on hierarchical density estimates. In: Pei, J., Tseng, V.S., Cao, L., Motoda, H., Xu, G. (eds.) PAKDD 2013. LNCS (LNAI), vol. 7819, pp. 160–172. Springer, Heidelberg (2013). https://doi.org/10.1007/978-3-642-37456-2_14

4. Cherman, E.A., Monard, M.C., Metz, J.: Multi-label problem transformation methods: a case study. CLEI Electron. J. **14**(1), 4–4 (2011)

5. Ganda, D., Buch, R.: A survey on multi label classification. Recent Trends Program. Lang. **5**(1), 19–23 (2018)

6. Godbole, S., Sarawagi, S.: Discriminative methods for multi-labeled classification. In: Dai, H., Srikant, R., Zhang, C. (eds.) PAKDD 2004. LNCS (LNAI), vol. 3056, pp. 22–30. Springer, Heidelberg (2004). https://doi.org/10.1007/978-3-540-24775-3_5

7. Goldberger, J., Hinton, G.E., Roweis, S.T., Salakhutdinov, R.R.: Neighbourhood components analysis. Adv. Neural Info. Process. Syst. **17**, 513–520 (2005)

8. Gouk, H., Pfahringer, B., Cree, M.: Learning distance metrics for multi-label classification. In: Asian Conference on Machine Learning, pp. 318–333. PMLR (2016)

9. Gupta, P., Anand, A.: Multi label classification using label clustering. In: Appearing in Proceedings of the 1st Indian Workshop on Machine Learning, IIT Kanpur, India (2013)

10. Katakis, I., Tsoumakas, G., Vlahavas, I.: Multilabel text classification for automated tag suggestion. In: Proceedings of the ECML/PKDD, vol. 18, p. 5 (2008)

11. Liu, W., Tsang, I.W.: Large margin metric learning for multi-label prediction. In: Proceedings of the National Conference on Artificial Intelligence (2015)

12. Madjarov, G., Kocev, D., Gjorgjevikj, D., Džeroski, S.: An extensive experimental comparison of methods for multi-label learning. Pattern Recogn. **45**(9), 3084–3104 (2012)

13. Nam, J., Kim, J., Loza Mencía, E., Gurevych, I., Fürnkranz, J.: Large-scale multi-label text classification — revisiting neural networks. In: Calders, T., Esposito, F., Hüllermeier, E., Meo, R. (eds.) ECML PKDD 2014. LNCS (LNAI), vol. 8725, pp. 437–452. Springer, Heidelberg (2014). https://doi.org/10.1007/978-3-662-44851-9_28

14. Nasierding, G., Tsoumakas, G., Kouzani, A.Z.: Clustering based multi-label classification for image annotation and retrieval. In: IEEE SMC, pp. 4514–4519 (2009)

15. Read, J., Pfahringer, B., Holmes, G.: Multi-label classification using ensembles of pruned sets. In: IEEE ICDM, pp. 995–1000. IEEE (2008)

16. Sugiyama, M.: Dimensionality reduction of multimodal labeled data by local fisher discriminant analysis. J. Mach. Learn. Res. **8**, 1027–1061 (2007)

17. Tsoumakas, G., Vlahavas, I.: Random k-labelsets: an ensemble method for multilabel classification. In: Kok, J.N., Koronacki, J., Mantaras, R.L., Matwin, S., Mladenič, D., Skowron, A. (eds.) ECML 2007. LNCS (LNAI), vol. 4701, pp. 406–417. Springer, Heidelberg (2007). https://doi.org/10.1007/978-3-540-74958-5_38

18. Wang, J., Yang, Y., Mao, J., Huang, Z., Huang, C., Xu, W.: CNN-RNN: a unified framework for multi-label image classification. In: IEEE CVPR (2016)

19. Weinberger, K.Q., Saul, L.K.: Distance metric learning for large margin nearest neighbor classification. J. Mach. Learn. Res. **10**(2), 1–38 (2009)

20. Yu, Z., Wang, Q., Fan, Y., Dai, H., Qiu, M.: An improved classifier chain algorithm for multi-label classification of big data analysis. In: 2015 IEEE 17th HPCC, 2015 IEEE 7th CSS, and 2015 IEEE 12th ICESS, pp. 1298–1301 (2015)

21. Zhang, M.L., Zhou, Z.H.: ML-KNN: a lazy learning approach to multi-label learning. Pattern Recogn. **40**(7), 2038–2048 (2007)

An Alternative Exploitation of Isolation Forests for Outlier Detection

Antonella Mensi[1]([✉])(iD), Alessio Franzoni[1], David M. J. Tax[2](iD),
and Manuele Bicego[1](iD)

[1] Department of Computer Science, University of Verona, Verona, Italy
{antonella.mensi,manuele.bicego}@univr.it,
alessio.franzoni_01@studenti.univr.it
[2] Faculty of Electrical Engineering, Mathematics and Computer Science, TU Delft,
Delft, The Netherlands
D.M.J.Tax@tudelft.nl

Abstract. Isolation Forests are one of the most successful outlier detection techniques: they isolate outliers by performing random splits in each node. It has been recently shown that a trained Random Forest-based model can also be used to define and extract informative distance measures between objects. Although their success has been shown mainly in the clustering field, we propose to extract these pairwise distances between the objects from an Isolation Forest and use them as input to a distance or density-based outlier detector. We show that the extracted distances from Isolation Forests are able to describe outliers meaningfully. We evaluate our technique on ten benchmark datasets for outlier detection: we employ three different distance measures and evaluate the obtained representation using a density-based classifier, the Local Outlier Factor. We also compare the methodology to the standard Isolation Forests scheme.

Keywords: Outlier detection · Isolation forests · Random forest-based similarity

1 Introduction

Isolation Forests (IF) [16,18] represent a Random Forest-based technique for outlier detection, which success have been assessed in many different contexts: for example, in the comparative analysis shown in [9], they were proven to be the most successful methodology to solve this task. In contrast to other Random Forests approaches for outlier detection [7,23], which are based on a standard classification Random Forest trained on normal data and artificially generated outliers, Isolation Forests use trees in which splits are performed completely at random (similarly to the Extremely Randomized Trees [10]). Given the trees, IFs solve outlier detection using the concept of "isolation", which encodes the fact that outliers are probably well separated from the rest, thus being able to be

© Springer Nature Switzerland AG 2021
A. Torsello et al. (Eds.): S+SSPR 2020, LNCS 12644, pp. 34–44, 2021.
https://doi.org/10.1007/978-3-030-73973-7_4

"isolated" from the remainder of the data within the early splits of the tree. Thus, the anomaly degree of a given point can be detected by looking at the depth of the leaf it reaches. Isolation Forests have been extensively employed, extended and improved in many different aspects [8,11,13,14,17,19,24]: most of these extensions [8,11,13,14,17,24] were devoted to improve the training stage, for example by defining novel ways to split a node; few of them focus on improving the testing phase, i.e. the anomaly score [13,19].

In this paper we propose and investigate an alternative exploitation of the Isolation Forests for outlier detection: instead of employing the isolation concept, we investigate the possibility of exploiting the IF to derive pairwise distances between objects, to be then used as input for a distance or density-based outlier detection classifier.

The proposed approach starts from the following observation: Random Forests (RF) are not used solely for classification or regression, but also as a valid and flexible data description tool. For example, in the field of clustering, there are different approaches which exploit the concept that the intrinsic nature of Random Forests allows to describe data in a meaningful way. In all these techniques –the so-called *distance-based RF clustering* methods [3,4,23,26,27]– the idea is to exploit RFs to derive a dissimilarity measure between points, to be subsequently used as input to a distance-based classifier. These measures have been proven to be more descriptive than standard geometric-based distances such as the Euclidean distance, and have been successfully applied in many different domains [1,12,21,22]. In almost all these methods the trained forests are standard binary classification RFs, built using the points to be clustered and a synthetically generated negative class. Very recently [3], however, other learning schemes have been investigated, able to work without generating a synthetic negative class that tends to hide the true nature and complexity of the data. Among other learning strategies, those based on random mechanisms were shown to perform surprisingly well, permitting to derive meaningful and informative distances.

Following these findings, we propose an alternative IF-based outlier detection scheme, in which we exploit Isolation Forests to derive dissimilarities to be used inside a distance-based outlier detector. In the paper we investigated three different strategies for computing the dissimilarity, based on different intuitions [23,27]. To investigate the suitability of the proposed framework we employed ten different benchmark outlier detection datasets, evaluating the different dissimilarities also in comparison with the standard Isolation Forest scheme. Results were encouraging, confirming the richness of the information that can be extracted from this particular type of Random Forests.

The remainder of the paper is divided as follows: in Sect. 2 we present the Isolation Forests in detail; in Sect. 3 we describe the proposed methodology and then we test it in Sect. 4. In Sect. 5 we make some conclusions.

2   Isolation Forests

The most successful and used Random Forest-based technique for outlier detection is called Isolation Forest, or IF [16,18]. Differently from other RF-based methodologies for outlier detection, which create artificial outliers in order to employ RF for classification [7,23], IFs work in a completely unsupervised way. They aim at separating each object from the rest of the dataset, independently of the class it belongs to. The success of the IFs can be attributed to the way in which they are built –the training phase– and secondly, by how the score of each object traversing the forest is computed –the testing one. In the two following Subsections we illustrate in detail such procedures.

2.1   Training Phase

An Isolation Forest is composed of several Isolation Trees (iTrees), which are built using a random subsample of the training set drawn without replacement. Each iTree is built recursively by partitioning each node into two children nodes in a completely random way, inspired by the Extremely Randomized Trees [10]. An axis-parallel split is performed in the following way: a feature is chosen completely at random, and then a random choice is made also for the value along which to split, in the domain of the selected feature. The tree is built until a stopping criterion is met: either we have reached the maximum established depth or it is impossible to split the node.

This tree structure is able to well differentiate outliers from inliers due to the fact that the former are usually fewer, different and heterogeneous with respect to the rest of the dataset. Indeed early splits will have a higher probability to separate outliers from the rest of the data due to the nature of outliers. Therefore we can infer that on average outliers will tend to end up in leaves that have a smaller depth than those that inliers will reach.

2.2   Testing Phase

In the testing phase an object x traverses each tree of a trained IF and a score is inferred, indicating the probability of x being an outlier. The definition of anomaly score $s(x)$, given by Liu et al. [16,18], is as follows:

$$s(x, S) = 2^{-\frac{E(h(x))}{c(S)}} \tag{1}$$

where S is the number of training samples used to build a tree, $c(S)$ is a normalization factor needed for comparing differently built forests and $E(h(x))$ is the average path length across all trees –for a more detailed explanation please refer to [16,18]. The score, which varies in the range between 0 and 1 behaves as expected: a smaller average depth will lead to a higher score which increases the probability of a point to be an outlier.

3   Methodology

The proposed methodology consists of three steps:

1. Train an Isolation Forest model $\mathcal{F}$.
2. Extract from $\mathcal{F}$ a distance matrix $\mathbf{D}$ which contains in cell (x, y) the pairwise distance between the xth and y^{th} object. We call it the IF-distance.
3. Classify the objects using an outlier detector that takes $\mathbf{D}$ as input.

Step 1: Training of IF

The first step represents the standard training of Isolation Forests, as described in Sect. 2. We train a forest $\mathcal{F}$ composed of T trees. Each tree t has been built using S samples drawn without replacement from the training set. The recursive building procedure continues until a maximum depth D is reached. Within each tree t we define the following elements: (i) $root$ is the root node of the tree; (ii) n is either an internal node of the tree, i.e. a node which can be split and is not the root, or a leaf node. Each node n contains $< S$ objects: we indicate this quantity with $|n|$ and (iii) $d()$ is the depth function which retrieves the depth of each node, where $d(root) = 0$.

Step 2: Derivation of IF-distance

First, we introduce some useful notation. When objects x and y are traversing a tree t, we define: i) $l_t(x)$ is the leaf node reached by x which has depth $d_t(x) = d(l_t(x))$; ii) $\mathcal{P}_t^x = \{n_1, n_2, \ldots n_{d_t(x)}\}$ is the path traversed by x in t in terms of set of nodes, excluding the root –since it is traversed by all objects. Note that $d_t(x) = |\mathcal{P}_t^x|$. iii) $LCA_t(x, y)$ is the lowest common ancestor of x and y, i.e. the last node in which x and y are together. The split defined in this node will separate x from y; iv) $\lambda_t(x, y) = d(LCA_t(x, y))$ and v) $\mathcal{P}_t^{(x,y)} = \{n_1, ..., LCA_t(x, y)\}$ is the path traversed by both objects, i.e. the subset of nodes traversed by both x and y. Note that $\lambda_t(x, y) = |\mathcal{P}_t^{(x,y)}|$.

The IF-distance $\mathbf{D}$ has been computed using three different proposals, widely and successfully employed in the clustering scenario [23,27].

1. In [4,23] two objects in a tree t are similar if they end up in the same leaf. Therefore, in a forest, two objects are more similar if they reach the same leaf in a greater number of trees. Formally, given objects x and y the Shi similarity between the two objects is defined as:

$$SimShi(x, y) = \frac{\sum_{t \in \mathcal{F}} \mathbb{1}(l_t(x) = l_t(y))}{T} \tag{2}$$

where $\mathbb{1}$ is the indicator function that returns 1 if the two leaves are equal and T is the number of trees in $\mathcal{F}$. This measure is then transformed into a distance in the following way:

$$Shi(x, y) = \sqrt{1 - SimShi(x, y)}. \tag{3}$$

The other two measures are defined by [27]. The authors generalize the concept introduced by [22]: objects which do not arrive at the same leaf may share some similarity, that can be measured via the length of their common path. The novel measures introduced in [27] are *ClustRF-Strct-Unfm* and *ClustRF-Strct-Adpt* which we will call *SimZhu2* and *SimZhu3* for the sake of simplicity:

2. Given two objects x and y that traverse a tree t, $SimZhu2_t$ is defined as:

$$SimZhu2_t(x,y) = \frac{\lambda_t(x,y)}{\max\{|\mathcal{P}_t^x|, |\mathcal{P}_t^y|\}}. \tag{4}$$

The length of the common path is divided by the length of the longest path: this is necessary since, given a fixed λ, the similarity between x and y should be higher if the denominator is closer to λ. The measure is extended to $\mathcal{F}$ in the following way:

$$SimZhu2(x,y) = \frac{\sum_{t \in \mathcal{F}} SimZhu2_t(x,y)}{T} \tag{5}$$

which is simply the average similarity between the two objects. We transform the similarity into a distance as follows:

$$Zhu2(x,y) = 1 - SimZhu2(x,y). \tag{6}$$

3. The variant called *SimZhu3* is a weighted version of *SimZhu2*. Each node is considered to have a depth-based importance since objects which are together in a very deep node are more similar than objects which are together only, for example, in the root. To account for this, in [27] they define the weight of a node k to be $\frac{1}{|k|}$ since smaller nodes are usually deeper in a tree. Therefore given objects x and y the similarity $SimZhu3_t$ in a tree t is:

$$SimZhu3_t(x,y) = \frac{\sum_{k \in \mathcal{P}_t^{(x,y)}} \frac{1}{|k|}}{\sum_{k \in \mathcal{P}_t^b} \frac{1}{|k|} + \frac{1}{|l_t(b)|}} \tag{7}$$

where $b = \underset{x,y}{\arg\max} |\mathcal{P}_t^b|$. The measure is extended to $\mathcal{F}$ in the following way:

$$SimZhu3(x,y) = \frac{\sum_{t \in \mathcal{F}} SimZhu3_t(x,y)}{T}. \tag{8}$$

We transform the similarity into a distance as follows:

$$Zhu3(x,y) = 1 - SimZhu3(x,y). \tag{9}$$

Step 3: Distance-based outlier detection

After having computed $\mathbf{D}$, we can apply any distance-based outlier detection method. Different techniques exist in the literature –for a detailed explanation please refer to [6]. The most simple methods exploit the distance to the kth

neighbor in different ways: an example is *NNd* [25]. NNd states that if the distance between an object and its nearest neighbor is greater than the distance between the latter and its nearest neighbor, then the object under analysis has an increased probability of being an outlier.

Then there are more refined techniques which employ an estimation of the relative density to solve the task, such as the *Local Outlier Factor* (LOF) [5]. LOF works by comparing the neighborhood density of the object under analysis with that of its neighbors. The object has a higher probability of being an outlier if at least one of the neighbors has a denser neighborhood than its own. The classifier has only one parameter to set: K, the neighborhood size. In our work we employ LOF since it is more sophisticated than NNd.

4   Experimental Evaluation

In this Section we first describe the datasets and some experimental details and then we present the obtained results and compare the methodology to the IFs.

4.1   Experimental Details

We evaluate the methodology on 10 UCI ML datasets[1] which were transformed into outlier detection datasets: in all of them nominal attributes were removed. Then the outlier and inlier classes are defined based on previous works: for Breastw, Ionosphere, Pima and Satellite see [16], for Glass and WBC see [15], for Arrhythmia and Wilt refer to [11], for Musk follow [2] and for Letter refer to [20] –as to this dataset further modifications were made other than defining the classes[2]. In Table 1, datasets are described in terms of number of objects, number of features and percentage of outliers. These datasets cover a large range of situations: they differ greatly in the number of features (from 5 up to 164), in the outlier percentage (from 3.17% up to 45.80%) and in the size (the smallest one has 213 samples while the biggest 6435).

After a preliminary evaluation –not shown here–, we chose the following parameters for the IF training: $S = 256, D = \log_2(S), T = 150$. The parameters of the methodology are very easy to set, as shown in [16,18]: indeed we only varied the forest size with respect to the default parametrizations since it shows better performances. Each experiment was repeated 20 times. For each iteration 50% of the objects was randomly assigned to the training set and the other 50% to the testing set, where the former did not contain any outlier.

Given the trained forests, we computed the IF-distances with the three variants described in Sect. 3: *Shi*, *Zhu2* and *Zhu3*. As to the chosen classifier, LOF, after preliminary analyses not shown here, we set $K = 14$ since it allows to achieve the best performances on average. As accuracy measure, as often done in outlier detection, we use the Area under the ROC Curve (AUC).

---

[1] Available at https://archive.ics.uci.edu/ml/index.php.

[2] All datasets adequately processed can be found at http://odds.cs.stonybrook.edu/, except for Arrhythmia for which we use a different version [11].

Table 1. Overview of the 10 datasets used for the experimental evaluation.

Datasets	Nr. of objects	Nr. of features	Outlier %
Arrhythmia	452	164	45.80%
Breastw	683	9	34.99%
Glass	214	9	4.21%
Ionosphere	351	32	35.90%
Letter	1600	32	6.25 %
Musk	3062	166	3.17%
Pima	768	8	34.90%
Satellite	6435	36	31.64%
WBC	378	30	5.56 %
Wilt	4839	5	5.39%

4.2   Results

The first analysis compares the three IF-distance measures we can compute from the Isolation Forests. In Figs. 1 (a), (b) and (c) we present a pairwise comparison between the distances: for each dataset we count for how many experiments the first named measure is better than the second (blue bar), the second is better than the first (orange bar) and for how many experiments the two distance measures perform the same (yellow bar). The green line represents the maximum number of experiments per dataset, which is 20. In Figs. 1 (a) and (c) we compare *Zhu2* with the other distances: its superiority is straightforward. Indeed for all datasets except two it is better than the other distance measures and for one, Musk, the performances are equal. Comparing instead *Zhu3* and *Shi*, in Fig. 1 (b), we can observe that in many cases *Zhu3* is the better choice, except for Breastw, Pima and Wilt which are all rather small; for Musk we observe again that the performances are independent of the used distance measure.

The second analysis compares the proposed methodology with the IF: we present the results in Table 2. For each dataset we report the median across the 20 repetitions. In detail, as to the proposed methodology we report the accuracy achieved with the best distance measure, which is indicated between parenthesis–if *All* is present, it means that all distances lead to the same accuracy. In addition we performed a Wilcoxon signed-rank test to assess whether the differences between the methodologies are statistically significant. The scores in bold are the best ones, and if a * is present, then the difference with the other methodology is statistically significant. From Table 2 we can observe that on seven datasets the best accuracy is reached when using the proposed methodology. In detail in four cases it is achieved when using *Zhu2* as distance measure and for four out of these seven datasets the difference is statistically significant –for Wilt and Letter the improvement is remarkable. In addition even though for the remaining datasets IF is significantly better, only for Breastw the proposed methodology actually fails. Finally if we observe the average results across all

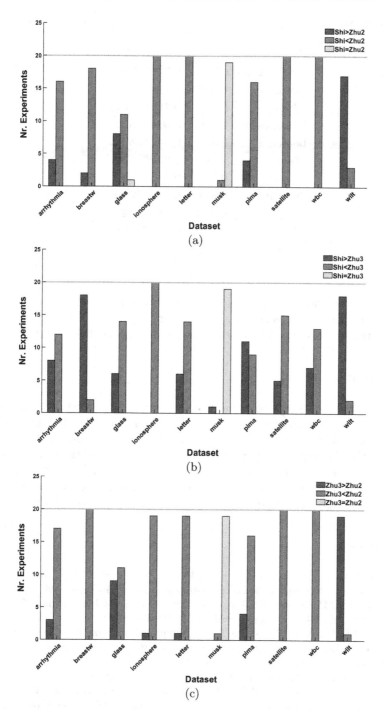

Fig. 1. Comparison between the proposed distances. Respectively each figure compares (a) Shi with Zhu2 (b) Shi with Zhu3 and (c) Zhu3 with Zhu2.

Table 2. Accuracy comparison between the IF and the proposed methodology.

Dataset	IF	LOF (Best Dist.)
Arrhythmia	0.773	**0.778(Shi)**
Breastw	**0.995***	0.582(Zhu2)
Glass	0.729	**0.733(Zhu2)**
Ionosphere	0.894	**0.906(Zhu2)**
Letter	0.641	**0.861*(Zhu2)**
Musk	0.988	**1.000*(All)**
Pima	**0.738***	0.696(Zhu2)
Satellite	0.810	**0.840*(Zhu2)**
Wbc	**0.954***	0.938(Zhu2)
Wilt	0.516	**0.903*(Shi)**
Average	0.804	**0.824**

the datasets the maximum accuracy is reached with the proposed technique. We can thus conclude it is advantageous to employ the IF-distance: this is particularly true if the dataset is big enough, i.e. if it has > 1000 objects.

5   Conclusions

In this paper we propose a novel methodology for outlier detection that exploits Isolation Forests. From the latter we extract a distance matrix which is then input to an outlier detector: the novel representation should be able to meaningfully describe the objects and identify the outliers, thanks to the intrinsic nature of the trees composing the forest. We employed different RF-based distance measures and evaluate the methodology on ten datasets: the proposed technique has been proven to be advantageous with respect to using Isolation Forests alone.

References

1. Abba, M.C., et al.: Breast cancer molecular signatures as determined by sage: correlation with lymph node status. Mol. Cancer Res. **5**(9), 881–890 (2007)
2. Aggarwal, C.C., Sathe, S.: Theoretical foundations and algorithms for outlier ensembles. SIGKDD Explor. Newsl. **17**(1), 24–47 (2015)
3. Bicego, M., Escolano, F.: On learning random forests for random forest-clustering. In: Proceedings of the 25th International Conference on Pattern Recognition, Forthcoming (2021)
4. Breiman, L.: Random forests. Mach. Learn. **45**(1), 5–32 (2001)
5. Breunig, M.M., Kriegel, H.P., Ng, R.T., Sander, J.: LOF: identifying density-based local outliers. In: Proceedings of SIGMOD International Conference on Managing Data, pp. 93–104 (2000)

6. Chandola, V., Banerjee, A., Kumar, V.: Anomaly detection: a survey. ACM Comput. Surv. **41**(3), 1–58 (2009)
7. Désir, C., Bernard, S., Petitjean, C., Heutte, L.: One class random forests. Pattern Recogn. **46**, 3490–3506 (2013)
8. Ding, Z., Fei, M.: An anomaly detection approach based on isolation forest algorithm for streaming data using sliding window. IFAC Proc. **46**(20), 12–17 (2013)
9. Emmott, A.F., Das, S., Dietterich, T., Fern, A., Wong, W.K.: Systematic construction of anomaly detection benchmarks from real data. In: Proceedings of SIGKDD Workshop Outlier Detection and Description, pp. 16–21 (2013)
10. Geurts, P., Ernst, D., Wehenkel, L.: Extremely randomized trees. Mach. Learn. **63**(1), 3–42 (2006)
11. Goix, N., Drougard, N., Brault, R., Chiapino, M.: One class splitting criteria for random forests. In: Proceedings of 9th Asian Conference Machine Learning, vol. 77, pp. 343–358 (2017)
12. Gray, K.R., Aljabar, P., Heckemann, R.A., Hammers, A., Rueckert, D.: Random forest-based similarity measures for multi-modal classification of Alzheimer's disease. NeuroImage **65**, 167–175 (2013)
13. Guha, S., Mishra, N., Roy, G., Schrijvers, O.: Robust random cut forest based anomaly detection on streams. In: Proceedings of the 33rd International Conference on Machine Learning, vol. 48, pp. 2712–2721 (2016)
14. Hariri, S., Kind, M.C., Brunner, R.J.: Extended isolation forest (2018). arXiv:1811.02141
15. Keller, F., Muller, E., Bohm, K.: HICS: high contrast subspaces for density-based outlier ranking. In: IEEE International Conference on Data Engineering, pp. 1037–1048. IEEE (2012)
16. Liu, F.T., Ting, K.M., Zhou, Z.H.: Isolation forest. In: IEEE International Conference on Data Mining, pp. 413–422 (2008)
17. Liu, F.T., Ting, K.M., Zhou, Z.H.: On detecting clustered anomalies using sciforest. In: ECML PKDD, pp. 274–290 (2010)
18. Liu, F.T., Ting, K.M., Zhou, Z.H.: Isolation-based anomaly detection. ACM Trans. Knowl. Discov. Data **6**(1), 1–39 (2012)
19. Mensi, A., Bicego, M.: A novel anomaly score for isolation forests. In: International Conference on Image Analysis and Processing, pp. 152–163 (2019)
20. Micenková, B., McWilliams, B., Assent, I.: Learning outlier ensembles: the best of both worlds-supervised and unsupervised. In: Proceedings of SIGKDD Workshop on Outlier Detection and Description, pp. 51–54 (2014)
21. Rennard, S., et al.: Identification of five chronic obstructive pulmonary disease subgroups with different prognoses in the eclipse cohort using cluster analysis. Ann. Am. Thorac. Soc. **12**(3), 303–312 (2015)
22. Shi, T., Seligson, D., Belldegrun, A., Palotie, A., Horvath, S.: Tumor classification by tissue microarray profiling: random forest clustering applied to renal cell carcinoma. Modern Pathol. **18**, 547–557 (2005)
23. Shi, T., Horvath, S.: Unsupervised learning with random forest predictors. J. Comput. Graph. Stat. **15**, 1–21 (2006)
24. Susto, G.A., Beghi, A., McLoone, S.: Anomaly detection through on-line isolation forest: an application to plasma etching. In: Annual SEMI Advanced Semiconductor Manufacturing Conference (2017)
25. Tax, D.: One-class classification; concept-learning in the absence of counterexamples. Ph.D. thesis, Delft University of Technology (2001)

26. Ting, K., Zhu, Y., Carman, M., Zhu, Y., Zhou, Z.H.: Overcoming key weaknesses of distance-based neighbourhood methods using a data dependent dissimilarity measure. In: Proceedings of International Conference on Knowledge Discovery and Data Mining, pp. 1205–1214 (2016)
27. Zhu, X., Loy, C., Gong, S.: Constructing robust affinity graphs for spectral clustering. In: Proceedings of International Conference on Computer Vision and Pattern Recognition, pp. 1450–1457 (2014)

Exponential Weighted Moving Average of Time Series in Arbitrary Spaces with Application to Strings

Alexander Welsing, Andreas Nienkötter, and Xiaoyi Jiang[(✉)] [iD]

Faculty of Mathematics and Computer Science, University of Münster,
Einsteinstrasse 62, 48149 Münster, Germany
xjiang@uni-muenster.de

Abstract. The exponentially weighted moving average (EWMA) is an important tool in time series analysis. So far the research on EWMA is typically limited to the real (vector) space $\mathbb{R}^n$. In this work we present an extension of this concept to arbitrary spaces. It is based on an interpretation of EWMA as a special case of weighted mean computation. We develop three computation methods. In addition to the direct computation in the original space, we particularly study an approach to embedding the data items of a time series into vector space. The feasibility of our EWMA computation framework is exemplarily demonstrated on strings.

1 Introduction

Time series data are omnipresent in the world. The analysis of such data leads to unique problems and solutions [17] that, amongst others, should take the inherent correlation of adjacent data items into account. There are several needs to clean time series data, including dealing with data missing, data inconsistency [3] and data errors [19].

The focus of this work is time series data smoothing for the latter purpose. The weighted moving average gives different weights to data at different positions in the sample window. Mathematically, it corresponds to the convolution of the data items with a fixed weighting function. The exponentially weighted moving average (EWMA) [7] is a popular technique of this class with the weighting for each older datum decreasing exponentially. Given an original noisy time series s_t, the EWMA results in a smoothed time series v_t and is formally defined by:

$$v_{t+1} \;=\; \beta \cdot v_t + (1 - \beta) \cdot s_{t+1}, \quad \beta \in [0, 1] \tag{1}$$

where the parameter β rules the degree of weighting decrease and a lower β discounts older data items faster. Commonly, the computation is initialized by setting $v_1 = s_1$. The $(t+1)$-th element of the smoothed series depends on all the previous data items of the original series:

$$v_{t+1} \;=\; \beta^t s_1 + \sum_{k=0}^{t-1} \beta^k (1 - \beta) \cdot s_{t-k+1}$$

© Springer Nature Switzerland AG 2021
A. Torsello et al. (Eds.): S+SSPR 2020, LNCS 12644, pp. 45–54, 2021.
https://doi.org/10.1007/978-3-030-73973-7_5

where β^n $(n = t, k)$ represents the nth power of β. EWMA has a broad range of applications. An important example is the momentum for gradient descent optimization [13], which is also beneficial for training neural networks.

So far the research on EWMA is typically limited to the real (vector) space $\mathbb{R}^n$. In this work we study the general case of EWMA in arbitrary spaces. Based on an interpretation of Eq. (1) as a special case of the so-called weighted mean (to be detailed in Sect. 2), we develop three methods for computing the related weighted mean. The feasibility of our EWMA computation framework is exemplarily demonstrated on strings. Potential applications include bioinformatics ("denoising" sequences of proteins belonging to the same family) and video analysis (e.g. smoothing sequences of gradually changing actions).

The remainder of the paper is organized as follows. We first formally define the problem of EWMA in arbitrary spaces in the next section. Then, Sect. 3 presents the fundamentals for the validation of our approach on strings. Section 4 describes three EWMA computation methods. For each method we present the related experimental results. We choose this somewhat unconventional structure (instead of presenting all experimental results in a single section) since our second (third) algorithmic variant is motivated by the experimental results and discussion of the first (second) method, respectively. For this reason we also describe the test data used in our experiments at the very beginning in Sect. 3. Finally, Sect. 5 concludes the paper.

2 Formal Problem Definition

Given an arbitrary space $\mathcal{O}$ with a distance function $\delta(o_i, o_j)$, the fundamental concept of weighted mean $\tilde{o}$ between objects $o_1, o_2 \in \mathcal{O}$ with ratio $\alpha \in [0, 1]$ is defined by:

$$\delta(o_1, \tilde{o}) = \alpha \cdot \delta(o_1, o_2), \qquad \delta(\tilde{o}, o_2) = (1 - \alpha) \cdot \delta(o_1, o_2) \qquad (2)$$

In other words, the weighted mean $\tilde{o}$ is a linear interpolation between objects o_1 and o_2. In many cases, this weighted mean can be derived from the distance function δ. Examples include strings [2], graphs [1] (adaptable to trees), graph correspondences [12], clustering [6] and bi-clusterings [14]. Note that although the relation between distances is theoretically exact in Eq. (2), it is not always possible to compute $\tilde{o}$ with these exact distances. One example is the Levenshtein string edit distance [18] (see Sect. 3.1) that usually returns a natural number (in case of integer costs of edit operations), while α is typically a rational number. In these cases, a string is returned that approximates this equation.

By replacing o_1 with v_t and o_2 with s_{t+1} in Eq. (2), it becomes clear that the EWMA v_{t+1} in Eq. (1) is simply the weighted mean of v_t and s_{t+1} with $\alpha = 1 - \beta$,

$$\delta(v_t, v_{t+1}) = (1 - \beta) \cdot \delta(v_t, s_{t+1}), \qquad \delta(v_{t+1}, s_{t+1}) = \beta \cdot \delta(v_t, s_{t+1})$$

Thus, the general case of EWMA in arbitrary spaces, i.e. computing the smoothed series v_{t+1}, is converted to that of weighted mean computation.

Weighted mean computation [1, 2, 6, 12] belongs to those efforts that extend popular concepts from vector space to arbitrary spaces. Another such example is the generalized median computation [14]. Our current work follows this research direction and studies times series smoothing.

3   Fundamentals of This Paper

3.1   String Edit Distance and Weighted Mean Computation

Strings are a fundamental representation in structural pattern recognition. In order to apply our approach to strings two requirements have to be fulfilled. First, a suitable distance function is needed to compare strings. Here, we use the popular Levenshtein string edit distance [18].

Secondly, the weighted mean based on the edit distance has to be determined. Given the optimal edit sequence from two strings x to z, the fundamental idea behind the algorithm [2] for computing the weighted mean y with the desired fraction a of the total distance $\delta(x, z)$ is as follows. Select a subset of all the edit operations so that the sum of their costs (approximately) amounts to a. The selected subset is applied to x (the remaining cost $\delta(x, z) - a$ is then realized by the remaining edit operations for obtaining z from y). Note that the weighted mean is generally not unique. However, it can be made unique if we select a subset of total cost a based on the optimal edit operations between x and z in their natural order from the edit distance computation.

3.2   Test Data

We use synthetic data to evaluate our algorithmic variants (and in fact to motivate the algorithmic development). Two strings g_1 and g_n ($n = 200$) of length 500 each were chosen at random from the public domain book "Robinson Crusoe" by Daniel Defoe. String g_1 was then gradually transformed into g_n using weighted mean computation and equally sampled α values. The series of strings $g_1, g_2, \ldots, g_n$ are guaranteed to be gradually changing and considered as ground truth (GT). A second, distorted series of strings $s_1, s_2, \ldots, s_n$ was created. For this, each symbol of each string was changed with a probability of 12%. A change had a 9% chance to be a deletion, a 4% chance to be an insertion and an 87% chance to be a substitution. The purpose of EWMA is to smooth this series of strings to achieve an output series of strings with smaller edit distances to GT compared to the input series of strings. In total we generated 20 such GT series and their distorted version. Experiments were done for the 20 distorted series and average results are reported later in the paper.

4   EWMA Computation Methods

4.1   Direct Method

The formal definition presented in Sect. 2 immediately leads to a direct EWMA computation method. Given a time series s_t in space $\mathcal{O}$ with distance function

Algorithm 1. EWMA computation: Direct method

Input: Distorted series $s_1, s_2, \ldots, s_n$ ($s_k \in$ arbitrary space $\mathcal{O}$)
Output: Smoothed series $v_1, v_2, \ldots, v_n$
1: $v_1 = s_1$
2: **for** $t = 1$ to $n - 1$ **do**
3:    Compute v_{t+1} as weighted mean of v_t and s_{t+1} according to Eq. (1)
4: **end for**

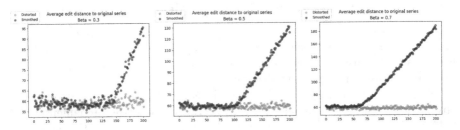

Fig. 1. Direct method: Average edit distance between distorted string s_t and GT g_t (red), average edit distance between smoothed string v_t and GT g_t (blue). (Color figure online)

$\delta(o_i, o_j)$, we determine the $(t + 1)$-th element of the smoothed series v_{t+1} by computing the weighted mean between v_t and s_{t+1} using some pre-specified parameter β. The pseudo-code of this method is given in Algorithm 1.

The results on the synthetic data as described in Sect. 3.2 are presented in Fig. 1 for three β values 0.3, 0.5 and 0.7. For each input string s_t, $t = 1, 2, \ldots, 200$, the figure shows the edit distance $\delta(s_t, g_t)$ (between the distorted string s_i and the related GT g_t) in red and $\delta(v_t, g_t)$ (between the smoothed string v_t and GT g_t) in blue. In each case the average over the 20 distorted series (of the same length of 200) is reported.

These results indicate that the direct method may be ineffective. The edit distance between the smoothed strings and GT sharply increases at a certain point, after initially being slightly higher than the distorted series. For $\beta = 0.7$, for instance, after about 30% of the steps has been smoothed, the quality of the smoothed series deteriorates quickly.

The reason for this is the way how the weighted mean of strings is generated. The original time series is created by increasing α. Each smoothed string v_t for $t > 1$ is created by using the first β percent of the symbols from s_t and the last $(1 - \beta)$ percent of symbols from v_{t-1}, meaning that the last $(1 - \beta)$ of s_n are never used in the smoothed string. After β percent has been smoothed, these symbols make up an increasing part of each step s_t, thus increasing the edit distance between the smoothed and original strings.

4.2   EWMA Computation: Vector Space Embedding Method

In contrast to general complex spaces, the vector space is much easier to handle, where many operations have simple solutions. This observation motivates a

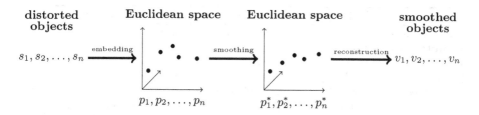

Fig. 2. Overview of vector space embedding.

universal vector space embedding approach consisting of three steps (see Fig. 2 for an illustration):

- Embedding: The input data items are embedded to points in a vector space.
- Operation: The operation at hand is performed on the embedded points.
- Reconstruction: The result is transformed back into the original space.

In structural pattern recognition, this problem-solving strategy has been used to solve a variety of instances of high-complex, partly NP-hard, generalized median problems [5,9,14]. This is due to the fact that the generalized median in vector space can be efficiently determined by applying the well-known Welszfeld algorithm [20].

Also the problem of EWMA computation is easily solvable in vector space based on the computation scheme in Eq. (1). Thus, we study this problem-solving strategy by embedding the distorted objects into vector space. The details of the two key steps embedding and reconstruction are described as follows. The pseudo-code of this method is presented in Algorithm 2.

Vector Space Embedding. In the past this embedding is commonly performed by the prototype embedding technique [5,9] using selected prototype objects. Intuitively, one would expect that the embedding should be done in a distance-preserving manner, i.e. find points $x_1, \ldots, x_n \in \mathbb{R}^d$ such that

$$\delta_e(x_i, x_j) = c \cdot \delta(s_i, s_j), \ \forall \ 1 \leq i, j \leq n, \ c > 0$$

with δ_e being the Euclidean distance between two points. Ideally, the scaling factor $c \in \mathbb{R}$ should be 1 to ensure an exact representation of the objects in vector space. The EWMA problem (likewise the generalized median), however, is scale invariant. Thus, we allow a constant scaling factor c in the formulation above. The rationale behind the distance-preserving requirement is to preserve the original object configuration in vector space. Distance distortion in the embedding process leads to distorted configuration in vector space and thus a "biased" solution vector, which in turn will result in a similarly "biased" operation (e.g. EWMA computation) in the original space. The recent work [14] shows that the commonly used prototype embedding technique does not satisfy the distance-preserving requirement well. Several distance-preserving embedding schemes were studied, which led to strong evidence of significantly improved

Algorithm 2. EWMA computation: Vector space embedding method

Input: Distorted series $s_1, s_2, \ldots, s_n$ ($s_k \in$ arbitrary space $\mathcal{O}$)
Output: Smoothed series $v_1, v_2, \ldots, v_n$
  1: $v_1 = s_1$
  2: **for** $t = 1$ to $n-1$ **do**
  3:    Embed $s_1, \ldots, s_{t+1}$ to $p_1, \ldots, p_{t+1}$ in vector space
  4:    Compute the smoothed series $p_1^*, \ldots, p_{t+1}^*$ in vector space directly using Eq. (1)
  5:    Reconstruct v_{t+1} by transforming p_{t+1}^* back to the original space
  6: **end for**

---

quality of generalized median computation using distance-preserving embedding. In particular, the curvilinear component analysis [4] turns out to be an excellent choice. Thus, we will use this embedding method for our work here. Note that in the formulation in Algorithm 2, all data items $s_1, \ldots, s_{t+1}$ are used for the embedding. This reflects our current implementation, but can be modified to only incrementally embed the new data item s_{t+1} while freezing the embedding of the previous data items $s_1, \ldots, s_t$.

Reconstruction. The last step is to transform the smoothed points from vector space back into the original space. Several inverse transformation methods have been studied before [5,14]. Although this was done in the context of generalized median computation, these transformation methods are of rather general nature and can be applied for other tasks. They need a number of corresponding pairs (s_i, p_i), where p_i represents the embedded point of distorted object s_i in vector space. In our case such pairs are given by the input distorted objects and their embedded points and can be used for transforming the smoothed points back to the original space. For the experimental work we apply the triangular reconstruction method. We refer to [5,14] for technical details of this and other reconstruction methods. In particular, we use the publicly available toolbox[1] implementing these reconstruction methods as described in [14] for our work.

With increasing time t, the number of corresponding pairs (s_i, p_i) used for the reconstruction increases accordingly. To further augment such pairs, we also study an additional variant of the vector space embedding method by including the pairs (v_i, p_i^*) resulting from the reconstructions so far. Here p_i^* represents the smoothed version of some embedding p_i and v_i its reconstruction.

In the following we report the experimental results of the vector space embedding method using the same test data as in Sect. 4.1. The dimension of the embedded vector space is set to $\lfloor 0.7 \cdot \min(|s_1|, |s_2|, \ldots, |s_n|) \rfloor$, where $|s_k|$ represents the length of string s_k. The study [14] indicated that a factor of 0.7 to 0.9 is a good choice for the dimension of embedding space. We thus used 0.7 in this work. For visualization purposes, we show the first two dimensions of the embeddings in Fig. 3 (1st row). The decrease in distortion for the smoothed embedded series is clearly visible in this figure with the smoothed points being much closer to the original series. The 2nd row shows the Euclidean distance between the

---
[1] https://www.uni-muenster.de/PRIA/forschung/dpe.html.

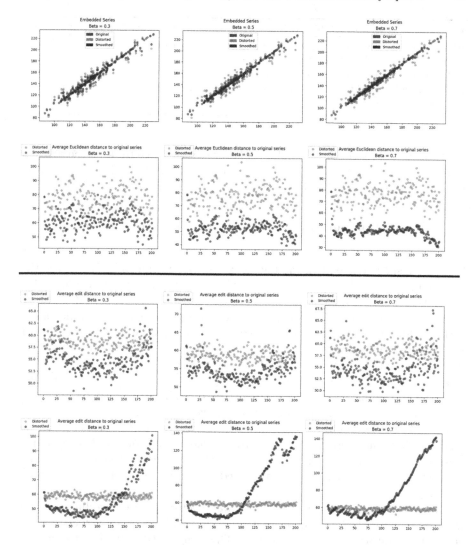

Fig. 3. Vector space embedding method. 1st row: visualization of two dimensions of the embeddings. 2nd row: average Euclidean distance between embedding of distorted string s_t and GT g_t (red), average Euclidean distance between embedding of smoothed string v_t and GT g_t (blue). 3rd/4th row: average edit distance between distorted string s_t and GT g_t (red), average edit distance between smoothed string v_t and GT g_t (blue) for the standard variant and the additional variant with (v_i, p_i^*). (Color figure online)

different series, where we can clearly see clusters separating the embedding of distorted and smoothed strings. Similar separation can be observed in the 3rd row in the original string space, although somewhat weaker than in the vector space in terms of the Euclidean distance for $\beta = 0.7$. For the additional variant with (v_i, p_i^*) the behavior differs considerably (4th row). For most β values

---

Algorithm 3. EWMA computation: Two-way vector space embedding method

---

Input: Distorted series $s_1, s_2, \ldots, s_n$ ($s_k \in$ arbitrary space $\mathcal{O}$)
Output: Smoothed series $v_1, v_2, \ldots, v_n$
 1: Compute the forward smoothed series $f_1, f_2, \ldots, f_n$ from $s_1, s_2, \ldots, s_n$
 2: Compute the backward smoothed series $b_n, \ldots, b_2, b_1$ from $s_n, \ldots, s_2, s_1$
 3: $v_1, v_2, \ldots . v_n = (f_1, \ldots, f_{n/2}, b_{n/2+1}, \ldots, b_n)$

---

the reconstructed series initially show a lower edit distance towards the original series than the standard variant. From the middle of the series, however, an increase in edit distance similar to the direct method in Fig. 1 is observable.

4.3   EWMA Computation: Two-Way Vector Space Embedding Method

The vector space embedding method is certainly an improvement against the direct method. Looking at the results of the variant with (v_i, p_i^*) in Fig. 3 (4th row) reveals that approximately the first half is better reconstructed than the second half. The reason for this phenomenon, in particular the distance spike in the middle, is related to the reconstruction and rather complex (see [21] for more details). This motivates us to suggest a two-way vector space embedding method. In contrast to the methods before, this is an off-line approach. That is, we assume to know the complete series $s_1, s_2, \ldots, s_n$. In a first step, we process the series to obtain a forward smoothed series. Similarly, we also consider the reversed series $s_n, \ldots, s_2, s_1$ to obtain a backward smoothed series. Then, we concatenate the first half of the forward smoothed series and the first half of the backward smoothed series (in reverse order). The pseudo-code of this method is presented in Algorithm 3 for even n. For uneven n, the element in the middle of either the forward or backward smoothed series can be inserted into the final smoothed series.

The experimental results of this two-way vector space embedding method are shown in Fig. 4. The two series of distorted and smoothed strings form a clear separation with almost no overlap. The steep increase in edit distance as observed before is no longer visible. A spike in edit distance, however, is notable in the middle of the series.

4.4   Discussion

All the embedding methods discussed before (prototype embedding approach [5] and distance-preserving embedding methods studied in [14]) are based on explicit transformations into the vector space. In contrast, it is also possible to use implicit transformations in terms of kernel functions. Kernel functions are well-known from their application in kernel machine [11], support vector machine, clustering, principal component analysis, etc. [16]. The development of kernel functions on non-vectorial data, in particular structural data (strings,

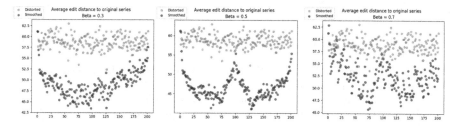

Fig. 4. Two-way vector space embedding method: Average edit distance between distorted string s_t and GT g_t (red), average edit distance between smoothed string v_t and GT g_t (blue). (Color figure online)

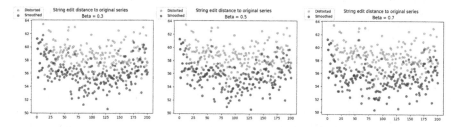

Fig. 5. Kernel method: Average edit distance between distorted string s_t and GT g_t (red), average edit distance between smoothed string v_t and GT g_t (blue). (Color figure online)

trees, graphs) [8, 10, 22], allows investigating kernel-based EWMA computation. We also studied this idea [21], which is based on the - rather complex - technique recently developed in [15]. The preliminary results as reported in [21] are comparable with the direct method, see Fig. 5. This is a direction worth further study in future.

5 Conclusion

So far the research on EWMA is typically limited to the real (vector) space $\mathbb{R}^n$. In this paper we have extended the EWMA computation to arbitrary spaces. To our knowledge this is the first extension of this kind reported in the literature. It is based on weighted mean computation. In addition to the direct computation in the original space, we particularly studied the vector space embedding approach. The feasibility of our EWMA computation framework was exemplarily demonstrated on strings.

The current experimental work provides an indication of the general feasibility of our approach to smoothing time series in arbitrary spaces. Future research will extend the experimental basis, especially on additional spaces.

References

1. Bunke, H., Günter, S.: Weighted mean of a pair of graphs. Computing **67**(3), 209–224 (2001)
2. Bunke, H., Jiang, X., Abegglen, K., Kandel, A.: On the weighted mean of a pair of strings. Pattern Anal. Appl. **5**(1), 23–30 (2002)
3. Chu, X., Ilyas, I.F., Krishnan, S., Wang, J.: Data cleaning: overview and emerging challenges. In: Proceedings of International Conference on Management of Data (SIGMOD), pp. 2201–2206 (2016)
4. Demartines, P., Hérault, J.: Curvilinear component analysis: a self-organizing neural network for nonlinear mapping of data sets. IEEE Trans. Neural Netw. **8**(1), 148–154 (1997)
5. Ferrer, M., Valveny, E., Serratosa, F., Riesen, K., Bunke, H.: Generalized median graph computation by means of graph embedding in vector spaces. Pattern Recogn. **43**(4), 1642–1655 (2010)
6. Franek, L., Jiang, X., He, C.: Weighted mean of a pair of clusterings. Pattern Anal. Appl. **17**(1), 153–166 (2014)
7. Gardner, E.S.: Exponential smoothing: the state of the art - Part II. Int. J. Forecast. **22**(4), 637–666 (2006)
8. Gärtner, T.: Kernels for Structured Data. World Scientific, Singapore (2008)
9. Jiang, X., Wentker, J., Ferrer, M.: Generalized median string computation by means of string embedding in vector spaces. Pattern Recogn. Lett. **33**(7), 842–852 (2012)
10. Kriege, N.M., Johansson, F.D., Morris, C.: A survey on graph kernels. Appl. Netw. Sci. **5**(1), 6 (2020)
11. Ma, S., Belkin, M.: Kernel machines that adapt to GPUs for effective large batch training. In: Proceedings of 2nd SysML Conference (2019)
12. Moreno-García, C.F., Serratosa, F., Jiang, X.: Correspondence edit distance to obtain a set of weighted means of graph correspondences. Pattern Recogn. Lett. **134**, 29–36 (2020)
13. Nakerst, G., Brennan, J., Haque, M.: Gradient descent with momentum - to accelerate or to super-accelerate? arXiv abs/2001.06472 (2020)
14. Nienkötter, A., Jiang, X.: Distance-preserving vector space embedding for consensus learning. IEEE Trans. Syst. Man Cybern.: Syst. **51**(2), 1244–1257 (2021)
15. Nienkötter, A., Jiang, X.: Kernel-based generalized median computation for consensus learning (2021). Under review
16. Shawe-Taylor, J., Cristianini, N.: Kernel Methods for Pattern Analysis. Cambridge University Press, Cambridge (2004)
17. Shumway, R.H., Stoffer, D.S.: Time Series Analysis and Its Applications: With R Examples. Springer, Heidelberg (2017). https://doi.org/10.1007/978-3-319-52452-8
18. Wagner, R.A., Fischer, M.J.: The string-to-string correction problem. J. ACM **21**(1), 168–173 (1974)
19. Wang, X., Wang, C.: Time series data cleaning: a survey. IEEE Access **8**, 1866–1881 (2020)
20. Weiszfeld, E., Plastria, F.: On the point for which the sum of the distances to n given points is minimum. Ann. Oper. Res. **167**(1), 7–41 (2009)
21. Welsing, A.: Moving average for time series of strings. Bachelor Thesis, University of Münster (2020)
22. Xu, L., Bai, L., Jiang, X., Tan, M., Zhang, D., Luo, B.: Deep Rényi entropy graph kernel. Pattern Recogn. **111**, 107668 (2021)

Experimental Analysis of Bidirectional Pairwise Ordinal Classifier Cascades

Peter Bellmann[1(✉)], Ludwig Lausser[2], Hans A. Kestler[2],
and Friedhelm Schwenker[1]

[1] Institute of Neural Information Processing, Ulm University,
James-Franck-Ring, 89081 Ulm, Germany
{peter.bellmann,friedhelm.schwenker}@uni-ulm.de
[2] Institute of Medical Systems Biology, Ulm University, Albert-Einstein-Allee 11,
89081 Ulm, Germany
{ludwig.lausser,hans.kestler}@uni-ulm.de

Abstract. Ordinal classifier cascades (OCCs) are basic machine learning tools in the field of ordinal classification (OC) that consist of a sequence of classification models (CMs). Each of the CMs is trained in combination with a specific subtask of the initial OC task. OCC architectures make use of a data set's ordinal class structure by simply arranging the CMs with respect to the corresponding class order (e.g., small - medium - large). Recently, we proposed bidirectional OCC (bOCC) architectures that combine two basic one-directional OCCs, based on a person-independent pain intensity recognition scenario, in combination with support vector machines. In the current study, we further analyse the effectiveness of bOCC architectures. To this end, we evaluate our proposed approach based on different OC benchmark data sets. Additionally, we analyse the proposed bOCCs in combination with two different classification models. Our outcomes indicate that it seems to be beneficial to replace basic pairwise one-directional OCCs by the pairwise bOCC architecture, in general.

Keywords: Ordinal classification · Ordinal classifier cascades

1 Introduction

Ordinal classification (OC) is an important field of supervised learning. In OC tasks, it is assumed that the classes constitute an ordinal structure. For instance, different medical applications represent interesting examples of OC tasks, such as the classification of pain intensities. More precisely, based on the pain intensity recognition example, the classes could constitute the following structure, *no pain ≺ low pain ≺ intermediate pain ≺ severe pain* [17].

Ordinal classifier cascades (OCCs) constitute basic tools for OC tasks [11]. An OCC system is a classification architecture that consists of a *sequence* of classification models (CMs). Each input sample is processed in sequential manner. The architecture's final decision is based on the prediction of one single CM

© Springer Nature Switzerland AG 2021
A. Torsello et al. (Eds.): S+SSPR 2020, LNCS 12644, pp. 55–64, 2021.
https://doi.org/10.1007/978-3-030-73973-7_6

from the sequence (see Sect. 2, for details). Thus, OCCs belong to the category of *selection ensembles* [12].

Basic OCCs are simple methods that benefit from the class structure. There is no need for an additional combination rule, as it is the case in common *decision fusion*-based classifier ensembles [5,10]. However, there exist strongly performing alternatives, such as the *error correcting output codes* (ECOC) [8], including the *one-versus-one* approach. Therefore, OCC architectures became popular tools for the *detection* of ordinal class structures [13,16], and substructures [15].

Aiming to improve the classification performance of basic OCCs, recently, we proposed a straightforward combination of OCC architectures [3]. Based on the BioVid Heat Pain Database [19], in [3], we provided a short ablation study on basic OCC architectures, in combination with support vector machines [18]. Our outcomes showed that the addition of one single classification model that is used to combine two OCCs significantly improved the corresponding person-independent pain intensity classification performance, with respect to the basic one-directional OCCs.

In the current study, we want to further analyse our recently proposed OCC modification. To this end, we provide an experimental evaluation in combination with different publicly available OC benchmark data sets. Moreover, we extend the numerical analysis by additionally evaluating each of the OCC architectures in combination with a decision tree classifier [6].

The remainder of the current work is organised as follows. In Sect. 2, we briefly summarise the functionality of OCC architectures. Subsequently, we provide our proposed design for the combination of basic OCCs. The data sets that will be evaluated in the experiments are described in Sect. 3. In Sect. 4, we first provide the details on the evaluation protocol, including the choice of performance measures and implementation software. Subsequently, we provide and discuss the experimental outcomes. Finally, in Sect. 5, we conclude the current study.

2   Ordinal Classifier Cascades

In the current section, we first provide the formalisation, followed by a brief introduction of different OCC approaches. Subsequently, we recap our proposed design of simply combined OCC architectures.

2.1   Formalisation

Let $\Omega = \{\omega_1, \dots, \omega_c\}$, $c \in \mathbb{N}$, $c > 2$, be the class label set of an ordinal c-class classification task. Without loss of generality, we assume that the class labels constitute an ordinal class structure with respect to the order $\omega_1 \prec \dots \prec \omega_c$. We denote the class labels ω_1 and ω_c as *edge classes*, or simply *edges*.

By $X \subset \mathbb{R}^d$, $d \in \mathbb{N}$, we denote the d-dimensional training set of the corresponding classification task. By $l(x)$, we denote the true label of data point

$x \in X$. Moreover, by $\mathrm{CM}_{i,j}$, we denote the classification model that is trained in combination with the set $X_{i,j} \subset X$, which is defined as follows,

$$X_{i,j} := \{x \in X : l(x) = \omega_i \vee l(x) = \omega_j\}, \quad \forall i,j \in \{1,\dots,c\}, \ (i \neq j).$$

Note that in the current work, we assume *symmetrical* classification models, i.e. it holds, $\mathrm{CM}_{i,j} = \mathrm{CM}_{j,i}$, for all $i,j \in \{1,\dots,c\}$.

Ordinal classifier cascades consist of a sequence of $c-1$ classification models. Each of the models is trained in combination with a binary subtask of the initial multi-class task. A model's output either provides the architecture's final decision, or indicates that the subsequent model has to process the current input sample. Note that the last model in the sequence always provides the architecture's final decision.

2.2 Current vs. Next and Current vs. Previous Ordinal Cascades

In *pairwise* OCCs, each model $\mathrm{CM}_{i,j}$ is trained in combination with one *pair* of *neighbouring* classes. In [3], we confirmed that the *direction* of an OCC can have a significant impact on the architecture's output distribution. Therefore, we divide the pairwise OCC approach into the following two architectures. In the *current vs. next* (CvsN) OCC architecture, the i-th model in the sequence is trained to separate the classes ω_i and ω_{i+1}. In the *current vs. previous* (CvsP) OCC architecture, the i-th model in the sequence is trained to separate the classes ω_{c-i+1} and ω_{c-i}. Therefore, the CvsN architecture consists of the model sequence $\mathrm{CM}_{1,2} \rightarrow \mathrm{CM}_{2,3} \rightarrow \dots \rightarrow \mathrm{CM}_{c-1,c}$. Analogously, the CvsP architecture consists of the model sequence $\mathrm{CM}_{c,c-1} \rightarrow \mathrm{CM}_{c-1,c-2} \rightarrow \dots \rightarrow \mathrm{CM}_{2,1}$. Both architectures are depicted in Fig. 1.

Note that both architectures, i.e. CvsN and CvsP, consist of the same set of classification models, due to the *symmetry* of the models, as discussed above. Therefore, the only difference between both models, which can lead to significant performance changes, is based on the architecture's direction, i.e. the choice of the *starting edge*.

2.3 Bidirectional Pairwise Ordinal Classifier Cascades

In [3], we proposed to complement the set of classification models of pairwise OCCs by a *selector component* that is trained in combination with the two edge classes, i.e. $\mathrm{CM}_{1,c}$. Each input sample, $z \in \mathbb{R}^d$, is first processed by the selector model. In the case that the output $\mathrm{CM}_{1,c}(z)$ is equal to ω_1, the input is moved to the CvsN sequence. In contrast, if the output $\mathrm{CM}_{1,c}(z)$ is equal to ω_c, the current input is moved to the CvsP sequence. Figure 2 depicts the bidirectional OCC (bOCC) architecture for the pairwise approach.

Note that there also exist different non-pairwise OCC architectures such as the *current vs. higher* approach [3], in which the first CM separates the class ω_1 from the classes $\{\omega_2,\dots,\omega_c\}$. However, the outcomes in [13] show that pairwise OCC approaches outperform the non-pairwise OCCs, with respect to classification performance, in general. Thus, in accordance to the experiments in [3], in the current study, we focus on the pairwise (b)OCC architectures.

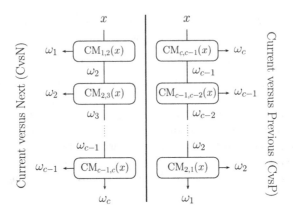

Fig. 1. Ordinal Classifier Cascades (Pairwise). Left: Current vs. Next (CvsN). Right: Current vs. Previous (CvsP). x: Input. c: Number of classes. ω_i: Class labels. $CM_{i,j}$: Classification model that separates the classes ω_i and ω_j. Arrowheads indicate that the corresponding output is taken as the architecture's final prediction.

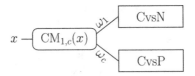

Fig. 2. Bidirectional Ordinal Classifier Cascade Architecture (Pairwise). x: Input. $CM_{1,c}$: Classification model that separates the edge classes ω_1 and ω_c. CvsN/CvsP: Current vs. Next and Current vs. Previous architectures (see Fig. 1).

3   Data Sets

In the current study, we include the following five publicly available data sets from the UCI machine learning repository [9] and the Weka website[1].

The **Contraceptive Method Choice (CMC)** data set is part of the National Indonesia Contraceptive Survey from the year 1987. The goal of this data set-specific classification task is to predict a married woman's contraceptive method. The class labels are denoted as follows, *no use* $\prec$ *short-term method* $\prec$ *long-term method*. The features describe the women's socio-economic and demographic characteristics, including *age*, *level of education*, and *husband's level of education*, amongst others.

The **Car Evaluation (Cars)** data set includes a set of class labels denoted by *unacceptable* $\prec$ *acceptable* $\prec$ *good* $\prec$ *very good*. The provided features are the *buying price*, the *maintenance price*, the *number of doors*, the *number of seats*, the *boot size*, as well as the car's estimated *safety level*.

---

[1] https://waikato.github.io/weka-wiki/datasets/.

The **Grub Damage (Grub)** data set represents a pasture damage estimation task. The class labels are denoted as *low* $\prec$ *average* $\prec$ *high* $\prec$ *very high*. The features consist of agriculture-specific characteristics. The Grub data set is available on the Weka website, included in the file denoted by *agridatasets.jar*.

The **Nursery** data set constitutes a job application evaluation task. Initially, this data set consists of five classes, including the class *recommended*. However, since this class contains only two data points, we focus on the remaining four classes, i.e. *not recommended* $\prec$ *very much recommended* $\prec$ *priority acceptance* $\prec$ *special priority acceptance*. The features describe the applicant's *housing*, as well as *social* and *health* conditions, amongst others.

The **Forest Type Mapping (Forests)** data set covers four types of forests. Note that in [4], we identified the following ordinal structure of the class labels, *Hinoki* $\prec$ *Sugi* $\prec$ *Mixed Deciduous* $\prec$ *Non-Forest*. The provided features consist of spectral characteristics of visible-to-near infrared wavelengths.

The characteristics of all data sets are summarised in Table 1.

Table 1. Data Sets. #C: Number of classes. #F: Total number of features (number of categorical features). #S: Number of samples.

Data set	#C	#F	#S	Class distribution			
CMC	3	9 (7)	1473	629 −	511 −	333 −	0
Car Evaluation	4	6 (6)	1728	1210 −	384 −	69 −	65
Grub Damage	4	8 (6)	155	49 −	41 −	46 −	19
Forests	4	27 (0)	523	86 −	195 −	159 −	83
Nursery	4	8 (8)	12958	4320 −	328 −	4266 −	4044

4   Experimental Analysis

In the current section, we first provide the experimental settings, followed by the presentation and discussion of the results.

4.1   Experimental Settings

Let $z \in Z \subset \mathbb{R}^d$ be a sample from the test set Z. For input z, we denote the final output of a classification architecture (CA) simply by $\mathrm{CA}(z)$. Moreover, by $\Delta : \Omega \times \Omega \to \{0, \ldots, c-1\}$, we denote the absolute difference between $\mathrm{CA}(z)$ and the true label of z, $l(z)$, i.e.

$$\Delta(\omega_i, \omega_j) := |i - j|, \quad \forall i, j \in \{1, \ldots, c\}.$$

Additionally, by $Z_i \subset Z$, we denote the subset of the test set Z, which includes all samples from the class ω_i, i.e.

$$Z_i = \{z \in Z : l(z) = \omega_i\}, \quad \forall i \in \{1, \ldots, c\}.$$

Classification Models. We evaluate each architecture in combination with two popular, basic machine learning classification models, i.e. a Support Vector Machine (SVM) [1,18] with linear kernel, as well as an unpruned Decision Tree (DT) [6] with the Gini Index as the node split criterion. More precisely, for each OCC approach, each sequence model is either a linear SVM, or a single DT. As these classifiers have different characteristics (connected/disconnected decision regions) the corresponding OCCs can take advantage of different ordinal embeddings [14]. For the implementation, we use the Matlab[2] software, in combination with the default parameters.

Performance Measures. Note that most of the data sets are highly imbalanced (see Table 1). Therefore, as performance measures, we focus on the weighted accuracy (wACC), as well as the weighted mean absolute error (wMAE), i.e.

$$wACC := \frac{1}{c} \sum_{i=1}^{c} \frac{|\{z \in Z_i : CA(z) = l(z)\}|}{|Z_i|},$$

$$wMAE := \frac{1}{c} \sum_{i=1}^{c} \sum_{z \in Z_i} \frac{\Delta(CA(z), l(z))}{|Z_i|}.$$

Evaluation Approaches. For each of the data sets, we apply a 10-fold cross validation. Note that we ensure that the initial class distribution is reflected in each of the folds, for all data sets. Moreover, for the statistical analysis, we apply the two-sided Wilcoxon signed-rank test [20], at a significance level of 5%.

4.2 Results

Tables 2 and 3 depict the results for the SVM and DT-based OCC architectures, respectively. From Tables 2 and 3, we can make the following observations.

First, considering the choice of classification models, there is no clear preference. Both model types, i.e. SVM and DT-based, outperform each other on different data sets, in combination with each of the OCC architectures. This is a natural outcome in many machine learning tasks, which is supported by Wolpert's *No-Free-Lunch Theorem* [21].

Second, in combination with SVM models, our proposed bOCC architecture outperforms both of the basic OCC approaches statistically significant on three of the five data sets, with respect to both performance measures, wACC and wMAE. Based on the Cars data set, the bOCC architecture is tied with the CvsP approach, also with respect to both performance measures. Based on the Grub data set, the bOCC architecture is slightly outperformed by the CvsP approach, also with respect to both performance measures.

---

[2] www.mathworks.com.

Table 2. SVM Results. Averaged 10-fold cross validation performance values ($\pm$ standard deviations). CvsN: Current vs. Next. CvsP: Current vs. Previous. bOCC: Bidirectional Ordinal Classifier Cascade. The best performing method is underlined. An asterisk (*) indicates a statistically significant difference between the bOCC and the best of the remaining architectures, according to the two-sided Wilcoxon signed-rank test, at a significance level of 5%.

SVM	Weighted accuracy			Weighted mean absolute error		
Data set	CvsN	CvsP	bOCC	CvsN	CvsP	bOCC
CMC	45.8 ± 4.8	46.0 ± 6.0	$\underline{50.0 \pm 5.3}^*$	0.69 ± 0.06	0.68 ± 0.09	$\underline{0.60 \pm 0.07}^*$
Cars	86.4 ± 5.6	$\underline{89.0 \pm 5.2}$	89.0 ± 5.2	0.17 ± 0.08	$\underline{0.12 \pm 0.06}$	0.12 ± 0.06
Grub	44.6 ± 13	$\underline{54.6 \pm 14}$	52.9 ± 12	0.95 ± 0.32	$\underline{0.65 \pm 0.21}$	0.75 ± 0.20
Forests	84.3 ± 6.2	74.2 ± 9.5	$\underline{87.9 \pm 4.3}^*$	0.21 ± 0.09	0.50 ± 0.25	$\underline{0.13 \pm 0.04}^*$
Nursery	63.9 ± 2.4	63.3 ± 2.3	$\underline{88.3 \pm 2.3}^*$	0.77 ± 0.03	0.62 ± 0.02	$\underline{0.12 \pm 0.02}^*$

Table 3. DT Results. Averaged 10-fold cross validation performance values ($\pm$ standard deviations). CvsN: Current vs. Next. CvsP: Current vs. Previous. bOCC: Bidirectional Ordinal Classifier Cascade. The best performing method is underlined. An asterisk (*) indicates a statistically significant difference between the bOCC and the best of the remaining architectures, according to the two-sided Wilcoxon signed-rank test, at a significance level of 5%.

DT	Weighted accuracy			Weighted mean absolute error		
Data set	CvsN	CvsP	bOCC	CvsN	CvsP	bOCC
CMC	46.0 ± 6.2	45.6 ± 6.0	$\underline{47.9 \pm 6.5}^*$	0.71 ± 0.07	0.71 ± 0.07	$\underline{0.67 \pm 0.07}^*$
Cars	84.8 ± 6.4	81.2 ± 5.5	$\underline{92.1 \pm 4.8}^*$	0.25 ± 0.10	0.36 ± 0.08	$\underline{0.11 \pm 0.06}^*$
Grub	42.8 ± 9.3	46.9 ± 15	$\underline{55.1 \pm 12}^*$	1.00 ± 0.19	0.78 ± 0.29	$\underline{0.68 \pm 0.26}$
Forests	80.6 ± 5.2	75.5 ± 5.9	$\underline{84.7 \pm 4.5}^*$	0.32 ± 0.10	0.45 ± 0.16	$\underline{0.18 \pm 0.05}^*$
Nursery	72.7 ± 1.0	73.6 ± 0.8	$\underline{98.6 \pm 0.8}^*$	0.69 ± 0.02	0.63 ± 0.01	$\underline{0.01 \pm 0.01}^*$

Third, in combination with DT models, our proposed bOCC architecture outperforms both of the basic OCC approaches on all data sets, with respect to both performance measures, wACC and wMAE. Based on the wACC measure, the improvement is always statistically significant. In combination with the wMAE measure, only for the Grub data set, the improvement is not statistically significant.

Finally, the best overall performance values, for both measures, are obtained in combination with the proposed bOCC architecture. In almost all cases, the improvement against both of the basic OCC approaches is statistically significant, according to the two-sided Wilcoxon signed-rank test, at a significance level of 5%. The highest improvements in performance are observed for the Nursery data set.

Table 4. SVM and DT Results. Averaged 10-fold cross validation class-specific MAE values, based on the Nursery data set. CvsN: Current vs. Next. CvsP: Current vs. Previous. bOCC: Bidirectional Ordinal Classifier Cascade. wMAE: Weighted MAE.

Nursery data set	Support vector machines					Decision trees				
Approach	ω_1	ω_2	ω_3	ω_4	wMAE	ω_1	ω_2	ω_3	ω_4	wMAE
CvsN	0.00	0.28	0.94	1.88	0.77	0.00	0.02	0.88	1.84	0.69
CvsP	2.00	0.28	0.11	0.08	0.62	2.47	0.02	0.02	0.01	0.63
bOCC	0.00	0.28	0.11	0.08	0.12	0.00	0.02	0.02	0.01	0.01

4.3 Discussion

The obtained results indicate that the simple introduction of the selector component, with the purpose of combining two basic OCC architectures, in most cases, has a positive effect with respect to the classification performance. It seems that, in general, *choosing the appropriate direction* merges the strengths of two one-directional OCCs.

Surprisingly, for the Nursery data set, the weighted accuracy values improved by more than 20%, in combination with both models, SVM and DT, respectively (see Tables 2 and 3). Most likely, this is due to the specific class distribution of the Nursery data set. This is the only data set, in which one of the classes constitutes the *minority class*, with respect to *both of the neighbouring classes*. More precisely, for the Nursery data set, the class distribution is equal to $4320-328-4266-4044$, with respect to the class order $\omega_1 \prec \omega_2 \prec \omega_3 \prec \omega_4$ (see Table 1). The number of samples in class ω_2 is *much lower* than the number of samples in the classes ω_1 and ω_3, respectively. This specific constellation seems to have a strong negative impact on both of the one-directional OCC architectures. Table 4 depicts the class-specific MAE values for the Nursery data set. From Table 4, we can observe that the CvsN architecture performs poorly specific to the classes ω_3 and ω_4. In contrast, the CvsP architecture performs poorly specific to the class ω_1. The combined bOCC architecture performs as well as the CvsN approach specific to the class ω_1, while performing as well as the CvsP approach specific to the classes ω_3 and ω_4, based on both models.

Note that, with respect to basic OCCs, the issue of class imbalance could be overcomed by applying some of the existing balancing approaches, such as the SMOTE [7] method, or the recently proposed *label manipulation* approach [2]. In contrast, our proposed bOCC architectures seem to automatically overcome the issue of class imbalances, as discussed above.

5   Conclusion

In the current work, we provided an extensive evaluation study for our recently proposed bOCC architectures. To this end, we analysed the effectiveness of pairwise bOCCs, in comparison to basic pairwise one-directional OCC approaches, based on a set of publicly available OC benchmark data sets. We conducted our experimental analysis in combination with two base classifiers, i.e. support vector machines and decision trees. Additionally, we evaluated the performance in combination with two measures, i.e. accuracy (ACC) and mean absolute error (MAE). To address the data sets' class imbalances, we focused on the *weighted* versions of the ACC and MAE measures. For the statistical analysis, we included the two-sided Wilcoxon signed-rank test. The obtained outcomes are discussed in detail, in Sect. 4. In most cases, it seems to be beneficial to extend pairwise OCC architectures by our proposed bOCC approach.

As future research, we aim at focusing on the analysis of our proposed *selector component* (SC). An optimised SC might further improve the classification performance of bOCCs. For instance, in the case of imbalanced edge classes, one could apply some of the existing balancing techniques. Alternatively, it might be beneficial to include the data from all available classes, during the training of the SC.

Acknowledgments. The work of Peter Bellmann and Friedhelm Schwenker is supported by the project *Multimodal recognition of affect over the course of a tutorial learning experiment* (SCHW623/7-1) funded by the German Research Foundation (DFG). We gratefully acknowledge the support of NVIDIA Corporation with the donation of the Tesla K40 GPU used for this research. Hans A. Kestler acknowledges funding from the German Science Foundation (DFG, 217328187 (SFB 1074) and 288342734 (GRK HEIST)). Hans A. Kestler also acknowledges funding from the German Federal Ministry of Education and Research (BMBF) e:MED confirm (id 01ZX1708C) and TRAN-SCAN VI - PMTR-pNET (id 01KT1901B).

References

1. Abe, S.: Support Vector Machines for Pattern Classification. Advances in Pattern Recognition. Springer, London (2005). https://doi.org/10.1007/1-84628-219-5
2. Bellmann, P., Hihn, H., Braun, D.A., Schwenker, F.: Binary classification: counterbalancing class imbalance by applying regression models in combination with one-sided label shifts. In: ICAART. SCITEPRESS (2021, to be published)
3. Bellmann, P., Lausser, L., Kestler, H.A., Schwenker, F.: Introducing bidirectional ordinal classifier cascades based on a pain intensity recognition scenario. In: Del Bimbo, A., et al. (eds.) ICPR 2021. LNCS, vol. 12666, pp. 773–787. Springer, Cham (2021). https://doi.org/10.1007/978-3-030-68780-9_58
4. Bellmann, P., Schwenker, F.: Ordinal classification: working definition and detection of ordinal structures. IEEE Access **8**, 164380–164391 (2020)
5. Bellmann, P., Thiam, P., Schwenker, F.: Multi-classifier-systems: architectures, algorithms and applications. In: Pedrycz, W., Chen, S.-M. (eds.) Computational Intelligence for Pattern Recognition. SCI, vol. 777, pp. 83–113. Springer, Cham (2018). https://doi.org/10.1007/978-3-319-89629-8_4
6. Breiman, L., Friedman, J.H., Olshen, R.A., Stone, C.J.: Classification and Regression Trees. Wadsworth, Belmont (1984)
7. Chawla, N.V., Bowyer, K.W., Hall, L.O., Kegelmeyer, W.P.: SMOTE: synthetic minority over-sampling technique. J. Artif. Intell. Res. **16**, 321–357 (2002)
8. Dietterich, T.G., Bakiri, G.: Error-correcting output codes: a general method for improving multiclass inductive learning programs. In: AAAI, pp. 572–577. AAAI Press/The MIT Press (1991)

9. Dua, D., Graff, C.: UCI machine learning repository (2017). http://archive.ics.uci. edu/ml

10. Hihn, H., Braun, D.A.: Specialization in hierarchical learning systems. Neural Process. Lett. **52**(3), 2319–2352 (2020). https://doi.org/10.1007/s11063-020-10351-3

11. Hühn, J.C., Hüllermeier, E.: Is an ordinal class structure useful in classifier learning? IJDMMM **1**(1), 45–67 (2008)

12. Kuncheva, L.I.: Combining Pattern Classifiers: Methods and Algorithms. Wiley, Hoboken (2014)

13. Lattke, R., Lausser, L., Müssel, C., Kestler, H.A.: Detecting ordinal class structures. In: Schwenker, F., Roli, F., Kittler, J. (eds.) MCS 2015. LNCS, vol. 9132, pp. 100–111. Springer, Cham (2015). https://doi.org/10.1007/978-3-319-20248-8_9

14. Lausser, L., Schäfer, L.M., Kestler, H.A.: Ordinal classifiers can fail on repetitive class structures. Arch. Data Sci. Ser. A **4**(1), 1–25 (2018)

15. Lausser, L., Schäfer, L.M., Kühlwein, S.D., Kestler, A.M.R., Kestler, H.A.: Detecting ordinal subcascades. Neural Process. Lett. **52**(3), 2583–2605 (2020). https:// doi.org/10.1007/s11063-020-10362-0

16. Lausser, L., Schäfer, L.M., Schirra, L.R., Szekely, R., Schmid, F., Kestler, H.A.: Assessing phenotype order in molecular data. Sci. Rep. **9**(1), 1–10 (2019)

17. Thiam, P., et al.: Multi-modal pain intensity recognition based on the senseemotion database. IEEE Trans. Affect. Comput. 1 (2019). https://doi.org/10.1109/taffc. 2019.2892090

18. Vapnik, V.: The Nature of Statistical Learning Theory. Springer, New York (2013)

19. Walter, S., et al.: The BioVid heat pain database data for the advancement and systematic validation of an automated pain recognition system. In: CYBCONF, pp. 128–131. IEEE (2013). https://doi.org/10.1109/CYBConf.2013.6617456

20. Wilcoxon, F.: Individual comparisons by ranking methods. Biom. Bull. **1**(6), 80–83 (1945)

21. Wolpert, D.H.: The lack of A priori distinctions between learning algorithms. Neural Comput. **8**(7), 1341–1390 (1996)

Deep Learning

On Calibration of Mixup Training for Deep Neural Networks

Juan Maroñas[1]([✉]), Daniel Ramos[2], and Roberto Paredes[1]

[1] PRHLT Research Center, Universitat Politècnica de València, Valencia, Spain
jmaronas@prhlt.upv.es
[2] AUDIAS - Audio, Data Intelligence and Speech, Universidad Autónoma de Madrid, Madrid, Spain

Abstract. Deep Neural Networks (DNN) represent the state of the art in many tasks. However, due to their overparameterization, their generalization capabilities are in doubt and still a field under study. Consequently, DNN can overfit and assign overconfident predictions – effects that have been shown to affect the calibration of the confidences assigned to unseen data. Data Augmentation (DA) strategies have been proposed to regularize these models, being Mixup one of the most popular due to its ability to improve the accuracy, the uncertainty quantification and the calibration of DNN. In this work however we argue and provide empirical evidence that, due to its fundamentals, Mixup does not necessarily improve calibration. Based on our observations we propose a new loss function that improves the calibration, and also sometimes the accuracy, of DNN trained with this DA technique. Our loss is inspired by Bayes decision theory and introduces a new training framework for designing losses for probabilistic modelling. We provide state-of-the-art accuracy with consistent improvements in calibration performance. (Appendix and code are provided here: GitHub link)

Keywords: Deep neural networks · Calibration · Data augmentation · Mixup training

1 Introduction

Deep Neural Networks (DNN) are probabilistic models (PM) that represent the state of the art in many tasks, either as end-to-end models [20], or as part of complex decision systems [6]. Many of the applications in which DNN has widely overcome previous approaches require that the parameterized probability distributions are interpretable. This means that both the prediction, (for instance the class selected in a classification problem), and the probability assigned to that prediction, are important for the correct performance of the whole system. Examples of these applications are medical diagnosis [3] or language recognition [1]. In all these problems, it is very different to decide towards an action with high probability, than doing it with a more moderated one. The ultimate consequences

© Springer Nature Switzerland AG 2021
A. Torsello et al. (Eds.): S+SSPR 2020, LNCS 12644, pp. 67–76, 2021.
https://doi.org/10.1007/978-3-030-73973-7_7

incurred in the decision process can be drastic if these probabilities are not reliable i.e., are not well-calibrated. In other words, our model is calibrated if the probabilities assigned reflect the uncertainty present in the data distribution. Moreover, our PM must be able to discriminate between the different classes, i.e. to separate them. Note that discrimination is inherent to the data distribution, which means that we cannot expect to separate our data if our data is not separable in its origins. Both good discrimination and a correct modelling of data uncertainty are mandatory to achieve optimal classification performance by the use of the Bayes Decision Rule (BDR).

The calibration and discrimination of a PM can be improved by optimizing the expected value of a proper scoring rule (PSR), a scalar obtained by the addition of both quantities [4]. For that reason, this optimization is not a guarantee of optimal calibration, as all the effort can be pushed into having better discriminative capabilities. This effect has been recently observed in the context of DNN where [7] showed that although these models are typically trained by optimizing the Negative Log-Likelihood (NLL), the calibration performance is compromised in the direction of over-confidence. This means that even though the accuracy provided by these models on several benchmarks are among the best published, the probabilities assigned are ultimately extreme and badly calibrated. One should not be surprised about this generalization limitation, as many theories that study the generalization capabilities of probabilistic models, such as the VC dimension [22] or the use of marginal likelihoods and Bayes rule for model selection [14], are instances of the Occam's Razor principle [14]. For instance a recent work [25] has shown that DNN can memorize the data input distribution and [18] has shown that many state of the art models overfit the test set.

For that reason, the community has been exploring different regularization techniques that can improve the generalization of these models, being Data Augmentation (DA) one of the gold standards. These techniques aim to increase the support on the input manifold, through transformations that are typically driven by expert knowledge, e.g. rotations or translations when the inputs are images. However, in many domains, it is not clear which kind of augmentations might be useful, which motivates the analysis of general-purpose DA techniques such as Mixup [26], whose fundamentals rely on empirical risk minimization (ERM) [23]. However, both Mixup and human-driven DA techniques share a common issue: they are not designed by analyzing the properties of the input distribution and the intersection of these with the PM; mainly because modern instances of these, such as DNN, are difficult to interpret. For that reason, the selection and performance of DA techniques depend, basically, on cross-validation; but there is no principled way to establish if a particular DA technique might boost the performance of a particular application or not.

Motivated by the fundamentals and good performance of Mixup, a very recent work [21] has studied how Mixup affects the uncertainty quantification and the calibration performance on DNN. They show that Mixup improves the calibration, and they attribute this fact to the smoothness that Mixup induces in the decision regions learned by a PM. Our work is built on top of this observation. We argue that the fundamentals of ERM and Mixup do not allow us to claim that learning smoother decision thresholds are a sufficient condition for having properly calibrated PM, because this decision is not based on the uncertainty of our input distribution. This also extrapolates to other strategies that have shown good regularization in terms of accuracy, uncertainty quantification or calibration, such as label smoothing [19] or more recently DA techniques [11,17].

In this work, we first provide empirical evidence that Mixup can degrade calibration. Secondly, we propose a new loss function to correct this calibration degradation by encouraging the PM to learn its discriminative capabilities, through the incorporation of a simple measure of data uncertainty. Thus, our loss function is inspired by how optimality is achieved in a BDR scenario, and we claim that this has to be done to achieve reliable probability distributions. Note that learning to assign $\{0,1\}$ probabilities only makes sense if the input distribution does not present any kind of overlapping, which is something really hard to assess. For that reason, it should not be surprising that a modern PM, such as a DNN, can have undesirables effects such as memorization [25] or overconfident-badly-calibrated probabilities [7] when forced to achieve this $\{0,1\}$ assignment, as it happens by learning through the categorical cross-entropy (CE). Note that a modern DNN, due to overparameterization, can successfully assign $\{0,1\}$ without any guarantees of generalization, and they typically rely on learning highly oscillating decision thresholds [24], which are also responsible for being vulnerable to adversarial attacks. The results of this work open new perspectives to design losses in this fashion, aiming at representing more sophisticated forms of data uncertainty.

2   Related Work

The first work that showed the badly calibrated probabilities of DNN is found in [7], where different classical calibration techniques are compared. The authors proposed Temperature Scaling. On top of this work [15] has shown how complex techniques can be employed for post-calibration if uncertainty is correctly incorporated, through the use of Bayesian Neural Networks. On the other hand, [9,10] has shown that using self-supervised learning and pre-trained models improves model robustness, uncertainty and calibration. Moreover, the same author has measured robustness against common perturbations [8], and [16] has measured the performance on calibration and uncertainty of several strategies under dataset shift. On the other hand, deep ensembles have also shown good performance for uncertainty quantification and calibration [13]. Finally, on the side of DA strategies, [21] measure the robustness and calibration of Mixup training and [17] propose On-Manifold Adversarial Data Augmentation, which attempts to generate challenging examples by following an on-manifold adversarial attack path in the latent space of a generative model. Moreover, [24] propose a similar technique to Mixup but on the hidden layers of a DNN, with good results in robustness against perturbations. Finally, Augmix has been proposed in [11] with good results in uncertainty quantification and robustness.

3   Background

In this section we describe calibration in the context of image classification and provide the fundamentals of Mixup before presenting our loss function in the next section. We are given N pairs of observed i.i.d. samples $\mathcal{O} = \{(x_n, t_n)\}_{n=1}^{N}$ drawn from some unknown joint probability $P(x, t)$. We then learn a categorical posterior distribution $p_\theta(t = k|x)$ by means of a function g_θ that maps input images x to class probabilities $\{k\}_{k=1}^{C}$ by maximum a posteriori. To make decisions we rely on BDR and chose the action α_i that minimize Bayes Risk:

$$R(\alpha_i|x) = \sum_{1 \leq k \leq C} \lambda_{ik} \cdot p(t = k|x)$$

$$\alpha_i = \underset{1 \leq i \leq C}{\operatorname{argmin}} R(\alpha_i|x)$$

(1)

where λ_{ik} represents the loss incurred when taking the action i if the ground truth is k. In this work we consider equal losses $\lambda_{ik} = 1, \lambda_{ii} = 0 \; \forall i, k$, which means that we choose the class with maximum posterior probability. This rule guarantees optimality when we plug in the data generating distribution [5]. In practice this distribution is substituted with the model $p_\theta(t|x)$ and thus, the lower the gap between the model and the data generating distribution, the closer we will be to an optimal decision.

In a classification scenario, we say that a model is calibrated if the confidences assigned by this model to a set of samples X towards class k are equal to the real proportion of samples in X that the model assigns to this class. This means that to be calibrated, a model should assign confidences considering the proportion of samples assigned to each of the classes. Moreover, in addition to calibration, a model should also present a sharpened probability distribution, a property known as discrimination or refinement [1,4]. With this property, we guarantee that our model can discriminate between classes. Thus, both good calibration and discrimination imply recovering how the data from the different classes is distributed or, in other words, good calibration and discrimination imply recovering data uncertainty. By doing so, our model will be forced to match the data generating distribution and this will guarantee asymptotic optimality in the decisions to be taken.

Note that the goal of a PM is to map any data distribution to a linear separable manifold. Thus, we can only achieve separability if: 1) the data is separable in its origins and 2) the model has enough capacity to do so. Thus, if 1) or 2) does not hold (which is something that we will not typically know), then it seems unreasonable to force the model to learn towards $\{0, 1\}$ probabilities; and we should expect an overparameterized model to experiment different pathologies such as overfitting [23], memorization [25] or bad calibration [7]. A very illustrative example of this pathology is: Why should we push probabilities towards 1.0 in a 1-dimensional input generative Gaussian classifier if Gaussians have support over $\mathbb{R}$? Based on this observation a training loss in a modern PM should somehow consider this inherent structure (uncertainty) in the data to reliably target the underlying distribution, and avoid the great ability of DNN to assign $\{0,1\}$ probabilities when we do not know if the distribution to be modelled is or can be linearly separated. This is the core idea of our proposed loss function, and we will further use it to justify why Mixup should not necessarily provide calibrated distributions.

Mixup has its fundamentals in vicinal risk minimization (VRM) [2][1], which is derived as a solution to the limitations present in ERM [22,23,26]. Contrary to other vicinal distributions, Mixup assumes that the samples in the vicinity distribution do not belong to the same class. For that reason, it is defined as the expected value of a linear interpolation between two input samples and their corresponding labels [26]. The interpolation is given by the coefficient γ, which is drawn from a beta distribution. An unbiased estimate of the empirical risk can be obtained by evaluating the average loss function on a set of samples drawn from this distribution as follows:

$$
\gamma \sim \text{Beta}(\beta, \beta)
$$
$$
\tilde{x} = \gamma \cdot x_1 + (1 - \gamma) \cdot x_2 \qquad (2)
$$
$$
\tilde{t} = \gamma \cdot t_1 + (1 - \gamma) \cdot t_2
$$

As a consequence, training with Mixup ensures a linear-soft transition between the confidence assigned by a model in the different parts of the input space. However, this only ensures smoothness in the confidence assigned to different regions of the input space, reducing the overconfidence but without any guarantee of an improved calibration, because the uncertainty is not considered at all. Note that Mixup just relies on an assumption on how the samples in the vicinity are distributed but do not take into consideration the proportion of samples present, which is at the core of a proper calibration.

As a consequence, only if the data distribution presents a linear relation between their corresponding classes, one could expect the ultimate distribution to be calibrated when applying this technique. In the experimental section, we show that some models trained with Mixup do not necessarily improve the calibration, as recently noted in [21]. In fact, we show that Mixup tends to worsen the calibration in many cases.

4   Proposed Loss: Auto-Regularized-Confidence

As illustrated in previous sections, our objective is to benefit from the improved accuracy of Mixup, but providing better calibrated distributions. To do so, we introduce a new loss function, which is a weighted combination of our proposed loss, named Auto-Regularized-Confidence (ARC), and the categorical cross entropy (CE). The ARC loss is inspired by the Expected Calibration Error (ECE) [7][2]. The idea, as argued in above sections, is to incorporate data uncertainty in the predictions. This is done by first partition the confidences p, assigned

---

[1] For unfamiliar readers we provide a wider description in appendix A.
[2] See appendix B for a detailed description of calibration metrics.

to a batch of samples X, into M bins B_i; and match these confidences to the accuracy μ_i in that bin, by means of any of these two variants:

$$
\begin{aligned}
\text{ARC\_V}_1 &= \frac{1}{M} \sum_{i=1}^{M} \left[\left(\frac{1}{|B_i|} \sum_{0 \le j \le |B_i|} p_{ij} \right) - \mu_i \right]^2 \\
\text{ARC\_V}_2 &= \frac{1}{M} \sum_{i=1}^{M} \left[\frac{1}{|B_i|} \sum_{0 \le j \le |B_i|} (p_{ij} - \mu_i)^2 \right]
\end{aligned}
\tag{3}
$$

The difference lies in whether the average confidence (ARC_V_1) or the individual confidences (ARC_V_2) are forced to match the accuracy. If we set $M = 1$ then our loss function is computed over the entire batch. We make the accuracy μ_i a constant value so learning gradients only depend on the confidence assigned by the model. Our loss is combined with the CE to avoid the local minimum in which the network parameterize a prior classifier (i.e., the one which assigns prior confidences to samples), as we found in our initial analysis. This is because a prior classifier is useless, but the trivial way of optimizing calibration. Thus the overall loss is given by:

$$
\text{L}(\theta) = \frac{1}{N} \sum_n \text{CE}(\theta, x_n, t_n) + \beta \cdot \text{ARC}(\theta, x_n, t_n)
\tag{4}
$$

where β is a hyperparameter that controls the relative importance given to each of the losses and is established with a validation set. As mentioned, this new loss targets the uncertainty of the learned representation, through the accuracy. The accuracy is used to summarize the proportion of samples from different classes that are being "mixed". So it somehow represents how the representations that the model can learn are distributed. It is clear that the accuracy is a very simple statistical summary of the data uncertainty and it is let to future work the search for other quantifiers that could encode more useful information such as how samples are distributed in the input space. Consequently, we can expect that by evaluating the CE loss on the Mixup image $\tilde{x}$, and the ARC loss on the mixing images x_1 and x_2, one can benefit from the improved discrimination as learned by the CE, but the ultimate confidences are assigned by how the classifier classifies samples x_1, x_2 from the generating distribution $p(x, t)$ and not those $\tilde{x}$ virtually generated by Mixup. It is then clear that ARC incorporates data uncertainty, which will improve the model representation of the underlying distribution, and thus its calibration. To validate this procedure, in our work we experiment with variants that compute ARC loss over x_1 and x_2; and we also compute ARC loss over $\tilde{x}$. In general, all datasets benefit more from the latter. A discussion is provided in the experimental section.

An additional analysis of this loss function is provided in appendix C and the experimental section. This includes the motivation beside experimenting with ARC_V_1 and ARC_V_2 and an analysis of why this loss might improve the accuracy, as we have found that some datasets improve this metric by applying the ARC loss.

Finally, we discuss one drawback of our proposal as being used as a general-purpose calibration tool. Note that, if applied on a DNN that presents near 100% accuracy on the training dataset (which is the case in many of the standard databases tested) then the ARC loss will provide the same learning signal as the CE, because it will for the average confidences to be 1.0. This means that it will not work in datasets where the training error is overfitted, as in CIFAR100. To solve this, we experiment with the following variant. We take a validation split from the training dataset where the DNN presents uncalibrated over-confidences. Let say that this validation set presents an 80% accuracy, with a 0.99 average confidence. Thus, we use the validation set to compute the ARC loss while the training dataset is only used for the CE.

5 Experiments

We perform several experiments that illustrate the main claims of this work. We show average results in the main work and provide specific results in Github, alongside code and details on loss hyperparameters (e.g. if the model uses ARC_V_1 or ARC_V_2). We evaluate a collection of classical benchmarks for this task: CIFAR100, CIFAR10, SVHN; and we also evaluate our model on more realistic problems such as the ones provided by Caltech Birds and Standford Cars, which contain bigger and more realistic images. Due to computational restrictions, we did not evaluate our model on ImageNet. We experiment with state-of-the-art configurations of computer vision DNN: Residual Networks, Wide Residual Networks and Densely Connected Neural Networks. Moreover, for each variant, we evaluate several configurations and models with and without dropout. We use pre-trained models on ImageNet for Birds and Cars. We evaluate different calibration metrics, detailed in appendix B. In the main work, we report the accuracy and ECE (with a partition of 15 bins) while the rest of the calibration metrics are reported in appendix D.1. We compare to a recent technique designed for implicitly calibrate a probabilistic DNN named MMCE over their best performing approach [12]. More details provided in appendix D.

For the sake of illustration, we provide average results of all the models in Table 1, and for the best-performing model per task in Table 2. First, as shown in rows B (Baseline) and B+M (Baseline+Mixup) in the tables, we see how Mixup degrades the calibration except in CIFAR100. By comparing with the results reported in [21], we can conclude that Mixup behaves particularly well in CIFAR100, probably because the intersection between classes can be explained through a linear relation. However, our tables demonstrate that this is not a general behaviour of Mixup as shown in the rest of datasets. It is surprising how Mixup degrades calibration in Birds and Cars, even though the DNN used for these datasets are pre-trained models which have been shown to provide better calibrated distributions [9]. In general, our results contrast with those reported in [21] where they provide general improvement in calibration performance due to Mixup. We can explain this difference with the fact that different models are used. For instance, while they use a VGG-16 and a ResNet-34, we are using much

Table 1. Table showing average accuracy and ECE in (%) of all the models considered in this work

	CIFAR10		CIFAR100		SVHN		Birds		Cars	
	ACC	ECE	ACC	ECE	ACC	ECE	ACC	ECE	ACC	ECE
Baseline (B)	94.76	3.41	77.21	11.57	96.32	1.90	78.51	2.39	86.74	2.06
Baseline + Mixup (B+M)	96.01	4.35	80.04	3.71	96.41	5.00	79.63	14.22	86.67	18.13
MMCE (M)	94.24	2.17	72.68	3.71	96.28	1.78	78.78	1.95	86.83	2.23
MMCE + Mixup (M+M)	91.90	5.69	78.52	5.48	96.59	2.83	79.99	12.37	86.03	13.07
ARC (A)	94.82	3.37	77.04	11.31	96.26	1.87	78.52	2.70	87.78	2.76
ARC + Mixup (A+M)	95.90	1.62	79.84	2.42	96.02	2.17	79.74	4.95	89.63	2.84

Table 2. Table showing the accuracy and ECE in (%) of the best model per task and technique.

	CIFAR10		CIFAR100		SVHN		Birds		Cars	
	ACC	ECE	ACC	ECE	ACC	ECE	ACC	ECE	ACC	ECE
Baseline (B)	95.35	2.97	79.79	5.06	97.07	0.50	80.31	4.34	89.13	2.57
Baseline + Mixup (B+M)	97.19	4.65	82.34	1.42	96.97	4.91	82.09	10.14	89.45	18.10
MMCE (M)	95.58	1.21	74.98	7.04	96.90	0.49	80.64	3.28	89.40	2.70
MMCE + Mixup (M+M)	97.02	1.11	81.31	4.46	97.17	3.69	82.41	10.93	88.47	11.56
ARC (A)	95.99	2.01	80.77	4.73	97.08	0.37	80.32	4.44	90.09	1.92
ARC + Mixup (A+M)	97.09	1.03	82.02	0.98	96.82	2.20	82.45	1.28	91.13	2.40

deeper models, such as a ResNet-101 or a DenseNet-121. The difference can be connected to the observation in [7] where they show that calibration is further degraded by deeper architectures. Moreover, we shall emphasize that our results on CIFAR10 are on the state-of-the-art ($\sim$97% ACC) and much better calibrated (1.03 top ECE and 1.62 average ECE) than in [21], while they report a 2.00 value of ECE.

Analyzing our loss function, we see how it can correct the miscalibration introduced by Mixup training. In CIFAR10 and CIFAR100 A+M is the best performing approach. In SVHN we see that A+M corrects the calibration error introduced in B+M, but the approach behaves similar to the others. SVHN is a dataset that presents good calibration in many models over the test set, as noted also in [7,15]. Finally, regarding Birds and Cars we see how our loss can highly correct the miscalibration introduced by Mixup. This means that our approach also performs well with pre-trained models on ImageNet. It should be noted that in this case, we do not achieve the same ECE error in Birds and Cars as with the baseline model. However, we have much better accuracy (over 3% on average results in Cars). In fact, our work reports nearly state of the art accuracy in Cars using a Dense-Net, where the best performing reported model has an accuracy only two points above but using much more complex architectures such as efficient net [20] or inception [19]. On the other hand, our method is better than the recently proposed MMCE [12]. We found this method to be unstable in some cases, as some models saturated during training or tended to degrade the accuracy, as shown in the tables.

Regarding the parameterization of the loss function, we found that most of the times the best configuration of hyperparameters was obtained with ARC_V$_1$. This can be explained by the fact that DNN typically learn invariant representations and thus, we avoid the pathological behaviour that ARC_V$_1$ can present, which is discussed in appendix C. Besides, we found that only in Birds and some CIFAR100 models, the ARC loss computed over the Mixup image $\tilde{x}$ worked better than when computed over x_1 and x_2, even though this configuration also improved the calibration. Thus, as we claim in Sect. 4, it seems reasonable that a loss function that takes into account, separately, the underlying structure present in the data can provide better calibrated uncertainties.

Finally, by looking at the results of applying ARC loss over the Baseline model (A in the tables) we see that the improvements in calibration are not significant, or at least not as when combined with Mixup. We have already argued the reason in Sect. 4. We mentioned that a possible solution could be to apply the ARC loss on a separate validation set. Surprisingly, the DNN learns to minimize the ARC loss by increasing the accuracy of this validation set rather than by relaxing the confidences assigned.

6 Conclusions and Future Work

This work has shown that Mixup does not ensure calibrated class distributions. The results and theory presented suggest that a similar analysis should be employed over different DA techniques, which is let for future work. We have also opened a new perspective to reduce overconfidence in DNN. As we cannot control how a model might overfit the dataset to achieve high discriminative performance, a good practice is to auto-regularize the model to incorporate the uncertainty of the learned representations. This work has shown a way of doing this on Mixup training, reporting state-of-the-art results in accuracy and calibration. Future work is concerned with the exploration of new loss functions for this purpose.

Acknowledgments. We gratefully acknowledge the support of Nvidia-corporation through the donation of two Nvidia TITAN XP. The research leading to these results has received funding from the European Union through Programa Operativo del Fondo Europeo de Desarrollo Regional (FEDER) from ComunitatValencia(2014–2020) under project Sistemas de frabricacióon inteligentes para la industria 4.0 (grant agreement IDIFEDER/2018/025). JM is supported by grant FPI-UPV under grant agreement 825111 Deep Health Project. DR and JM are supported by the Spanish National Ministry of Education through grant RTI2018-098091-B-I00.

References

1. Brümmer, N., et al.: On calibration of language recognition scores. In: Proceedings of Odyssey. San Juan, Puerto Rico (2006)

2. Chapelle, O., et al.: Vicinal risk minimization. In: Leen, T.K., et al. (eds.) NIPS, pp. 416–422. MIT Press (2001)
3. Crowson, C., et al.: Assessing calibration of prognostic risk scores. Stat. Methods Med. Res. **25**(4), 1692–1706 (2016)
4. deGroot, M.H., et al.: The comparison and evaluation of forecasters. The Statistician **32**, 12–22 (1983)
5. Duda, R.O., et al.: Pattern Classification, 2nd edn. Wiley, New York (2001)
6. Gulcehre, C., et al.: On integrating a language model into neural machine translation. Comput. Speech Lang. **45**(C), 137–148 (2017)
7. Guo, C., et al.: On calibration of modern neural networks. In: Precup, D., Teh, Y.W. (eds.) ICML, vol. 70, pp. 1321–1330 (2017)
8. Hendrycks, D., et al.: Benchmarking neural network robustness to common corruptions and perturbations. In: ICLR (2019)
9. Hendrycks, D., et al.: Using pre-training can improve model robustness and uncertainty. In: ICLR (2019)
10. Hendrycks, D., et al.: Using self-supervised learning can improve model robustness and uncertainty. In: Wallach, H., et al. (eds.) NIPS, pp. 15663–15674 (2019)
11. Hendrycks, D., et al.: AugMix: a simple data processing method to improve robustness and uncertainty. In: ICLR (2020)
12. Kumar, A., et al.: Trainable calibration measures for neural networks from kernel mean embeddings. In: Dy, J., Krause, A. (eds.) ICML, vol. 80, pp. 2805–2814 (2018)
13. Lakshminarayanan, B., et al.: Simple and scalable predictive uncertainty estimation using deep ensembles. In: Guyon, I., et al. (eds.) NIPS, pp. 6402–6413 (2017)
14. MacKay, D.: Bayesian methods for adaptive models. Ph.D. thesis, California Institute of Technology (1992)
15. Maroñas, J., et al.: Calibration of deep probabilistic models with decoupled Bayesian neural networks. Neurocomputing **407**, 194–205 (2020)
16. Ovadia, Y., et al.: Can you trust your model's uncertainty? Evaluating predictive uncertainty under dataset shift. In: Wallach, H., et al. (eds.) NIPS (2019)
17. Patel, K., et al.: On-manifold adversarial data augmentation improves uncertainty calibration (2019)
18. Recht, B., et al.: Do ImageNet classifiers generalize to ImageNet? In: Chaudhuri, K., Salakhutdinov, R. (eds.) ICML vol. 97, pp. 5389–5400 (2019)
19. Szegedy, C., et al.: Rethinking the inception architecture for computer vision. In: CVPR (2016)
20. Tan, M., et al.: EfficientNet: rethinking model scaling for convolutional neural networks. In: Chaudhuri, K., Salakhutdinov, R. (eds.) ICML, pp. 6105–6114 (2019)
21. Thulasidasan, S., et al.: On mixup training: Improved calibration and predictive uncertainty for deep neural networks. In: Wallach, H., et al. (eds.) NIPS (2019)
22. Vapnik, V., et al.: On the uniform convergence of relative frequencies of events to their probabilities (1971)
23. Vapnik, V.N.: Statistical Learning Theory. Wiley, New York (1998)
24. Verma, V., et al.: Manifold Mixup: better representations by interpolating hidden states. In: Chaudhuri, K., Salakhutdinov, R. (eds.) ICML, pp. 6438–6447 (2019)
25. Zhang, C., et al.: Understanding deep learning requires rethinking generalization. In: ICLR (2017)
26. Zhang, H., et al.: Mixup: beyond empirical risk minimization. In: ICLR (2018)

Augmenting Graph Convolutional Neural Networks with Highpass Filters

Fatemeh Ansarizadeh$^{(\boxtimes)}$ [ID], David B. Tay[ID], Dhananjay Thiruvady[ID],
and Antonio Robles-Kelly[ID]

School of IT, Deakin University, Waurn Ponds, VIC 3216, Australia
{f.ansarizadeh,david.tay,dhananjay.thiruvady,
antonio.robles-kelly}@deakin.edu.au

Abstract. In this paper, we propose a graph neural network that employs high-pass filters in the convolutional layers. To do this, we depart from a linear model for the convolutional layer and consider the case of directed graphs. This allows for graph spectral theory and the connections between eigenfunctions over the graph and Fourier analysis to employ graph signal processing to obtain an architecture that "concatenates" low and high-pass filters to process data on a connected graph. This yields a method that is quite general in nature applicable to directed and undirected graphs and with clear links to graph spectral methods, Fourier analysis and graph signal processing. Here, we illustrate the utility of our graph convolutional approach to the classification using citation datasets and knowledge graphs. The results show that our method provides a margin of improvement over the alternative.

Keywords: Citation graph · Graph convolutional neural networks · Knowledge graph

1 Introduction

Recent breakthroughs in machine learning techniques have resulted in substantial progress in a wide range of areas, from image classification to natural language understanding. Predominately, for convolutional neural networks (CNNs), the input data, *i.e. images*, can be viewed as structured in an Euclidean space and hence abstracted onto a planar graph on a lattice. However, in many fields of research, including social networks and brain connectomes, input instances can have non-planar structure and hence, be defined over non-Euclidean spaces. Indeed, developing efficient implementations of CNNs for high-dimensional, non-Euclidean domains such as non-planar graphs, polygonal meshes or manifolds [4] its not a straightforward task [5].

The recent advent of graph neural networks provides the ideal basis for applying machine learning algorithms to datasets whose instances are not structured as a lattice and that require the capacity to process more general graphs. In particular, applying CNNs on data architectures in non-Euclidean domains was

© Springer Nature Switzerland AG 2021
A. Torsello et al. (Eds.): S+SSPR 2020, LNCS 12644, pp. 77–86, 2021.
https://doi.org/10.1007/978-3-030-73973-7_8

initially conceived in 1998 [6] and further developed by Scarselli *et al.* [7]. The resulting approach – graph neural networks (GNNs) – provide an efficient way to model problems as a graph, which can be easily integrated with neural networks. GNNs generally pose the problem of learning at each vertex of a graph using its neighborhood information. Early approaches modelled this local support using an information-theoretic standpoint whereby a signal on a graph with N vertices can be viewed as the counterpart of a discrete-time signal with N samples in classical signal processing. This has given rise to graph signal processing, where the most serious hurdle on the application of classical signal processing methods on graphs concerns discrete-time signals whereby dependencies arising from the irregular data domain are difficult to detect [8].

Neural networks on graphs have been proposed elsewhere [2,5,7,12,22]. In this paper, we propose a layer-wise GCN which employs a localized first-order approximation of spectral graph convolution to process data on a connected graph. This sort of architectures were presented in [22] and later advanced by [12]. The main difference between these and that presented in this paper stems from the fact that our architecture not only contains the equivalent of low-pass filters, but augments these with high-pass ones. To combine low and high-pass filters, we depart from graph-spectral theory. This yields a method to "concatenate" the low and high pass filter responses on the graph.

2 Background

In this paper, the focus is on the connections between graph convolutional networks (GCNs) and spectral graph theory. Recall that graph-spectral methods [13] allow for a direct link to be made with the Fourier transform through Chebychev polynomials. Thus, spectral methods carry a natural notion of frequency conveyed through the natural relationship between eigenfuctions over the graph, their corresponding eigenvalues, and the natural frequencies or "modes" of connected graphs computed making use of the graph Laplacian. For example, eigenvalues that are small in value correspond to eigenfunctions on the graph that vary smoothly and change gradually and, hence, can be viewed as representing low frequencies. In contrast, large eigenvalues correspond to rapidly changing functions, *i.e.* high frequencies. This is analogous to the treatment of eigenvalues and eigenvectors in classical Fourier analysis.

Throughout the paper we will extensively use graphs. Formally, in a graph $\mathcal{G} = \{\mathcal{V}, \mathcal{E}\}$, $\mathcal{V}$ and $\mathcal{E}$ are the set of vertices and edges, respectively. For the purposes of inference and learning, instances are abstracted as vertices and the relationship between them correspond to edges in the graph. For instance, when applied to our sample citation datasets, the vertices $\mathcal{V} = \{v_1, v_2, \cdots, v_N\}$ represent documents under consideration and the edges $\mathcal{E} = \{e_1, e_2, \cdots, e_m\}$ stand for whether a document cites another or not, *i.e.* an edge $e_k = (v_i, v_j)$ implies that document v_j cites v_i.

As mentioned above, graph-spectral methods are based upon the Laplacian or adjacency matrix [15]. Recall that the graph connectivity can be described by the adjacency matrix $\boldsymbol{A} \in \mathbb{R}^{N \times N}$, where its element $a_{i,j}$ is the weight of the edge connecting nodes i and j, and a zero weight indicates no connection. A graph can have directed or undirected edges, whereby, for the latter case A is symmetric. The normalized graph Laplacian matrix defined as $L = I - D^{-\frac{1}{2}} A D^{-\frac{1}{2}}$ [9], where D is a diagonal matrix called the degree matrix (the ith diagonal element correspond to the degree of the node indexed i) and I is the $N \times N$ identity matrix. Thus, $[D]_{i,i} \equiv \sum_j a_{i,j}$ and $[D]_{i,i}$ represents the sum of the weights of all the edges incident to vertex i. An important property of the normalized Laplacian matrix (for undirected graphs) is that it is a real, symmetric, positive semidefinite matrix. As a result, the normalized Laplacian matrix can be factorized as $L = U \Lambda U^T$, where $U = [u_0|u_1|\cdots|u_{N-1}]$ is the orthogonal matrix of eigenvectors. These eigenvectors correspond to the eigenvalues given by the diagonal matrix $\Lambda \equiv \text{diag}(\lambda_0, \lambda_1, \cdots, \lambda_{N-1})$, $i.e.$ $\Lambda_{ii} = \lambda_i$. These eigenvalues define the spectrum of the graph and provide a frequency interpretation of the "modes" for the graph. Small eigenvalues correspond to low frequencies and vice-versa.

Many graph signal processing methods are spectrally-based [8] as it allows for the analysis of the signals in terms of their frequency content. In a connected graph, the eigenvalues of the normalized graph Laplacian, $\Lambda = [\lambda_0, \lambda_1, \cdots, \lambda_{N-1}]$, satisfy the inequality, $0 = \lambda_0 < \lambda_1 \leqslant \cdots \leqslant \lambda_{Max} \leqslant 2$. Moreover, the eigenvectors of the normalized Laplacian matrix form an orthonormal space, $i.e.$ $u_i^T u_j = \delta(i - j)$. Let a graph signal $x \in R^N$ represent a feature vector for all nodes of a graph, where x_i is the feature value for the ith node. The graph Fourier transform of the signal x is defined as $\hat{x} \equiv \mathcal{F}(x) \equiv U^T x$, where $\hat{x}$ is the transformed signal. The inverse graph Fourier transform is defined as $x = \mathcal{F}^{-1}(\hat{x}) \equiv U\hat{x}$.

The transformed signal $\hat{x}$ possesses the coordinates of the graph signal in the new orthogonal space U. Therefore, the input signal x can be expressed as $x = \sum_i \hat{x}_i u_i$, which is the inverse graph Fourier transform expression. Considering this, the spectral convolution in the Fourier domain between a signal and a filter $g_\theta \equiv \text{diag}(\theta)$, where $\theta \in R^N$, is given by $g_\theta \hat{x} = g_\theta(U^T x)$. This is equivalent to the multiplication of each diagonal element of g_θ with each element of $\hat{x}$. The filter output signal $z \in R^N$ in the vertex domain is then obtained by taking the inverse Fourier transform of $g_\theta \hat{x} = g_\theta(U^T x)$ as $z \equiv g_\theta * x = U g_\theta U^T x$.

The above definition of convolution is the basis of all spectral convolutions on graph neural networks. Note that the only distinction among these spectral convolutions relies on the properties of the filter g_θ. In a machine learning setting, θ is the vector of learnable parameters. Furthermore, this spans a matrix Θ where each column corresponds to a vector θ. This matrix can be viewed as the span of "channels", where each of these corresponds to a different filter g_θ. In practice, the computational complexity of the above graph convolution is of order $\mathcal{O}(N^3)$, which makes it impractical for large graphs. To circumvent this issue, Chebyshev polynomials are used to approximate the filter g_θ and thereby reducing the complexity to $\mathcal{O}(N)$ [5]. Thus, the filter g_θ is approximated by Chebyshev polynomials of the diagonal matrix of eigenvalues as $g_\theta = \sum_{i=0}^{K} \theta_i T_i(\hat{\Lambda})$, where $\hat{\Lambda} = 2\Lambda/\lambda_{max} - I$ contain the normalized eigenvalues in the interval $[-1, 1]$, and $T_i(x)$ are Chebyshev polynomials of the first kind.

By defining $\tilde{L} \equiv 2L/\lambda_{Max} - I$ and using the property $T_i(\tilde{L}) = UT_i(\tilde{\Lambda})U^T$, the filter output $g_\theta \hat{x} = g_\theta(U^T x)$ can be written as $z = g_\theta * x = \sum_{i=0}^{K} \theta_i T_i(\tilde{L})x$. Therefore, we can infer that approximating the filter by Chebyshev polynomials makes GCNs to be locally supported in the vertex space, *i.e.* only signal values in a neighbourhood of a given vertex are needed to compute the output value of the vertex [8]. Using a first-order approximation of $g_\theta * x = \sum_{i=0}^{K} \theta_i T_i(\tilde{L})x$, with $K = 1$ and $\lambda_{Max} = 2$, the simplified version of graph convolution becomes $g_\theta * x = \theta_0 x - \theta_1 D^{-\frac{1}{2}} A D^{-\frac{1}{2}} x$ [12]. To preclude over-fitting by limiting the number of parameters, we assume $\theta = \theta_0 = -\theta_1$. After substituting this equality in previous equation we get $g_\theta * x = \theta \left(I + D^{-\frac{1}{2}} A D^{-\frac{1}{2}} \right) x$.

We now consider multiple channels of input/output and introduce non-linearities. This yields

$$H = g_\theta * X = f\left(\Theta \left(I + D^{-\frac{1}{2}} A D^{-\frac{1}{2}} \right) X \right), \qquad (1)$$

where X is a matrix of node feature vectors X_i, and $f(\cdot)$ is called an activation function.

Since the numerical experiments show instability in the above GCN [9], we opt to solve the problem via re-normalization. To this end, $I + D^{-\frac{1}{2}} A D^{-\frac{1}{2}}$ is replaced with $\tilde{D}^{-\frac{1}{2}} \tilde{A} \tilde{D}^{-\frac{1}{2}}$, where $\tilde{A} \equiv A + I$ and $\tilde{D}$ is the diagonal degree matrix $\tilde{D}_{i,i} = \sum_j \tilde{A}_{(i,j)}$. As a result, the output of a single layer is then calculated using this equation

$$H = f\left(\Theta \tilde{D}^{-\frac{1}{2}} \tilde{A} \tilde{D}^{-\frac{1}{2}} X \right), \qquad (2)$$

Thus, the forward propagation rule applied in a multi-layer graph convolutional networks (GCNs) follows an iterative scheme given by:

$$H^{(l+1)} = \sigma\left(\tilde{D}^{-\frac{1}{2}} \tilde{A} \tilde{D}^{-\frac{1}{2}} H^{(l)} W^{(l)} \right). \qquad (3)$$

In Eq. (3), the filter θ and activation function $f(\cdot)$ are replaced by W and σ, respectively. Additionally, the state of the lth layer is represented by $H^{(l)}$, where the initial value of the state is $H^{(0)} = X$.

3   Convolutional Filters for GCNs

The filter in equation $g_\theta * x = \theta \left(I + D^{-\frac{1}{2}} A D^{-\frac{1}{2}} \right) x$ is essentially a low-pass one, which only captures low frequency components of a signal/feature[8]. This low-pass filter essentially computes a weighted average of the signal values in a localized neighbourhood about a given vertex. Here, by introducing high-pass filtering, we aim to capture those high frequency components, which represent variation of the signal values about the localized neighbourhood.

3.1   Layer-Wise Linear Model

To commence, we depart from a localized first-order approximation of spectral graph convolution, which comprises an input layer, two convolutional layers and one output layer. In previous section, the symmetric normalization in Eq. (3) is only valid for undirected graphs. For the more general case of directed graphs, left-normalization (also called row-normalization) is used [10]. Therefore, symmetric normalization in Eq. (3), $\tilde{D}^{-\frac{1}{2}}\tilde{A}\tilde{D}^{-\frac{1}{2}}$, is replaced by $\tilde{D}^{-1}\tilde{A}$. This yields the forward propagation rule on directed graphs as

$$H^{(l+1)} = \sigma\left(\tilde{D}^{-1}\tilde{A}H^{(l)}W^{(l)}\right), \tag{4}$$

where l represents the current layer index and $l+1$ corresponds to that for the layer immediately after. We further simplify this equation by replacing $\tilde{D}^{-1}\tilde{A}$ by $\hat{A}$, which is defined below. This gives the final propagation rule as

$$H^{(l+1)} = \sigma\left(\hat{A}H^{(l)}W^{(l)}\right). \tag{5}$$

3.2 Low and High-Pass Filters

We now turn our attention to the modification of $\tilde{A}$, so as to incorporate both low and high-pass filters, in each convolutional layer. Consider N instances in each dataset and let A to be a symmetric adjacency matrix. We now define the following matrices:

$$\tilde{A} = I + A \qquad \text{and} \qquad \tilde{B} = I - A \tag{6}$$

where I is the identity matrix. To understand these equations in terms of its spectral filtering characteristics, consider the normalized Laplacian matrix which can be written as $L = I - A$. The relationship between each eigenvalue of the Laplacian and the adjacency matrices is given by $\lambda = 1 - \mu$, [16], where λ and μ are the eigenvalues of Laplacian and adjacency matrices, respectively. The (diagonal) matrix form of $\lambda = 1 - \mu$ is $\Lambda = I - M$. Using the definition of eigenvalues and eigenvectors, we have $LU = \lambda U$, and furthermore, using $L = I - A$, we get:

$$(I - A)U = \Lambda U \implies AU = (I - \Lambda)U. \tag{7}$$

Recall that the spectrum of normalized Laplacian matrix lies in range $0 \leqslant \lambda \leqslant 2$. Therefore, since $\lambda = 1 - \mu$, the spectrum of the corresponding adjacency matrix is $-1 \leqslant \mu \leqslant 1$. From a graph signal processing viewpoint, the frequency response function $1 + \mu$ represents a low-pass filter corresponding to the matrix $\tilde{A}$. When $\lambda = 0$, corresponding to the lowest frequency, $\mu = 1$ and the filter function value is $1 + \mu = 2$, corresponding to a large gain. When $\lambda = 2$, corresponding to the highest frequency, $\mu = -1$ and the filter function value is $1 + \mu = 0$, corresponding to a zero gain. The opposite is true for the high-pass filter, with frequency response function $1 - \mu$. This function takes the lowest and highest values at the smallest $\lambda = 0$ and largest $\lambda = 2$, respectively, and corresponds to the matrix $\tilde{B}$.

3.3   Incorporating Filters into the Convolutional Layers

Here, we proceed to apply row normalization to $\tilde{A}$ and $\tilde{B}$ as expressed in Eqs. (6). To compute $\hat{A}$ and $\hat{B}$:

$$\hat{A} = \tilde{D}^{-1}\tilde{A} \qquad \text{and} \qquad \hat{B} = \tilde{D}^{-1}\tilde{B}, \qquad (8)$$

where $\tilde{D}$ is the degree matrix and $\tilde{D}_{i,i} = \sum_j \tilde{A}$. These two matrices, $\hat{A}$ and $\hat{B}$, are sparse since $\tilde{D}$ is a diagonal matrix. This sparsity is crucial since real-world datasets are often sparse and hence this approach prevents over-fitting [17]. Here, $\hat{A}$ and $\hat{B}$ represent the low-pass and high-pass filters, respectively. To combine the effect of both filters, we concatenate $\hat{A}$ and $\hat{B}$ horizontally. Hence, in $H^{(l+1)} = \sigma\left(\hat{A}H^{(l)}W^{(l)}\right)$, $\hat{A}$ is replaced by $\left[\hat{A} \vdots \hat{B}\right]$, which results in the rule $H^{(l+1)} = \sigma\left(\left[\hat{A} \vdots \hat{B}\right]H^{(l)}W^{(l)}\right)$. To weight each of the filter matrices separately, the state or feature matrix, H, is multiplied by $\hat{A}$ and $\hat{B}$ individually. This gives $\left[\hat{A}H^{(l)} \vdots \hat{B}H^{(l)}\right]$. The purpose here is to be able to modify the classic forward propagation rule so as to incorporate the low-pass and high-pass filters. Note that, up to this step, the propagation rule is as follows:

$$H^{(l+1)} = \sigma\left(\left[\hat{A}H^{(l)} \vdots \hat{B}H^{(l)}\right]W^{(l)}\right). \qquad (9)$$

Here, we initialize the weight matrix making use of the Glorot or Xavier method expressed in [18]. Since we are interested in analysing the influence of low-pass and high-pass filters, the weight matrices for each of them is initialized separately. Hence, two different weight matrices are allocated to the low-pass and high-pass filters. The weight matrix, W, is a concatenated matrix formed by two other weight matrices as Δ and Γ. To be mathematically compatible, these two sub-matrices must be concatenated vertically $W^{(l)} = \begin{bmatrix} \Delta^{(l)} \\ \Gamma^{(l)} \end{bmatrix}$. The final forward propagation rule can then be written as:

$$H^{(l+1)} = \sigma\left(\left[\hat{A}H^{(l)} \vdots \hat{B}H^{(l)}\right]\begin{bmatrix} \Delta^{(l)} \\ \Gamma^{(l)} \end{bmatrix}\right). \qquad (10)$$

Now $\Delta^{(l)}$ is multiplied by $\hat{A}H^{(l)}$ and $\hat{B}H^{(l)}$ is multiplied by $\Gamma^{(l)}$. By concatenating $\hat{A}H^{(l)}$ and $\hat{B}H^{(l)}$ horizontally, both the low and high-pass filters can simultaneously influence the output at each node. Note that the elements of the resulting matrix, inside the activation function, consists of terms arising from both the low and the high-pass filters. Furthermore, each layer's output forms the input of the next layer after passing through a nonlinear activation function, σ.

4   Implementation Issues

Figure 1 shows the schematic of our GCNN implementation. In our implementation, the feature matrix X, is provided at input. According to equation $g_\theta * x = \theta\left(I + D^{-\frac{1}{2}}AD^{-\frac{1}{2}}\right)x$, in the first layer, $H^{(0)} = X$, we have:

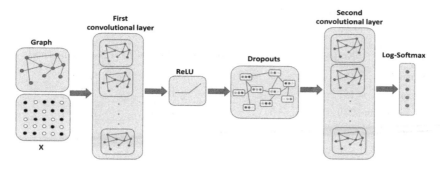

Fig. 1. A schematic representing our two-layer convolutional graph neural network. Each convolutional layer learns a hidden representation by aggregating feature information using local support. After the first layer, a ReLU function and dropout with probability of 0.5 is applied. In the output layer, a log-softmax is used for classification.

$$H^{(1)} = ReLU\left(\left[\widehat{A}X \vdots \widehat{B}X \right] \begin{bmatrix} \Delta^{(0)} \\ \Gamma^{(0)} \end{bmatrix} \right). \tag{11}$$

The output of Eq. (11) is the input to the next layer, which can be expressed as:

$$H^{(2)} = \left[\widehat{A}H^{(1)} \vdots \widehat{B}H^{(1)} \right] \begin{bmatrix} \Delta^{(1)} \\ \Gamma^{(1)} \end{bmatrix}. \tag{12}$$

In our network, multiplication of the combined low and high-pass filtering, expressed in $\left[\widehat{A} \vdots \widehat{B} \right]$, with the features matrix, is followed by the first convolutional layer. The process of convolution entails multiplication by the weight matrix. After convolution in the first layer, a *ReLU* activation function is applied. To prevent over-fitting in the model, after compiling the first layer, a dropout is introduced to the architecture. After the second convolutional layer, a log-Softmax loss is used to perform classification for mutually exclusive classes in the output layer.

5   Experiments

We now illustrate the utility of our GCNN for purposes of classification. To this end, we show results and provide comparison with the method in [12]. Our choice of alternative is based upon the notion that, as mentioned earlier, our

Table 1. Statistics of the citation network datasets used and results yielded by our method and the alternative in [12]

Dataset	Nodes	Edges	Classes	Features	Method in [12]	Our method
Cora	2078	5429	7	1433	80.1%	82.5%
Citeseer	3327	4732	6	3703	67.9%	69%

Table 2. Statistics of the knowledge graph dataset WebKB used in our experiments and results yielded by our method and the alternative in [12]

Dataset	Nodes	Edges	Classes	Features	Method in [12]	Our method
Cornell	195	304	5	1703	90%	90%
Texas	187	328	5	1703	42%	47.4%
Washington	230	446	5	1703	55.6%	77.8%
Wisconsin	265	530	5	1703	56%	60%

architecture is based on first order graph convolution approximation first presented in [22] and later advanced by [12].

All the datasets used to investigate the performance of our proposed approach were trained for 200 epochs using the Adam optimizer, which involves a first-order gradient-based optimization. Also, for purpose of providing a fair comparison, we have set all hyper-parameters of our approach to the same values as those reported in [12]. We implemented our network in Python with all layers defined according to Fig. 1 with a learning rate of 0.01 and a dropout rate of 0.5. In order to avoid over-training the model, a L-2 regularisation method with the term coefficient 5×10^{-4} is used in all our experiments.

5.1 Datasets

For our experiments, we have used three widely available datasets. These are the Cora [19][1], Citeseer [20][2] and WebKB[3] datasets. Here, and for the sake of consistency in our comparison with the alternative in [12], both the Citeseer and WebKB datasets were pre-processed before feeding them to the GCNs. A data cleaning was applied to the CiteSeer dataset as described in [21]. In case of the Cora dataset, we have used the same dataset splits as those employed in [12]. Regarding the other two datasets, we have used random 60%-20%-20-% dataset splits for training, validation and testing. All necessary matrices are constructed according to the metadata files provided with each dataset, which contain the information on nodes and edges necessary for constructing the adjacency matrices for the graph neural network. The number of nodes, features, classes and other information for the datasets have been summarised in Tables 1 and 2.

5.2 Classification Results

The results of our proposed approach are summarized and compared with the method in [12] in Table 1 and Table 2, which show the average accuracy and improvement over 10 runs, where each run has a unique random initialization. From these tables, we can appreciate improvements across datasets as compared

---

[1] The dataset can be accessed at https://relational.fit.cvut.cz/dataset/CORA.

[2] Widely available at http://networkrepository.com/citeseer.php.

[3] For more information on WebKB, go to http://www.cs.cmu.edu/~webkb/.

to the approach in the alternative, which only employs low-pass filters. The WebKB-Cornell dataset is the only case where our approach shows the same level of accuracy as the alternative. Breaking down the results, for the Cora and Citeseer datasets, we find slightly improved accuracy whereas some of the differences in the WebKB dataset are substantial. The reasons for more improvement in accuracy by adding highpass filters in these datasets can be attributed to the structure of the datasets. In particular, we see that the WebKB datasets consist of a smaller number of nodes, edges and classes. Moreover, the proportion of the number of edges to nodes are: (1) Cora = 2.6, (2) Citeseer = 1.3, (3) WebKB – Cornell = 1.5, (4) WebKB – Texas = 1.7, (5) WebKB – Washington = 1.94 and (6) WebKB – Wisconsin = 2. This hints that, increasing the density of the graph also implies increasing the classification accuracy. This is consistent with the results, where large gains can be seen with small problem instances whose graph structure is denser. Another point of interest is the level of accuracy across these datasets. We notice that the accuracy in the Cora dataset is higher as compared to that yielded for the Citeseer dataset. This is consistent for both our method and the alternative. Furthermore, we see that the range is large for the WebKB datasets (47.4% to 90%), which is not dissimilar to previous results. This is also consistent to the intuitive notion that, in a dense graph, the local support is provided by more edges within a particular neighbourhood, leading to improved classification accuracy.

6   Conclusions

In this paper, we have presented a method to integrate low and high-pass filters into convolutional layers in GCNNs. Our method is quite general in nature and applies to both, directed and undirected graphs. We used concepts from spectral graph theory, and exploited the relationship between eigenvalues and the modes of the graph to incorporate high-frequency information in the learning process. We illustrated the utility of our method for classification making use of widely available citation datasets and compared our results against those yielded by an alternative. In our experiments, our method outperforms the alternative, providing a margin of improvement in the classification accuracy.

References

1. Dou, W., Zhang, X., Liu, J., Chen, J.: HireSome-II: towards privacy-aware cross-cloud service composition for big data applications. IEEE Trans. Parallel Distrib. Syst. **2**(26), 455–466 (2013)
2. Li, Y., Tarlow, D., Brockschmidt, M., Zemel, R.: Gated graph sequence neural networks. In: International Conference on Learning Representations (2016)
3. Fukushima, K., Miyake, S., Ito, T.: Neocognitron: a neural network model for a mechanism of visual pattern recognition. IEEE Trans. Syst. Man Cybern. B Cybern. **2**(5), 826–834 (1983)
4. Taubiǹ, G.: Geometric signal processing on polygonal meshes. In: Proceedings of EUROGRAPHICS 2000: state of the art report (2000)

5. Defferrard, M., Bresson, X., Vandergheynst, P.: Convolutional neural networks on graphs with fast localized spectral filtering. In: Advances in Neural Information Processing Systems, pp. 3844–3852 (2016)
6. LeCun, Y., Bottou, L., Bengio, Y., Haffner, P.: Gradient-based learning applied to document recognition. In: Proceedings of the IEEE, pp. 2278–2324. IEEE (1998)
7. Scarselli, F., Gori, M., Tsoi, A., Hagenbuchner, M., Monfardini, G.: The graph neural network model. IEEE Trans. Neural Netw. **2**(20), 61–80 (2008)
8. Shuman, D.I., Narang, S.K., Frossard, P., Ortega, A., Vandergheynst, P.: The emerging field of signal processing on graphs: extending high-dimensional data analysis to networks and other irregular domains. IEEE Signal Process. Mag. **2**(30), 83–98 (2013)
9. Wu, Z., Pan, S., Chen, F., Long, G., Zhang, C., Philip, S.Y.: A comprehensive survey on graph neural networks. IEEE Trans. Neural Netw. Learn. Syst. **32**, 4–24 (2020)
10. Zhou, J., Cui, G., Zhang, Z., Yang, C., Liu, Z., Wang, L., Li, C., Sun, M.: Graph neural networks: a review of methods and applications. ArXiv preprint arXiv:1812.08434 (2018)
11. Spielman, D.A.: Algorithms, graph theory, and linear equations in Laplacian matrices. In: 4th Proceedings of the International Congress of Mathematicians, pp. 2698–2722. Plenary Lectures and Ceremonies Vols. II-IV: Invited Lectures, World Scientific (2010)
12. Kipf, T.N., Welling, M.: Semi-supervised classification with graph convolutional networks. In: International Conference on Learning Representations (2017)
13. Bruna, J., Zaremba, W., Szlam, A., LeCun, Y.: Spectral networks and locally connected networks on graphs. arXiv preprint arXiv:1312.6203 (2013)
14. Sarkar, S., Boyer, K.L.: Quantitative measures of change based on feature organization: Eigenvalues and Eigenvectors. Comput. Vis. Image Underst. **17**(1), 110–136 (1998)
15. Narang, S.K., Ortega, A.: Perfect reconstruction two-channel wavelet filter banks for graph structured data. IEEE Trans. Signal Process. **60**(6), 2786–2799 (2012)
16. Weiss, Y.: Segmentation using eigenvectors: a unifying view. In: International Conference on Computer Vision, pp. 975–982 (1999)
17. Xiang, G., Wei, H., Ongming, G.: Exploring structure-adaptive graph learning for robust semi-supervised classification. In: 2020 IEEE International Conference on Multimedia and Expo (ICME), pp. 1–6 (2020)
18. Glorot, X., Bengio, Y.: Understanding the difficulty of training deep feedforward neural networks. In: Proceedings of the Thirteenth International Conference on Artificial Intelligence and Statistics, pp. 249–256 (2010)
19. McCallum, A.K., Nigam, K., Rennie, J., Seymore, K.: Automating the construction of internet portals with machine learning. Inf. Retrieval **3**(2), 127–163 (2000)
20. Giles, C.L., Bollacker, K.D., Lawrence, S.: CiteSeer: an automatic citation indexing system. In: Proceedings of the third ACM conference on Digital libraries, pp. 89–98 (1998)
21. Wang, Y., et al.: A data cleaning method for citeseer dataset. In: Cellary, W., Mokbel, M.F., Wang, J., Wang, H., Zhou, R., Zhang, Y. (eds.) WISE 2016. LNCS, vol. 10041, pp. 35–49. Springer, Cham (2016). https://doi.org/10.1007/978-3-319-48740-3_3
22. Bruna, J., Zaremba, W., Szlam, A., LeCun, Y.: Spectral networks and locally connected networks on graphs. In: International Conference on Learning Representations (2014)

Selecting Features from Time Series Using Attention-Based Recurrent Neural Networks

Michal Myller[1,2] , Michal Kawulok[1,2] , and Jakub Nalepa[1,2(✉)]

[1] Silesian University of Technology, Gliwice, Poland
jnalepa@ieee.org
[2] KP Labs, Gliwice, Poland

Abstract. Capturing, storing, and analyzing high-dimensional time series data are important challenges that need to be effectively tackled nowadays, as the extremely large amounts of such data are being generated every second. In this paper, we introduce the recurrent neural networks equipped with attention modules that quantify the importance of features, hence can be employed to select only an informative subset of all available features. Additionally, our models are trained in an end-to-end fashion, hence are directly applicable to infer over the unseen data. Our experiments included datasets from various domains and showed that the proposed technique is data-driven, easily applicable to new use cases, and competitive to other dimensionality reduction algorithms.

Keywords: Attention · RNN · Feature selection · Time series analysis

1 Introduction

The amount of generated data grows rapidly in many domains, hence collecting, storing, and processing such massive datasets are critical problems that need to be tackled in order to extract value from big data. Additionally, high data dimensionality can easily lead to the problem of the curse of dimensionality which refers to various phenomena that arise while analyzing such data that do not occur in low-dimensional spaces. Since the number of required examples grows exponentially with the number of features, capturing ground-truth sets which could be used for training supervised learners is often infeasible in practice.

To circumvent the problem of high data dimensionality, we exploit various dimensionality reduction techniques, encompassing *feature selection* and *feature extraction*. The former methods elaborate a subset of informative features that are already present in the data, whereas those features that are not relevant or

This work was co-financed by the Silesian University of Technology grant for maintaining and developing research potential, and by the Polish National Centre for Research and Development under Grant POIR.01.01.01-00-0853/19. JN was supported by the Silesian University of Technology funds (02/080/BKM20/0012).

A. Torsello et al. (Eds.): S+SSPR 2020, LNCS 12644, pp. 87–97, 2021.
https://doi.org/10.1007/978-3-030-73973-7_9

are redundant should be filtered out. Such approaches include similarity-based algorithms [11] (which often fail to handle feature redundancy, as they repeatedly find correlated features), the techniques built upon the information theory and statistical analysis [7] that require ground truth, sparse learning approaches that are embedded into the learning process [6], and recent deep networks [4]. On the other hand, feature extraction builds new low-dimensional data representations that should retain the underlying data properties and characteristics [2]. Despite the fact that many dimensionality reduction approaches do not consider the temporal nature of data, they are commonly applied to time series with highly correlated data points captured in consecutive time points.

We introduce a deep learning-powered feature selection method that is specifically designed to capture temporal characteristics of time series. We build upon our previous work concerning the attention-based convolutional neural networks [9], and propose recurrent neural networks (RNNs) equipped with attention modules (Sect. 2) that quantify the importance of specific features in the time series data. Such networks may be used not only for reducing the dimensionality of time series via selecting informative features, but also to elaborate fully functional models that are ready to be deployed over the unseen data. In the experimental study (Sect. 3), we verified the impact of attaching attention modules to long short-term memory (LSTM) networks, and confronted our method with well-established feature selection and extraction algorithms.

2   Method

We present our model equipped with attention that determines the most informative features in the time series data. First, we discuss the baseline model (Sect. 2.1), and then our attention-based models (Sect. 2.2). This section is concluded with an illustrative example of our feature selection (Sect. 2.3).

2.1   Baseline Model

In our baseline recurrent model (Fig. 1), the time series sample of size T is processed by the two-layered LSTM network and it is encoded into a single latent vector, which is essentially the last hidden state of the second layer of the RNN. Then, such vector is transferred to the *classifier* part of the model, which is a simple multi-layer perceptron with one input, one hidden, and one output layer combined with the parametric ReLU activations. The classifier contains only one hidden layer, because—as the *universal approximation theorem* [5] states—a feed-forward network with a single hidden layer can approximate any continuous function for inputs within a specific range. The number of neurons n in the input and hidden layers was kept small to minimize the number of trainable parameters and ultimately avoid overfitting. Moreover, a larger number of neurons would increase the processing time, and—with datasets containing a large number of samples—that would hamper the practical deployment of such models. The number of neurons in the last layer is dependent on the number of classes C in the dataset. However, for binary classification, the number of output

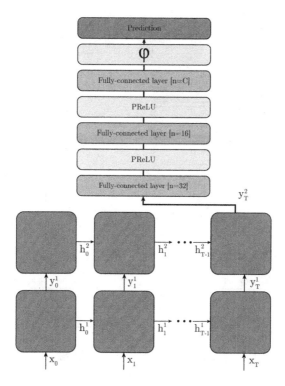

Fig. 1. The baseline model. The violet cells are the LSTM cells of the RNN, light gray indicates activations, and dark gray are the fully-connected layers. (Color figure online)

neurons is set to one. The same stands true for the last activation function φ, where *softmax* is used for multiclass scenario, and *sigmoid* for the binary one.

2.2 The Model with Attention Modules

The model with an additional *attention module* and *attention classifier* is presented in Fig. 2. Similarly to the baseline model, the time series sample of size T is processed through the two-layered LSTM network. The original classifier receives the last hidden state of the LSTM, and elaborates its own prediction.

The evaluation of features' importance is performed by the *attention module*. It accepts a hidden state y_t^2 at each time step t, processes it, and using the *sigmoid* activation returns the attention score for each feature. When all hidden states are evaluated in such a manner, the scores are collected into a vector, which is then passed through the *softmax* activation to normalize the output into a probability distribution. Such vector H_T can be referred to as a *heatmap* indicating the most important features. The larger the value in the heatmap is, the more contribution it had into the model's predictions.

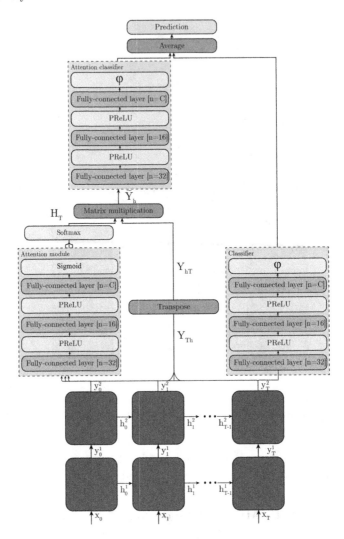

Fig. 2. The model with attention (n is the number of neurons in each fully-connected layer). The violet shows the LSTM cells, light gray indicates activations, dark gray are the fully-connected layers, and light orange marks arbitrary mathematical operations. (Color figure online)

To take advantage of the heatmap, the matrix multiplication is performed:

$$\tilde{Y}_h = Y_{Th}^{\mathsf{T}} H_T, \text{ where}$$
$$\tilde{Y}_h \in \mathbb{R}^m, Y_{Th} \in \mathbb{R}^{n \times m} \text{ and } H_T \in \mathbb{R}^n. \tag{1}$$

Here, Y_{Th} is the matrix containing all the hidden states concatenated from the second layer of the LSTM network, n is the size of a time step sample, and m is the size of the hidden state. This process weights each hidden state based on its importance captured in the heatmap. Such weighted vector $\tilde{Y}_h$ is passed to the *attention classifier* which evaluates it. Predictions from the standard classifier and from the attention classifier are averaged to form the final prediction.

It should be pointed out that each component, i.e., *classifier, attention module*, and *attention classifier* can be easily replaced, meaning that they could be built out of any kind of a layer. For example, they could exploit convolutional layers, which would take advantage of spatial relations within the data.

2.3   Illustrative Example

During the training phase, the heatmap is constantly updated as the attention module learns to output higher values for entries which are more important. When the training process is finished, cells with the higher values represent features with higher importance. To better understand the process of *feature selection*, Fig. 3 presents an example of the heatmap extraction. The vectors y_T, where $T = 4$, represent time steps (features) returned by the last layer of the LSTM model. Each of them is passed to the attention module. Such values are concatenated for all vectors, and processed by *softmax* to form the final heatmap. In the example, the second feature holds the highest score, so it should be picked during the final selection. The first one achieved the lowest score, meaning that attention considers it irrelevant—the higher is the value in the final heatmap, the more important the feature becomes.

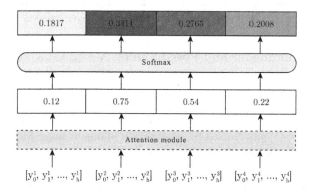

Fig. 3. The heatmap extraction example—the darker the cell of the heatmap becomes, the more important the corresponding feature is.

3   Experiments

The objectives of the experiments are multifold. First, we compare the attention-based RNNs with other dimensionality reduction approaches, and confront our method with principal component analysis (PCA), and the mutual information feature selection (MI) [1]. For all of the algorithms, we elaborate the reduced representations of sizes $\mathcal{F} \in \{2, 5, 10, 15, 20, 25\}$. Additionally, we investigate the classification abilities of the models applied over the full and reduced datasets to verify if there are any differences in the overall performance of the models. To quantify the performance of the classifiers, we report their overall accuracy (OA), balanced accuracy (BA), being the recall averaged across all classes (we tackle both binary and multiclass problems), F_1 score (a harmonic mean of the precision and recall—we report F_1 for the binary problems), specificity (Spec.), and the Cohen's kappa (κ) for the multiclass problems [8]. The κ and F_1 metrics range from 0 to 1, where one means the perfect score, whereas OA, BA, and Spec. are reported in %. We consider sets from various domains to show that the attention RNNs are data-driven and can be applied to any time series data.

All the models were implemented in `Python 3.7` with `PyTorch`, and trained using ADAM with default parameters: the learning rate of 0.001, $\beta_1 = 0.9$, and $\beta_2 = 0.999$. Each model was trained for 200 epochs, unless the stop criterion was met (OA over the validation set V did not improve for 15 epochs).

3.1   Datasets

In our study, we focus on three experiments: (i) classifying smiles as posed and spontaneous based on image sequences, (ii) segmenting hyperspectral images, and (iii) classifying heartbeat electrocardiogram sequences. The following subsections discuss the corresponding datasets, alongside the followed setup for dividing them into the training (T), validation (V), and test (Ψ) subsets. The T's are used for training the supervised models, whereas the V's are utilized to monitor the training process. The final results are quantified over unseen Ψ's.

Classifying Spontaneous and Posed Smiles. We exploit a database of RGB image sequences (UvA-NEMO), where people either fake their smile, or smile genuinely. It contains 1240 sequences (1920 × 1080 resolution, 50 fps, 44–606 frames), where 597 of them are spontaneous and 643 are posed, collected from 400 subjects (age of 8–76). We extract 17 Action Units (AU) [10] that encode various facial movements caused by contraction of one or more facial muscles for each frame. As the proposed attention-based model operates on samples having the equal number of time steps, the frames count with their respective AUs in each sequence have been interpolated to 44 to match the shortest one. We utilize the two level 10-fold cross-validation, where each time a test fold is separated, 9-fold cross-validation is used to train/validate the model. During the 9-fold cross-validation, one fold is separated for validation where the rest is used for training. Hence, we execute 90 runs, and average the results.

Segmentation of Hyperspectral Images. Hyperspectral cubes capture multiple images acquired across a continuous range of wavelengths to reflect subtle properties of the objects in the scene. Even though hyperspectral images are not time series, they indeed could be treated as such. By considering each pixel's spectral dimension as a sequence, each consecutive band acts as a time step in a time series—the consecutive bands in the hyperspectral image are highly-correlated. In this experiment, we use two benchmark scenes: Pavia University (PU) and Salinas Valley (SV). The PU scene (of size 340×610, 103 bands, 430–850 nm, 9 classes of urban objects) was captured over the Pavia University in Lombardy, Italy using the Reflective Optics System Imaging Spectrometer sensor. SV (217×512, 204 bands, 400–2450 nm, 16 vegetation classes) was acquired over Salinas Valley in California, USA, using the NASA Airborne Visible/Infrared Imaging Spectrometer sensor. The Monte Carlo cross-validation was used as the validation technique for both scenes—we divide them 30 times into T, V, and Ψ with 80%, 10%, and 10% of all pixels, and average the results.

Classification of Heartbeat Electrocardiogram Sequences. This dataset consists of 5000 time series, each with 140 time steps. Each sample corresponds to a heartbeat from a single patient, obtained using the electrocardiogram (ECG). The set originally contains five classes, but because of the extreme imbalance ratio between them, we merge all diseaseous classes into one, hence we tackle binary classification with 2919 healthy and 2081 diseaseous samples. The sequences are randomly divided into T and Ψ with 500 and 4500 examples [3]. Our models were trained 30× with different V's (10% of random T sequences).

Table 1. The quantitative results (best are boldfaced) obtained using all algorithms.

Alg.	$\mathcal{F}$	UvA-NEMO				PU			SV			ECG			
		OA	BA	F_1	Spec.	OA	BA	κ	OA	BA	κ	OA	BA	F_1	Spec.
B	All	67.7	67.9	**0.70**	65.3	**90.9**	**88.7**	**0.88**	81.5	81.9	**0.93**	0.79	91.8	0.90	**94.2**
B(A)	All	**68.1**	**68.6**	**0.70**	**65.8**	88.7	86.5	0.85	85.1	**88.4**	0.92	0.83	90.6	0.90	93.0
A	25	64.3	64.7	0.67	62.5	78.6	72.7	0.70	84.6	88.7	0.84	0.83	82.8	0.80	82.8
PCA	25	53.7	29.1	0.69	4.5	78.9	70.6	0.70	83.2	85.5	0.76	0.81	72.9	0.73	65.9
MI	25	66.5	67.0	0.68	63.9	83.7	79.9	0.78	81.3	83.4	0.78	0.79	71.5	0.66	77.9
A	20	65.6	65.9	0.68	63.3	73.9	66.2	0.64	84.1	88.1	0.84	0.82	82.7	0.79	79.8
PCA	20	53.4	38.7	0.66	22.8	83.5	78.0	0.77	88.5	92.7	0.82	0.87	80.9	0.80	75.4
MI	20	66.1	66.4	0.68	63.6	81.6	76.2	0.75	78.7	79.8	0.84	0.76	79.6	0.76	83.8
A	15	65.3	65.6	0.68	63.1	78.4	71.8	0.70	84.3	85.7	0.85	0.83	79.9	0.81	78.4
PCA	15	53.2	42.0	0.63	28.9	84.4	79.7	0.79	89.4	93.6	0.88	0.88	88.4	0.89	83.9
MI	15	65.2	65.6	0.67	62.9	81.0	75.5	0.74	83.2	86.9	0.83	0.81	80.0	0.77	83.1
A	10	65.6	66.0	0.68	63.7	72.6	64.0	0.62	81.9	82.0	0.88	0.80	86.6	0.86	84.5
PCA	10	52.9	47.4	0.59	39.3	87.3	83.3	0.83	**90.2**	**94.5**	0.92	**0.89**	92.1	**0.93**	90.7
MI	10	65.6	66.1	0.68	64.1	70.1	57.2	0.59	83.7	87.8	0.76	0.82	68.5	0.62	74.2
A	5	63.8	63.9	0.67	61.5	62.9	52.6	0.48	76.7	76.4	0.81	0.74	79.5	0.80	76.4
PCA	5	53.7	50.5	0.59	44.4	84.4	80.8	0.79	88.7	92.9	**0.93**	0.87	**92.4**	0.92	93.8
MI	5	66.7	67.1	0.69	64.8	64.0	43.9	0.49	76.6	77.9	0.84	0.74	81.5	0.79	87.0
A	2	63.5	63.7	0.66	61.3	63.8	53.2	0.49	62.5	55.3	0.73	0.58	71.6	0.70	65.1
PCA	2	52.8	49.0	0.61	43.2	80.0	73.3	0.72	83.6	87.7	0.88	0.82	86.8	0.85	90.8
MI	2	64.0	64.4	0.68	62.4	64.6	44.4	0.50	64.8	56.0	0.78	0.61	72.6	0.68	79.2

3.2   The Results

In Table 1, we gather the experimental results obtained over all datasets. We can appreciate that attaching attention modules to the RNNs (A) not only allows us to perform feature selection which does not significantly deteriorate the classification in most cases, when compared to the full data, but also helps improve the abilities of the baseline model (B) without attention—see B and B(A). Additionally, incorporating attention into the model has little impact on the number of its trainable parameters ($11.0 \cdot 10^4$ vs. $13.7 \cdot 10^4$ for UvA-NEMO, and $0.94 \cdot 10^4$ vs. $12.1 \cdot 10^4$ for all other sets without and with attention, respectively), and the corresponding floating point operations per seconds (883 vs. 1002, 1733 vs.

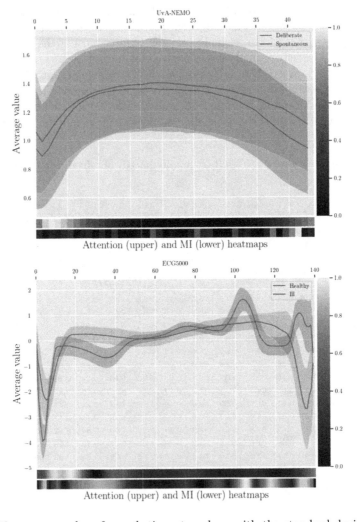

Fig. 4. The average values for each time step along with the standard deviation for each class in UvA-NEMO (upper plot) and ECG5000 (bottom plot). The heatmaps are displayed at the bottom—lighter areas indicate more important features.

2009, 3430 vs. 3974, and 2355 vs. 2729 kilo FLOPs for UvA-NEMO, PU, SV, and ECG without and with attention). The two-tailed Wilcoxon tests (at $p < 0.01$) revealed that only for $\mathcal{F} = 2$ selected features the differences were significantly worse than those elaborated over a larger subset of features using attention RNNs for almost all sets, with UvA-NEMO being an exception (the differences were not statistically important for $\mathcal{F} = \{2, 5, 15, 25\}$). Overall, B(A) delivered the best OA across all sets and investigated techniques (20 variants in total, see Table 1) with the average rank of 3.0, and the fourth best BA with the average rank of 6.25 (it was outperformed by PCA with 10, 15, and 5 principal components that obtained the average rank of 2.50 and 4.50, and 5.00, respectively).

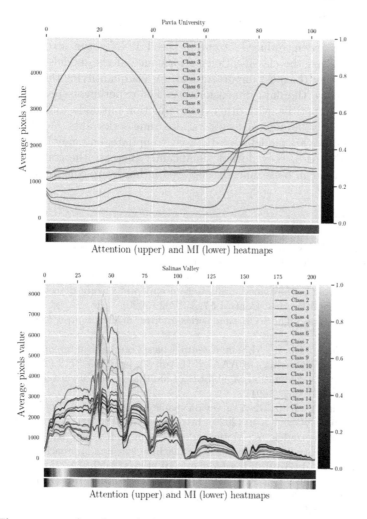

Fig. 5. The average values for each time step and for each class in Pavia University (upper plot) and Salinas Valley (bottom plot). The heatmaps are displayed at the bottom—lighter areas indicate more important features.

Although the feature selection results are comparable with other data reduction techniques, the qualitative evaluation presented in Figs. 4 and 5 shows that our approach can accurately capture the important information across the entire feature space. In Fig. 5, we can observe that the most important part of the electromagnetic spectrum is determined with our attention-based models, indicating those hyperspectral bands that convey the most distinguishable information about the underlying objects. Similarly, the most informative features in UvA-NEMO and ECG reflect those parts of the signal that can be sufficient to classify examples into two classes (see the initial parts of the signals in Fig. 4).

4   Conclusion

In this paper, we introduced attention-based RNNs—the attention modules not only improved the classification performance of the models, but were also exploited to select the most important features via quantifying their importance, hence to decrease the data dimensionality. The experiments revealed that our technique is competitive with other data reduction methods, and it is very flexible, as can easily be deployed over various time series data. Additionally, the attention modules are model-agnostic, easy to update, and may be incorporated into any RNN. Also, they do not significantly increase the complexity of the model, therefore can be seamlessly deployed in hardware-constrained settings.

References

1. Battiti, R.: Using mutual information for selecting features in supervised neural net learning. IEEE Trans. Neural Netw. **5**(4), 537–550 (1994)
2. Boonyakitanont, P., Lek-uthai, A., Chomtho, K., Songsiri, J.: A review of feature extraction and performance evaluation in epileptic seizure detection using EEG. Biomed. Signal Process. Control **57**, 101702 (2020)
3. Dau, H.A., et al.: The UCR time series archive. arXiv preprint arXiv:1810.07758 (2019)
4. Gui, N., Ge, D., Hu, Z.: AFS: an attention-based mechanism for supervised feature selection. In: Proceedings of AAAI CAI, vol. 33, pp. 3705–3713 (2019)
5. Hornik, K.: Approximation capabilities of multilayer feedforward networks. Neural Netw. **4**(2), 251–257 (1991)
6. Li, X., Wang, Y., Ruiz, R.: A survey on sparse learning models for feature selection. IEEE Trans. Cybern., 1–19 (2020). https://doi.org/10.1109/TCYB.2020.2982445
7. Masoudi-Sobhanzadeh, Y., Motieghader, H., Masoudi-Nejad, A.: FeatureSelect: a software for feature selection based on machine learning approaches. BMC Bioinform. **20**(1), 170 (2019). https://doi.org/10.1186/s12859-019-2754-0
8. Nalepa, J., Myller, M., Kawulok, M.: Transfer learning for segmenting dimensionally reduced hyperspectral images. IEEE Geosci. Remote Sens. Lett. **17**(7), 1228–1232 (2020)
9. Ribalta Lorenzo, P., Tulczyjew, L., Marcinkiewicz, M., Nalepa, J.: Hyperspectral band selection using attention-based convolutional neural networks. IEEE Access **8**, 42384–42403 (2020)

10. Zhi, R., Liu, M., Zhang, D.: A comprehensive survey on automatic facial action unit analysis. Visual Comput. **36**(5), 1067–1093 (2019). https://doi.org/10.1007/s00371-019-01707-5
11. Zhu, X., Wang, Y., Li, Y., Tan, Y., Wang, G., Song, Q.: A new unsupervised feature selection algorithm using similarity-based feature clustering. Comput. Intell. **35**(1), 2–22 (2019)

Feature Extraction Functions for Neural Logic Rule Learning

Shashank Gupta[✉], Antonio Robles-Kelly, and Mohamed Reda Bouadjenek

School of IT, Deakin University, Waurn Ponds, VIC 3216, Australia
guptashas@deakin.edu.au

Abstract. Combining symbolic human knowledge with neural networks provides a rule-based ante-hoc explanation of the output. In this paper, we propose feature extracting functions for integrating human knowledge abstracted as logic rules into the predictive behaviour of a neural network. These functions are embodied as programming functions, which represent the applicable domain knowledge as a set of logical instructions and provide a modified distribution of independent features on input data. Unlike other existing neural logic approaches, the programmatic nature of these functions implies that they do not require any kind of special mathematical encoding, which makes our method very general and flexible in nature. We illustrate the performance of our approach for sentiment classification and compare our results to those obtained using two baselines.

Keywords: Neural logic · Feature extracting functions · Rule learning

1 Introduction

Deep Neural Networks tend to suffer from the Black Box problem, mainly because their training is often purely data-driven, with no direct or indirect human intervention [17]. As a result, the interpretation of the input-output mapping is often challenging, if not almost intractable. Moreover, they do not have an inherent representation of causality or logical rule application. Indeed, previous work has shown that supervision purely in the form of data can lead a model to learn some unwanted patterns and provide misleading and incorrect predictions [13, 19]. These drawbacks hinder their applications in a wide range of domains such as cyber-security, healthcare, food safety, power generation and environmental management, which require a level of trust or confidence associated with the output of the network [18].

A common approach to make the predictions of a Neural Network explainable is to encode the intended rules or patterns derived from human domain knowledge in its trainable parameters [22]. This can be viewed as the process of combining structured logical knowledge representing high-level cognition with neural systems [3]. Indeed, logic rules provide a way to represent human knowledge in a structured format. However, logic rules need to be translated from natural language to logical representations. Moreover, they require a suitable encoding format, which is not a straightforward task because in most cases this encoding is application-specific.

© Springer Nature Switzerland AG 2021
A. Torsello et al. (Eds.): S+SSPR 2020, LNCS 12644, pp. 98–107, 2021.
https://doi.org/10.1007/978-3-030-73973-7_10

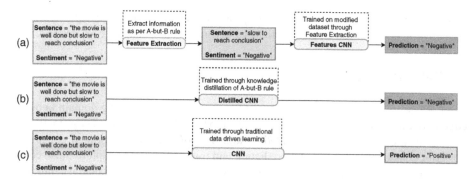

Fig. 1. (a) Overview of our proposed approach, which abstracts domain knowledge into the sentiment prediction of a neural network using feature extraction on the input data instead of distillation (bottom-way prediction). This achieves an ante-hoc rule based explanation of Neural Network inferential process as compared to (b), a distillation approach (middle-way prediction), which encodes knowledge into the network parameters or to (c), a straight application of a CNN to the sentence-sentiment tuple (top-way prediction) devoid of neural logic.

One way to efficiently encode human knowledge abstracted as first order logic rules into the parameters of a neural network is to use the iterative-knowledge distillation method [7], a process summarized in Fig. 2a. Briefly, iterative-knowledge distillation consists of representing structured human knowledge as a set of declarative first-order logic rules using soft-logic [1], then, encoding these rules into the parameters of the network via indirect supervision making use of knowledge distillation [5] at each training iteration. However, while iterative knowledge distillation makes the network to learn from both data and rules, we find that it implicitly makes an assumption of knowledge to remain static and true for every data point in the data set. Also, it imprints the knowledge into the network parameters permanently through distillation and do not provide any mechanism to accommodate for any change in the existing rules or addition of new ones. Thus, updating the rules requires to re-train the whole network. This can sometimes lead to a decrease in performance as shown in our experimental results.

To overcome the aforementioned issues, we propose to construct feature-extracting functions instead of logic rules from human knowledge as summarized in Fig. 1. These functions are analogous to decision rules [2] but modified to provide supervision similar to logic rules [7]. They are directly applied on the data so as to transfer the human knowledge into a distribution of the input data and influence the output of the network. We do this by viewing each function as a mini-batch processing step during each iteration. Since the functions are applied directly to the data, we do not need to compute the probability distributions nor construct a teacher network. This effectively reduces the complexity of our method. Also, these feature-extracting functions can be modified at any time during the training process, thus providing a lot of flexibility in adapting to qualitative and quantitative characteristics of the data under consideration. This is consistent with the well known properties of feature-extracting functions to express natural language [11], exploiting these traits for the training of deep networks to provide a more direct nature of supervision based upon the input data. Our method is quite general in nature, being a flexible manner of providing human knowledge supervision to the network, hence, it can be applied to tasks beyond Natural Language Processing.

2   Related Work

A lot of research has been done in the past few years for incorporating domain knowledge about a problem into machine learning models [4,7,12,20,21]. These methods essentially use knowledge represented in logical and/or symbolic form to construct posterior constraints on the model prediction and train the model to capture those constraints. Iterative Knowledge Distillation [7] sets itself apart from other neural symbolic methods as it provides a very flexible framework for integrating knowledge represented as first order logic rules with general purpose neural networks such as CNNs and RNNs.

A recent paper [9] gives a detailed analysis on the methodology used in [7], comparing its performance to other neural symbolic methods and arguing that it is not very effective in transferring knowledge to the neural network model (student network). Our work is consistent with this finding, achieving better performance by representing knowledge purely in terms of data, which is directly given as input for training a neural network. Moreover, since we have used the same data sets as those in [7], we employ an identical type of supervision as that used for the iterative knowledge distillation.

Finally, the authors in [9] suggest using a deep contextualized word representation model such as ELMo (Embeddings from Language Models [16]) and feed the embedding to the neural network to better capture the rule knowledge. However, this still fails to accommodate the dynamic nature of rules acquired from domain knowledge and its only limited to Language-related tasks.

3   Feature-Extracting Functions and Neural Logic

In our approach, we develop feature-extracting functions from human knowledge instead of constructing logic rules, which are expressed as programming functions and take the data instances in terms of independent features as input. They enforce the knowledge directly upon the neural network during training. This eliminates the need for constructing a teacher network and provide the flexibility to allow these functions to be applied either during the training process or during the pre-processing phase of the data. Figure 2b summarizes our approach.

3.1   Distillation vs Feature Extraction

In iterative distillation, a parametric baseline neural network is used as a "student", which needs to be provided with logical knowledge by a non-parametric "teacher" network. The teacher network is a projection of the student network over a regularized sub-space whereby the training data is constrained by logical rules. These logic rules are encoded using soft-logic [1] for the sake of constructing soft-boundaries and for calculating rule-regularized distributions. Thus, the training data comprises a set $D = \{(x_n, y_n)\}_{n=1}^{N}$ of N tuples (x_i, y_i), where x_i is an input instance (an independent

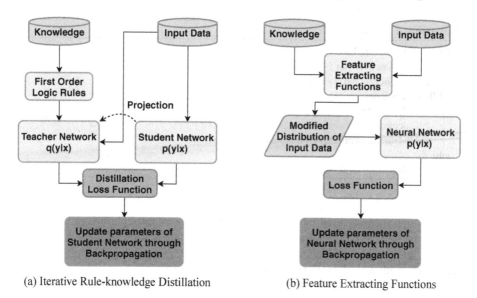

(a) Iterative Rule-knowledge Distillation   (b) Feature Extracting Functions

Fig. 2. (a) shows an overview of the iterative knowledge distillation framework in [7]; (b) shows an overview of the method we propose.

variable or a set of independent variables) and y is the corresponding target. The set of logic rules are expressed as $R = \{(R_l, \lambda_l)\}_{l=1}^{L}$ where R_l is the lth rule constructed from human knowledge over D and λ_l is the corresponding confidence value. A logic rule can be made up of several conditions or logic expressions. Each logic expression when instantiated on D produces a set of groundings as $\{(r_{l_g}(D))\}_{g=1}^{G_l}$ and thus, represent a rule as a set of ground expressions on D where each r_{l_g} is g^{th} grounding of the lth rule. The combined set of D and R is called *learning resources*.

For example, consider a set of movie reviews in which x comprises a set of tokens and the target y represents the sentiment value (0 for negative and 1 for positive reviews). From human knowledge, we know that, if a sentence has a syntactic structure of "A-but-B", then the sentiment of the sentence should be consistent with that of "B" component. Therefore, we can express the "A-but-B" statement as a logic rule stated as R_1 with an assumption that at least one ground expression will evaluate to "True" ($\lambda_1 = 1$). To encode this formally, we define a Boolean random variable $r_{l_g}(x, y) =$ "has an *A-but-B* structure", then apply an expectation operator on it to calculate sets of valid distributions in D, which will be further used to construct a "teacher network". This process is complex and time consuming which is not applicable to different types of datasets from different domains.

To tackle this drawback, our method combines the input and human knowledge to provide a pre-processed data set, which can be used for training the neural network. For the sake of consistency, we denote the input data $D = \{(x_n, y_n)\}_{n=1}^{N}$ as a set of N tuples (x, y), where x_i is a set of input independent variables and its corresponding target y_i, and the human knowledge $F = \{(F_l(D)\}_{l=1}^{L}$ as a set of L feature extracting

---

Algorithm 1: Training process.

Input: The training batch set $D = \{(x_n, y_n)\}_{n=1}^{N}$,
    The functions set $F = \{(F_l(D)\}_{l=1}^{L}$
Initialize the neural network parameters θ
while <u>Iteration</u> **do**
| 1: Calculate $D^* = \{(x_n^*, y_n)\}_{n=1}^{N}$
| 2: Calculate the probability distribution $p_\theta(Y|X^*)$
| 3: Update the parameters θ using objective function in Eq.(2)
end
Output: Trained neural network

---

functions, which are applied on D. Revisiting the previous example, instead of using soft-logic using auxiliary random variables, for the "A-but-B" rule we write a function $F_l = A - but - B(x, y)$, which outputs $(x*, y)$, where $x*$ has only 'B' features to be is consistent with $\lambda_l = 1$ as presented above.

3.2 Feature-Extracting Functions

Consider the conditional probability distribution $p_\theta(y_i|x_i)$ with parameter set θ as the softmax output of a Neural Network. Here, inspired by the labeling functions used by Ratner *et al.* [2], we use the input instance x_i to compute a post-processed instance x_i^*. We can view the post-processed instance x_i^* as an explicit representation of the domain knowledge, expressed in the rule under consideration and mapped onto the input instance x_i. This is an important observation since it hints at a minimisation problem on the cumulative output on the feature extracting functions so as to obtain the parameter set θ which can be expressed formally as follows:

$$\theta = \arg\min_{\theta \in \Theta} \frac{1}{N} \Sigma_{n=1}^{N} L(y_n, p_\theta(Y|X^*)) \tag{1}$$

where $L(\cdot)$ is the loss function of choice and $p_\theta(Y|X)$ is the conditional probability distribution of the target set Y given the set X^* of all the post-processed instances x_i^*. Since the information is purely present in the modified feature-set, the feature extracting functions become a post-processed input data for the network.

The treatment above also has the advantage of ease of implementation. We summarise the training and testing process of our method in Algorithm 1. Note that at each training iteration, we calculate the post-processed data set $D^* = \{(x_n^*, y_n)\}_{n=1}^{N}$ using the feature extracting functions $F_l \in F$ as applied on the input batch $D = \{(x_n, y_n)\}_{n=1}^{N}$. These are passed to the neural network to calculate the conditional probability $p_\theta(y_i|x_i^*)$ for each $(x_i^*, y_i) \in D^*$.

4 Experiments

We performed sentence-level binary sentiment classification and compared our method (CNN-F) with a baseline network (CNN) devoid of knowledge support and its knowledge distilled version (CNN-rule) created from Iterative knowledge Distillation [7]. We used the same convolutional neural network architecture proposed in [8] employing it's "non-static" version with the exact same configuration as that presented by the authors. We have compared our method CNN-F against the non-static version of the CNN in [8] as published by the authors and the CNN-rule in [7], which is a knowledge distilled version of CNN. Also, we have initialised word vectors using word2vec [10] and used fine-tuning, training the neural network using stochastic gradient descent (SGD) with the AdaDelta optimizer [23].

Since contrasting senses are hard to capture, we define a linguistically motivated rule called "A-but-B" rule akin to that in [7]. It states that if a sentence has an "A-but-B" syntactic structure, the sentiment of the whole sentence will be consistent with the sentiment of it's "B" component. For example, for the sentence S = "you can taste it, but there's no fizz", its sentiment is decided by only the sentiment of its B component = "there's no fizz". From this rule, we can define a feature-extracting function $F_1 = A - but - B(x, y)$ on set D which takes the input pair of sentence-label (x, y) and outputs $(x*, y)$ where $x*$ is corresponding features of "B".

We evaluate our method on three public data-sets:

1. The Stanford sentiment tree bank dataset (SST2) [15], which contains 2 classes (negative and positive), and 6,920/872/1,821 sentences in the train/dev/test sets respectively. Following [8], we train the models on both, sentences and phrases.
2. The movie review one (MR) introduced in [14]. This data set consists of 10,662 one-sentence movie reviews with negative or positive sentiments.
3. The customer reviews of various products data set (CR) presented in [6], which contains 2 classes and 3,775 instances[1].

We also evaluate our method only on the sentences containing "A-but-B" structure in the test sets of all three data sets under study to show that better performance of our method CNN-F on the whole test set is indeed attributed to the better performance on sentences having "A-but-B" structure. SST2 test set has a total of 1,821 instances out of which 210 instances exhibit the "A-but-B" structure. For MR data, it has a total of 10,662 instances out of which 1603 instances are found to have "A-but-B" structures. Finally, the CR data set has a total of 3,775 instances out of which 413 instances contain sentences with "A-but-B" structures. For the MR and CR dataset, we use nested 10-fold cross validation and report mean ±95% confidence interval for all performance metrics over the ten trails corresponding to the 10-fold cross validation. For these results, we have used the models of CNN, CNN-rule and CNN-F trained using the whole data sets.

---

[1] As we present our method as an alternative to the iterative-knowledge distillation [7], a direct comparison was necessary in terms of results and thus, we adopted the same methodology to produce results as in [7]. The authors in [7] also employ 10-fold cross validation for the MR and CR data sets.

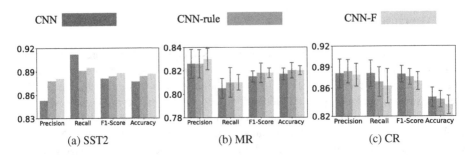

Fig. 3. Performance obtained using our method (CNN-F), the method in [7] (CNN-rule) and that in [8] (CNN) on the data sets under study. Errors bars denote 95% confidence intervals around the mean.

In Table 1 and Fig. 3, we show the precision, recall, F-1 score and accuracy for the positive sentiment class yielded by our method (CNN-F), the method in [7] (CNN-rule) and that in [8] (CNN). From the experimental results, we observe that our method outperforms the two methods on both the SST2 and MR data sets for all measures. Note that the performance decreases on the CR data set for both CNN-rule and CNN-F, which indicates that the "A-but-B" rule cannot be generalized to data points coming from similar distributions.

In Table 2 and Fig. 4, we show the results obtained only on sentences having the "A-but-B" syntactic structure. At first glance, we note that our method works as intended and is quite competitive, outperforming the two baselines despite using only one rule for comparison. Since our method represents knowledge purely in terms of a distribution on input data, we can argue that it was bound to perform better than iterative-knowledge distillation [7] since the neural network will only process the input features that are consistent with the human knowledge. Also, a decrease in performance is observed on the CR data set for both CNN-rule and CNN-F, which is consistent with the fact that the "A-but-B" rule cannot be generalized for every sentence and should not be encoded in the parameters permanently.

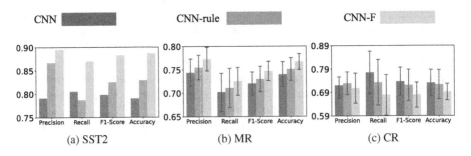

Fig. 4. Performance obtained using our method (CNN-F), the method in [7] (CNN-rule) and that in [8] (CNN) on the data sets under study making use only of sentences containing **with A-but-B** structures. Errors bars denote 95% confidence intervals around the mean.

Table 1. Performance obtained using our method (CNN-F), the method in [7] (CNN-rule) and that in [8] (CNN) on the data sets under study.

Method	SST2			
	Precision	Recall	F-1 Score	Accuracy
CNN	0.853	**0.912**	0.881	0.877
CNN-rule	0.878	0.891	0.884	0.884
CNN-F	**0.881**	0.895	**0.888**	**0.887**
	MR			
	Precision	Recall	F-1 Score	Accuracy
CNN	0.826 ± 0.012	0.805 ± 0.008	0.815 ± 0.005	0.817 ± 0.005
CNN-rule	0.826 ± 0.012	0.810 ± 0.012	0.818 ± 0.008	0.820 ± 0.006
CNN-F	$\mathbf{0.830 \pm 0.009}$	$\mathbf{0.810 \pm 0.007}$	$\mathbf{0.818 \pm 0.004}$	$\mathbf{0.820 \pm 0.004}$
	CR			
	Precision	Recall	F-1 Score	Accuracy
CNN	0.881 ± 0.020	$\mathbf{0.881 \pm 0.018}$	$\mathbf{0.880 \pm 0.012}$	$\mathbf{0.847 \pm 0.014}$
CNN-rule	$\mathbf{0.884 \pm 0.017}$	0.869 ± 0.020	0.876 ± 0.011	0.844 ± 0.012
CNN-F	0.879 ± 0.016	0.863 ± 0.024	0.870 ± 0.013	0.836 ± 0.013

Table 2. Performance obtained using our method (CNN-F), the method in [7] (CNN-rule) and that in [8] (CNN) on the data sets under study making use only of sentences containing **with A-but-B** structures.

Method	SST2			
	Precision	Recall	F-1 Score	Accuracy
CNN	0.791	0.805	0.798	0.790
CNN-rule	0.867	0.787	0.825	0.829
CNN-F	**0.895**	**0.870**	**0.883**	**0.887**
	MR			
	Precision	Recall	F-1 Score	Accuracy
CNN	0.744 ± 0.029	0.702 ± 0.041	0.720 ± 0.025	0.740 ± 0.027
CNN-rule	0.750 ± 0.026	0.711 ± 0.042	0.730 ± 0.027	0.751 ± 0.024
CNN-F	$\mathbf{0.773 \pm 0.025}$	$\mathbf{0.725 \pm 0.030}$	$\mathbf{0.747 \pm 0.020}$	$\mathbf{0.767 \pm 0.017}$
	CR			
	Precision	Recall	F-1 Score	Accuracy
CNN	0.720 ± 0.055	$\mathbf{0.775 \pm 0.091}$	$\mathbf{0.737 \pm 0.060}$	$\mathbf{0.731 \pm 0.057}$
CNN-rule	$\mathbf{0.729 \pm 0.049}$	0.733 ± 0.098	0.721 ± 0.067	0.724 ± 0.063
CNN-F	0.708 ± 0.064	0.679 ± 0.087	0.679 ± 0.054	0.692 ± 0.036

Finally, the results in both tables indicate that when there is a performance gain on datasets SST2 and MR by incorporating A-but-B rule, it is best for CNN-F and when there is a performance drop, it is worst for CNN-F. This suggests that the feature extracting functions not only can be used as an alternative but also in conjunction with iterative knowledge distillation in order to provide a "Maximum Performance Gain or Drop value" from the constructed logic rules. This value can be used to select the best combination of these rules via cross validation when using distillation. This also provides a mechanism to quantitatively evaluate how effectively the rule knowledge was distilled into the parameters of the neural network when applied to distillation approaches.

5 Conclusion

In this paper, we have shown how feature extracting functions can be employed to learn logic rules for sentiment analysis. This provides a means to representing human knowledge in neural networks via programmable feature extracting functions. Moreover, we have shown that, using these feature extracting functions, we can obtain a model whose posterior output can be influenced by domain knowledge expressed in terms of logic rules without the need of transferring these into the network parameters. The approach presented here is quite general in nature, being applicable to a wide variety of logic rules that can be expressed using rule-to-knowledge conditional probability distributions. We have illustrated the utility of our method for textual sentiment analysis and compared our results with those obtained using two baselines. In our experiments, our method was quite competitive, outperforming the alternatives.

References

1. Bach, S.H., Broecheler, M., Huang, B., Getoor, L.: Hinge-loss Markov random fields and probabilistic soft logic. CoRR abs/1505.04406 (2015)
2. Bach, S.H., et al.: Snorkel drybell: a case study in deploying weak supervision at industrial scale. In: Proceedings of the 2019 International Conference on Management of Data, pp. 362–375 (2019)
3. Gabbay, A., Garcez, A., Broda, K., Gabbay, D.M., Gabbay, P.: Neural-Symbolic Learning Systems: Foundations and Applications. Springer, London (2002). https://doi.org/10.1007/978-1-4471-0211-3
4. Ganchev, K., Graça, J., Gillenwater, J., Taskar, B.: Posterior regularization for structured latent variable models. J. Mach. Learn. Res. 11, 2001–2049 (2010)
5. Hinton, G., Vinyals, O., Dean, J.: Distilling the knowledge in a neural network. In: NIPS Deep Learning and Representation Learning Workshop (2015)
6. Hu, M., Liu, B.: Mining and summarizing customer reviews. In: Proceedings of the Tenth ACM SIGKDD International Conference on Knowledge Discovery and Data Mining, pp. 168–177 (2004)
7. Hu, Z., Ma, X., Liu, Z., Hovy, E., Xing, E.: Harnessing deep neural networks with logic rules. In: Proceedings of the 54th Annual Meeting of the Association for Computational Linguistics (Volume 1: Long Papers), Berlin, Germany, pp. 2410–2420. Association for Computational Linguistics (2016)
8. Kim, Y.: Convolutional neural networks for sentence classification. CoRR abs/1408.5882 (2014)

9. Krishna, K., Jyothi, P., Iyyer, M.: Revisiting the importance of encoding logic rules in sentiment classification. In: Proceedings of the 2018 Conference on Empirical Methods in Natural Language Processing, pp. 4743–4751 (2018)
10. Le, Q.V., Mikolov, T.: Distributed representations of sentences and documents. CoRR abs/1405.4053 (2014)
11. Lewis, D.D.: Feature selection and feature extraction for text categorization. In: Proceedings of the Workshop on Speech and Natural Language, pp. 212–217 (1992)
12. Liang, X., Hu, Z., Zhang, H., Lin, L., Xing, E.P.: Symbolic graph reasoning meets convolutions. In: Advances in Neural Information Processing Systems, pp. 1853–1863 (2018)
13. Nguyen, A.M., Yosinski, J., Clune, J.: Deep neural networks are easily fooled: high confidence predictions for unrecognizable images. CoRR abs/1412.1897 (2014)
14. Pang, B., Lee, L.: Seeing stars: exploiting class relationships for sentiment categorization with respect to rating scales. In: Proceedings of the 43rd Annual Meeting of the Association for Computational Linguistics (ACL 2005) (2005)
15. Pennington, J., Socher, R., Manning, C.: GloVe: global vectors for word representation. In: Proceedings of the 2014 Conference on Empirical Methods in Natural Language Processing (EMNLP), pp. 1532–1543 (2014)
16. Peters, M., Neumann, M., Iyyer, M., Gardner, M., Clark, C., Lee, K., Zettlemoyer, L.: Deep contextualized word representations. In: Proceedings of the 2018 Conference of the North American Chapter of the Association for Computational Linguistics: Human Language Technologies, pp. 2227–2237 (2018)
17. Rai, A.: Explainable AI: from black box to glass box. J. Acad. Mark. Sci. **8**(1), 137–141 (2020). https://doi.org/10.1007/s11747-019-00710-5
18. Ribeiro, M.T., Singh, S., Guestrin, C.: "Why should I trust you?": Explaining the predictions of any classifier. CoRR abs/1602.04938 (2016)
19. Szegedy, C., et al.: Intriguing properties of neural networks. In: International Conference on Learning Representations (2014)
20. Taskar, B., Guestrin, C., Koller, D.: Max-margin markov networks. In: NIPS, pp. 25–32 (2003)
21. Tran, S.N.: Unsupervised neural-symbolic integration. CoRR abs/1706.01991 (2017)
22. Vilone, G., Longo, L.: Explainable artificial intelligence: a systematic review. arXiv preprint arXiv:2006.00093 (2020)
23. Zeiler, M.D.: ADADELTA: an adaptive learning rate method. CoRR abs/1212.5701 (2012)

Learning High-Resolution Domain-Specific Representations with a GAN Generator

Danil Galeev[(✉)], Konstantin Sofiiuk, Danila Rukhovich, Mikhail Romanov, Olga Barinova, and Anton Konushin

Samsung AI Center, Moscow, Russia
{d.galeev,k.sofiiuk,d.rukhovich,m.romanov,o.barinova,
a.konushin}@samsung.com

Abstract. In recent years generative models of visual data have made a great progress, and now they are able to produce images of high quality and diversity. In this work we study representations learnt by a GAN generator. First, we show that these representations can be easily projected onto semantic segmentation map using a lightweight decoder. We find that such semantic projection can be learnt from just a few annotated images. Based on this finding, we propose LayerMatch scheme for approximating the representation of a GAN generator that can be used for unsupervised domain-specific pretraining. We consider the semi-supervised learning scenario when a small amount of labeled data is available along with a large unlabeled dataset from the same domain. We find that the use of LayerMatch-pretrained backbone leads to superior accuracy compared to standard supervised pretraining on ImageNet. Moreover, this simple approach also outperforms recent semi-supervised semantic segmentation methods that use both labeled and unlabeled data during training.

Keywords: Semi-supervised learning · Adversarial learning · Transfer learning

1 Introduction

Generative models of visual data, and generative adversarial nets (GANs) in particular, have made remarkable progress in recent years [8,9], and now they are able to produce images of high quality and diversity. Generative models have long been considered as a means of representation learning, with common assumption that the ability to generate data from some domain implies understanding of the semantics of that domain. Thus, various ideas about using GANs for representation learning have been studied in the literature [17]. Most of these works are focused on producing universal feature representations by training a generative model on a large and diverse dataset [4,5].

© Springer Nature Switzerland AG 2021
A. Torsello et al. (Eds.): S+SSPR 2020, LNCS 12644, pp. 108–118, 2021.
https://doi.org/10.1007/978-3-030-73973-7_11

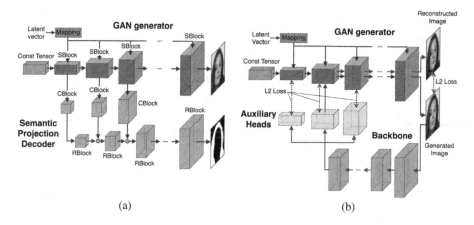

Fig. 1. (a) Semantic projection implemented by a decoder built on top of a style-based generator as described in Sect. 3.1; (b) LayerMatch scheme for pretraining a backbone to approximate the activations of a GAN model as described in Sect. 4.1.

In this work we consider the task of unsupervised *domain-specific pretraining*. Rather than trying to learn a universal representation on a diverse dataset, we focus on producing a specialized representation for a particular domain. Our intuition is that GAN generators are most efficient for learning *high-resolution representations*, as generating a realistically-looking image implies learning appearance and location of different semantic parts. Thus, we experiment with semantic segmentation as a target downstream task. To illustrate our idea, we perform experiments with *semantic projection* of GAN generator and show that it can be easily converted into a semantic segmentation model. Based on this finding, we introduce a novel *LayerMatch* scheme that trains a model to predict the activations of the internal layers of GAN generator. Since the proposed scheme is trained on synthetic data and requires only a trained generator model, it can be used for unsupervised domain-specific pretraining.

The rest of the paper is organized as follows. In Sect. 3.1 we explore semantic projections of GAN generator and describe our experiments. In Sect. 4.1 we introduce the LayerMatch scheme for unsupervised domain-specific pretraining that is based on inverting GAN generator. Section 5 describes our experiments with the models pretrained using LayerMatch. In Sect. 2 we discuss related works.

2 Related Work

Several works consider generative models for unsupervised pretraining [4,5,11, 15]. One of the approaches [17] uses representation learnt by a discriminator. Another line of research extends GAN to bidirectional framework (BiGAN) by introducing an auxiliary encoder branch that predicts the latent vector from a natural image [4,5]. The encoder learnt via BiGAN framework can be used as

---

Algorithm 1: Training semantic projection model

Input: GAN model (G, D)

Output: Semantic projection model P

1   Generate n images from the random latent vectors $I_i = G(l_i), l_i \sim \mathcal{N}(\mathbf{0}, \sigma)$,
     $i = 1 \dots n$ and store them along with their features $\{(I_i, \mathbf{\Phi}_i)\}$, $i = 1 \dots n$

2   Annotate the images and create semantic maps L_i, $i = 1 \dots n$

3   Train a decoder P on pairs $\{(\mathbf{\Phi}_i, L_i)\}$, $i = 1 \dots n$

---

feature extractor for downstream tasks [5]. The use of GANs as universal feature extractors has severe limitations. First, GANs are not always capable of learning a multimodal distribution as they tend to suffer from mode collapse. The trade-off between GAN precision and recall is still difficult to control [10]. Besides, training a GAN on a large dataset of high-resolution images requires an extremely large computational budget, which makes ImageNet-scale experiments prohibitively expensive. Our approach differs from this line of work, as we use a GAN to specialize a model to a particular domain rather than trying to obtain universal feature representations. We explore the representation of a GAN generator that, to the best of our knowledge, has not been previously considered for transfer learning.

While image-level classification has been extensively studied in a semi-supervised setting, dense pixel-level classification with limited data has only drawn attention recently. Most of the works on semi-supervised semantic segmentation borrow the ideas from semi-supervised image classification and generalize them on high-resolution tasks. [6] adopt an adversarial learning scheme and propose a fully convolutional discriminator that learns to differentiate ground truth label maps from probability maps of segmentation predictions. [16] use two network branches that link semi-supervised classification with semi-supervised segmentation including self-training.

3   Semantic Projection of a GAN Generator

Let us introduce the following notation. A typical GAN model consists of a jointly trained generator G and a discriminator D. Generator G transforms a random latent vector $l \in \mathbb{R}^k$ into an image $I_{gen} \in \mathbb{R}^{3 \times H \times W}$ and discriminator $D : \mathbb{R}^{3 \times H \times W} \to \mathbb{R}$ classifies whether an image is real or fake. Let us denote the activations of internal layers of G for latent vector l by $\mathbf{\Phi}(l)$.

Semantic projection of a generator P is a mapping of the features $\mathbf{\Phi}(l)$ onto the dense label map $L \in \{1, \dots, C\}^{W \times H}$ where C is the number of classes. It can be implemented as a decoder that takes the features from different layers of a generator and outputs the semantic segmentation result. An example of a decoder architecture built on top of a style-based generator is shown in Fig. 1 (a).

3.1 Converting Semantic Projection Model into a Segmentation Model

Training procedure of a semantic projection model is shown in Algorithm 1. First, we sample a few images using GAN generator G and store corresponding activations $\{\mathbf{\Phi}_i\}, \mathbf{\Phi}_i = (\phi_1^i, \ldots, \phi_k^i), i = 1 \ldots n$ of internal layers. The latent vectors are sampled from normal distribution. Then, we manually annotate a few generated images. The decoder is trained in a supervised manner using the segmentation masks from the previous step with corresponding intermediate generator features. We use cross-entropy between the predicted mask and ground truth as a loss function.

Once we have trained a semantic projection model P, we can obtain pixelwise annotation for generated images. For this purpose, we can apply P to the features produced by generator $L_{gen} = P(\mathbf{\Phi})$. However, the features of a generator are not available for real images. Since semantic projection alone does not allow obtaining semantic segmentation maps for real images, we propose Algorithm 2 for converting the semantic projection model into a semantic segmentation model applicable to real images. The intuition is that training on a large number of GAN-generated images along with accurate annotations provided by semantic projection should result in an accurate segmentation model.

Algorithm 2: Converting semantic projection into semantic segmentation model

---

Input: GAN model (G, D), semantic projection model P
Output: Semantic segmentation model S

1   Generate N images from the random latent vectors $I_i = G(l_i), l_i \sim \mathcal{N}(\mathbf{0}, \sigma)$ $i = 1 \ldots N$ and store them along with their features $\{(I_i, \mathbf{\Phi}_i)\}, i = 1 \ldots N$
2   Compute results of semantic projection $\{P(\mathbf{\Phi}_i)\}, i = 1 \ldots N$
3   Train semantic segmentation model S on pairs $\{(\mathbf{\Phi}_i, P(\mathbf{\Phi}_i))\}, i = 1 \ldots N$

3.2 Experiments with Semantic Projections

In this section we address the following questions: 1) Will a lightweight decoder be sufficient to implement an accurate semantic projection model? 2) How many images are required to train semantic projection to a reasonable accuracy? 3) Will the use of Algorithm 2 lead to improved performance on real images?

Experimental Protocol. We perform experiments with style-based generator [7] on two datasets (FFHQ and LSUN-cars). In both experiments, we manually annotate 20 randomly generated images for training the semantic projection models. For FFHQ experiment we use two classes: hair and background. For LSUN-cars we use car and background categories. We also train DeepLabV3+ [3] model using Algorithm 2 with semantic projection models trained on 20 images. In all experiments we use ResNet-50 as a backbone. For LSUN-cars we experiment with both ImageNet-pretrained and randomly initialized backbones.

(a)   (b)

Fig. 2. (a) - Evaluation results of the semantic projection model on two classes (background and hair) with respect to the number of images in training; (b) - outputs of semantic projection model for test images generated by StyleGAN. Note that while the model was trained just on 20 images, it provides quite accurate segmentation.

Table 1. Comparison of the segmentation models trained with equal amount of supervision. See text for more details.

Categories	Method	ImageNet-pretrained backbone	Accuracy	IoU
Hair/Background	Training on 20 labeled images	+	0.952	0.819
	Algorithm 2 with 20 labeled images		**0.968**	**0.876**
Car/Background	Training on 20 labeled images	−	0.859	0.698
	Algorithm 2 with 20 labeled images		0.979	0.941
	Training on 20 labeled images	+	0.964	0.905
	Algorithm 2 with 20 labeled images		**0.986**	**0.961**

For comparison we train a similar DeepLabV3+ model on 20 labeled real images. 80 annotated real images are used for testing the semantic segmentation models. Pixel accuracy and intersection-over-union (IoU) are measured for methods comparison.

Architecture of the Semantic Projection Model. The lightweight decoder architecture for semantic projection is shown in Fig. 1. It has an order of magnitude fewer parameters compared to the standard decoder architectures and 16 times fewer than DeepLabV3+ decoder. Each CBlock of the decoder takes the features from corresponding SBlock of StyleGAN as an input. CBlock consists of a 50% dropout, a convolutional and a batch normalization layers. Each RBlock of a decoder has one residual block with two convolutional layers. The number of feature maps in each convolutional layer of the decoder is set to 32, as wider feature maps resulted in just minor improvement in our experiments.

Results and Discussion. Figure 2 (b) shows outputs of a semantic projection model trained on 20 synthetic images using Algorithm 1. The results of varying the size of a training set from 1 to 15 synthetic images is shown in Fig. 2 (a). The test set in this experiment contains 30 manually annotated GAN-generated images. We observe that even with a single image in training set, the model achieves reasonable segmentation quality, and the quality grows quite slowly after 10 images.

Next, we compare two semantic segmentation models trained with equal amount of supervision. The first one uses Algorithm 2 with semantic projection model trained on 20 synthetic images. The second one uses ImageNet-pretrained backbone and is trained on 20 real images. Table 1 shows quantitative comparison of the two models. One can notice that in case when the backbone for DeepLabV3+ is randomly initialized, the model trained with Algorithm 2 is significantly more accurate compared to the baseline approach. When using ImageNet-pretrained backbones, Algorithm 2 leads to 6% improvement in terms of IoU for both datasets.

Our experiments of two datasets demonstrate that a lightweight decoder is sufficient to implement an accurate semantic projection model. We observe that just a few annotated images are enough to train semantic projection to a reasonable accuracy. The Algorithm 2 leads to improved accuracy on real images compared to simply training a similar model with the same number of annotated images.

4   Transfer Learning Using Generator Representation

Training a semantic projection model introduced in Sect. 3.1 requires manual annotation of GAN-generated images. Thus, we cannot use standard real-image datasets for comparison with other works. Real images could potentially be embedded into the GAN latent space, but in practice this approach has its own limitations [2]. Besides, some of the images produced by GAN generators can be hard to label.

A semantic segmentation network transforms an image $I \in \mathbb{R}^{3 \times H \times W}$ into a segmentation map. At the same time a GAN generator G transforms a random vector $l \in \mathbb{R}^k$ into an image $I_{gen} \in \mathbb{R}^{3 \times H \times W}$. Obviously, the input dimensions of these two types of models do not match. Therefore, the models trained for image generation cannot be directly applied to image segmentation. To overcome this issue one can think of inverting a generator. Inverted GAN generators have been widely used for the task of image manipulation [1,2]. For this purpose, an encoder model is usually trained to predict the latent vector from an image. Following [1,2] we train an encoder network, but predict the activations of a fixed GAN generator instead of the latent vector. The backbone of the trained encoder can then be used to initialize a semantic segmentation model.

4.1   Unsupervised Pretraining with LayerMatch

The scheme of the LayerMatch algorithm is shown in Fig. 1 (b). We can view generator G as a function of the latent vector l and all the intermediate activations: $G = G(l, \phi_1, \phi_2, .., \phi_n)$, where intermediate features themselves depend on the latent vector and all the previous features: $\phi_i = \phi_i(l, \phi_1, \phi_2, .., \phi_{i-1})$.

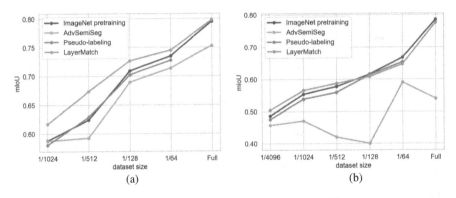

(a)   (b)

Fig. 3. Comparison of the models trained with Algorithm 3 to semi-supervised segmentation methods for a varying number of annotated samples. (a) - FFHQ+CelebA-HQ dataset. (b) - LSUN-church dataset

The generated image I_{gen} is fed to the encoder E, which tries to predict the n specified activation tensors: $\hat{\Phi} = E(I_{gen})$, where $\hat{\Phi} = (\hat{\phi}_1, \hat{\phi}_2, \ldots, \hat{\phi}_n)$.

The loss function for LayerMatch training consists of two terms:

$$\mathcal{L} = \mathcal{L}_{rec} + \mathcal{L}_{match}, \tag{1}$$

where matching loss $\mathcal{L}_{match}$ is the sum of the L2-losses between generated and predicted features that penalizes difference between the outputs of the encoder and the activations of the generator:

$$\mathcal{L}_{match} = \frac{1}{n} \sum_{i=1}^{n} \|\phi_i - \hat{\phi}_i\|_2^2 \tag{2}$$

Reconstructed image I_{rec} is obtained by replacing random feature ϕ_m with $\hat{\phi}_m$, where $1 \leq m \leq n$, and recalculating features:

$$\tilde{\phi}_j = \phi_j(l, \phi_1, \ldots, \phi_{m-1}, \hat{\phi}_m, \tilde{\phi}_{m+1} \ldots \tilde{\phi}_{j-1}), m < j \leq n \tag{3}$$

$$I_{rec} = G(l, \phi_1, \ldots, \phi_{m-1}, \hat{\phi}_m, \tilde{\phi}_{m+1}, \ldots, \tilde{\phi}_n) \tag{4}$$

Reconstruction loss $\mathcal{L}_{rec}$ is the L2-loss between the generated image and the reconstructed image. This loss controls that the generator produces an image which is close to the original one when generator activations are replaced with the outputs of the backbone:

$$\mathcal{L}_{rec} = \|I_{rec} - I_{gen}\|_2^2 \tag{5}$$

Algorithm 3: Unsupervised pretraining with LayerMatch

Input: GAN model (G, D) trained on a large unlabeled dataset, a small labeled dataset $(I_j, L_j), j = 1 \ldots n$

Output: Semantic segmentation model S

1   Generate N images from the random latent vectors $I_i = G(l_i), l_i \sim \mathcal{N}(\mathbf{0}, \sigma)$, $i = 1 \ldots N$ and store them along with their features $\{(I_i, \mathbf{\Phi}_i)\}$, $i = 1 \ldots N$

2   Train the backbone using LayerMatch using the pairs $\{(I_i, \mathbf{\Phi}_i)\}$, $i = 1 \ldots N$

3   Train a semantic segmentation model S on the labeled part of the data $\{(I_j, L_j)\}$, $j = 1 \ldots n$

---

5   Experiments with LayerMatch

Evaluation Protocol. The standard protocol for evaluation of unsupervised learning techniques proposed in [20] involves training a model on unlabeled ImageNet, freezing its learned representation, and then training a linear classifier on its outputs using all of the training set labels. This protocol is based on the assumption that the resulting representation is universal and applicable to different domains. We rather focus on domain-specific pretraining, or "specializing" the backbone to a particular domain. We aim at high-resolution tasks, *e.g.* semantic segmentation. Therefore, we apply a different evaluation protocol.

We assume that we have a high-quality GAN model trained on a large unlabeled dataset from the domain of interest along with a limited number of annotated images from the same domain. The unlabeled data is used for training a GAN model, which in turn is used for pretraining the backbone model using LayerMatch (see Algorithm 3). The pixelwise-annotated data is later used for training a fixed semantic segmentation network with the pretrained backbone using a standard cross-entropy loss. Then, we evaluate the resulting model on a test set across the standard semantic segmentation metrics such as mIOU and pixel accuracy. We perform experiments with varying fraction of labeled data. In all our experiments we initialize the networks with ImageNet-pretrained backbones.

Comparison with Prior Work. For all compared methods we use the same network architectures differing only in training procedure and loss functions used. The first baseline uses a standard ImageNet-pretrained backbone without domain-specific pretraining. The semantic segmentation model is trained using available annotated data and does not use the unlabeled data.

The other two baselines are recent semi-supervised segmentation methods using both labeled and unlabeled data during training. In the experiments with these methods we used exactly the same amount of both labeled and unlabeled data as for LayerMatch. Namely, for the experiments with Celeba-HQ we used both the unlabeled part of CelebA and the FFHQ dataset, that was used for GAN training. For the experiments with LSUN-church all the unlabeled data in LSUN-church dataset was used during training.

Table 2. Comparison of segmentation models trained on CelebA-HQ dataset with equal amount of supervision. Notice that LayerMatch provides better results for almost all categories and improves the IoU for eye glasses category by several times.

Method	pixAcc	mIoU	background	skin	nose	eye glasses	left eye	right eye	left brow	right brow	left ear	right ear	mouth	upper lip	lower lip	hair	hat	earrings	necklace	neck	cloth
ImageNet only	.92	.62	.89	.89	.87	.01	.75	.77	.67	.68	.71	.72	.76	.74	.79	.87	0	.34	0	.77	.62
AdvSemiSeg [6]	.91	.62	.89	.89	.85	.14	.75	.77	.66	.66	.71	.70	.77	.73	.78	.86	0	.29	0	.75	.59
Pseudo-labeling [13]	.92	.63	.89	.90	.86	.01	.77	.78	.71	.70	.73	.72	.79	.75	.79	.86	0	.33	0	.77	.58
LayerMatch	.93	.67	.90	.91	.87	.68	.78	.78	.70	.69	.75	.74	.79	.76	.80	.87	0	.34	0	.80	.64

The first semi-supervised segmentation method that we use for comparison is based on pseudo-labeling [13]. Unlabeled data is augmented by generating pseudo-labels using the network predictions. Only the pixels with high-confidence pseudo-labels are used as ground truth for training. The second one is an adversarial semi-supervised segmentation approach [6]. In our experiments we used the official implementation provided by the authors, and changed only the backbone.

Datasets. Celeba-HQ [12] contains 30,000 high-resolution face images selected from the CelebA dataset [14], each image having a segmentation mask with the resolution of 512×512 and 19 classes including all facial components and accessories. We use a StyleGAN2 model trained on FFHQ dataset provided in [9] that has a FID measure 3.31 and PPL 125. In the experiments with Celeba-HQ we vary the fraction of labeled data from 1/1024 to the full dataset.

LSUN-church contains 126,000 images of churches of 256×256 resolution. We have selected top 10 semantic categories that occupy more than 1% of image area, namely road, vegetation, building, sidewalk, car, sky, terrain, pole, fence, wall. We use a StyleGAN2 model provided in [9] that has a FID measure 3.86 and PPL 342. As LSUN dataset does not contain pixelwise annotation, we take the outputs of the Unified Scene Parsing Network [19] as ground truth in this experiment similarly to [2]. In the experiments with Celeba-HQ we vary the fraction of labeled data from 1/4096 to the full dataset.

Implementation Details. HRNet [18] is used as an encoder architecture. We add K auxiliary heads for each of K activations that we want to predict (see Fig. 1 (b)). After training, auxiliary heads are discarded and only the pretrained backbone is used for transfer learning, similar to ImageNet pretraining. For pretraining the encoder we use Adam optimizer with the learning rate 10^{-4} and the cosine learning rate decay. We use source code from HRNet repository for training semantic segmentation networks.

Results and Discussion. Figure 3 shows the comparison of the proposed LayerMatch pretraining scheme to 3 baseline methods across 2 datasets with varying fraction of annotated data. Pseudo-labeling is applicable in case when some part of the dataset is unlabelled.

One can see that LayerMatch pretraining shows significantly higher IoU compared to the baseline methods on Celeba-HQ (see Fig. 3 (a)) for any fraction of the labeled data. For LSUN-church it shows higher accuracy compared to other methods in cases when up to 1/512 of the data is annotated. Table 2 shows category-wise results for all four compared models trained with 1/512 of labeled data. LayerMatch pretraining leads to significant accuracy improvement for the eyeglasses category.

Overall, LayerMatch pretraining leads to improved results in semi-supervised learning scenario compared to both simple ImageNet pretraining and to semi-supervised segmentation methods. Lower accuracy for larger fraction of annotated datasets on LSUN-church can be attributed to lower quality of LSUN-church GAN generator compared to Celeba-HQ GAN generator. Another possible reason for this effect may be the imperfect annotation of both training and test data, which may lead to inaccuracies in evaluation.

6   Conclusion

We study the use of GAN generators for the task of learning domain-specific representations. We show that the representation of a GAN generator can be easily projected onto semantic segmentation map using a lightweight decoder. Then, we propose LayerMatch scheme for unsupervised domain-specific pretraining that is based on approximating the generator representation. We present experiments in semi-supervised learning scenario and compare to recent semi-supervised semantic segmentation methods.

References

1. Abdal, R., Qin, Y., Wonka, P.: Image2StyleGAN: how to embed images into the StyleGAN latent space? In: Proceedings of the IEEE International Conference on Computer Vision, pp. 4432–4441 (2019)
2. Bau, D., et al.: Seeing what a GAN cannot generate. In: Proceedings of the IEEE International Conference on Computer Vision, pp. 4502–4511 (2019)
3. Chen, L.C., Zhu, Y., Papandreou, G., Schroff, F., Adam, H.: Encoder-decoder with atrous separable convolution for semantic image segmentation. In: ECCV (2018)
4. Donahue, J., Krähenbühl, P., Darrell, T.: Adversarial feature learning. arXiv preprint arXiv:1605.09782 (2016)
5. Donahue, J., Simonyan, K.: Large scale adversarial representation learning. In: Advances in Neural Information Processing Systems, pp. 10541–10551 (2019)
6. Hung, W.C., Tsai, Y.H., Liou, Y.T., Lin, Y.Y., Yang, M.H.: Adversarial learning for semi-supervised semantic segmentation. arXiv preprint arXiv:1802.07934 (2018)
7. Karras, T., Laine, S., Aila, T.: A style-based generator architecture for generative adversarial networks. arXiv preprint arXiv:1812.04948 (2018)
8. Karras, T., Laine, S., Aila, T.: A style-based generator architecture for generative adversarial networks. In: Proceedings of the IEEE Conference on Computer Vision and Pattern Recognition, pp. 4401–4410 (2019)
9. Karras, T., Laine, S., Aittala, M., Hellsten, J., Lehtinen, J., Aila, T.: Analyzing and improving the image quality of stylegan. arXiv preprint arXiv:1912.04958 (2019)

10. Kynkäänniemi, T., Karras, T., Laine, S., Lehtinen, J., Aila, T.: Improved precision and recall metric for assessing generative models. In: Advances in Neural Information Processing Systems, pp. 3929–3938 (2019)
11. Larsen, A.B.L., Sønderby, S.K., Larochelle, H., Winther, O.: Autoencoding beyond pixels using a learned similarity metric. arXiv preprint arXiv:1512.09300 (2015)
12. Lee, C.H., Liu, Z., Wu, L., Luo, P.: MaskGAN: towards diverse and interactive facial image manipulation. arXiv preprint arXiv:1907.11922 (2019)
13. Lee, D.H.: Pseudo-label: the simple and efficient semi-supervised learning method for deep neural networks. In: Workshop on challenges in representation learning, ICML, vol. 3, p. 2 (2013)
14. Liu, Z., Luo, P., Wang, X., Tang, X.: Deep learning face attributes in the wild. In: Proceedings of International Conference on Computer Vision (ICCV), December 2015
15. Makhzani, A., Shlens, J., Jaitly, N., Goodfellow, I., Frey, B.: Adversarial autoencoders. arXiv preprint arXiv:1511.05644 (2015)
16. Mittal, S., Tatarchenko, M., Brox, T.: Semi-supervised semantic segmentation with high-and low-level consistency. IEEE Trans. Pattern Anal. Mach. Intell. **43**, 1369–1379 (2019)
17. Radford, A., Metz, L., Chintala, S.: Unsupervised representation learning with deep convolutional generative adversarial networks. arXiv preprint arXiv:1511.06434 (2015)
18. Sun, K., et al.: High-resolution representations for labeling pixels and regions. arXiv preprint arXiv:1904.04514 (2019)
19. Xiao, T., Liu, Y., Zhou, B., Jiang, Y., Sun, J.: Unified perceptual parsing for scene understanding. In: Proceedings of the European Conference on Computer Vision (ECCV), pp. 418–434 (2018)
20. Zhang, R., Isola, P., Efros, A.A.: Colorful image colorization. In: Leibe, B., Matas, J., Sebe, N., Welling, M. (eds.) ECCV 2016. LNCS, vol. 9907, pp. 649–666. Springer, Cham (2016). https://doi.org/10.1007/978-3-319-46487-9_40

Predicting Polypharmacy Side Effects Through a Relation-Wise Graph Attention Network

Vincenzo Carletti[⊠], Pasquale Foggia, Antonio Greco, Antonio Roberto, and Mario Vento

University of Salerno, Via Giovanni Paolo II, 132, 84084 Fisciano, SA, Italy
vcarletti@unisa.it

Abstract. Polypharmacy is the combined use of multiple drugs, widely adopted in medicine to treat patients that suffer of complex diseases. Therefore, it is important to have reliable tools able to predict if the activity of a drug could unfavorably change when combined with others. State-of-the-art methods face this problem as a link prediction task on a multilayer graph describing drug-drug interactions (DDI) and protein-protein interactions (PPI), since it has been demonstrated to be the most effective representation. Graph Convolutional Networks (GCN) are the method most commonly chosen in recent research for this problem. We propose to improve the performance of GCN on this link prediction task through the addition of a novel relation-wise Graph Attention Network (GAT), used to assign different weight to the different relationships in the multilayer graph. We experimentally demonstrate that the proposed GCN, compared with other recent methods, is able to achieve a state-of-the-art performance on a publicly available polypharmacy side effect network.

1 Introduction

The treatment of complex diseases often requires the combined use of multiple drugs, called a *polypharmacy therapy*, with the expectation to improve the therapeutic effect. The drawback of such a therapy is the possibility that interactions among the drugs can cause the onset of undesired side effects. Although many of the side effects are well-known in the scientific literature, several are still undiscovered, especially for new experimental therapies. Therefore, it is desirable to have tools to predict if the activity of a drug could unfavorably change when combined with others, possibly starting from already known interactions.

As for many problems in biology, graphs are the most natural way to represent the complexity of the structures and the interactions involved in biological processes [14]. Noteworthy examples of the usage of graphs are protein and molecule structures, protein interaction networks and recently graph representations of the human genome [2,4]. As described in [21], more complex representations, obtained through the combination of different kind of networks, are

A. Torsello et al. (Eds.): S+SSPR 2020, LNCS 12644, pp. 119–128, 2021.
https://doi.org/10.1007/978-3-030-73973-7_12

usually adopted to model the interactions among entities of different families, such as genes, proteins or drugs. Graph representing these heterogeneous networks are named *multilayer graphs* where each *layer* is a subnetwork composed of homogeneous entities. In particular, discovering polypharmacy side effects can be formalized as a *link prediction* task on a multilayer undirected graph [3,13,20], aiming to estimate the probability that a particular side-effect occurs due to the interaction a pair of drugs. The use of a multilayer graph is justified in [20] where the authors have shown that combining the known interactions among the drugs with those of the proteins targeted by the drugs can effectively improve the prediction accuracy. The proposed representation is a two-layer graph, shown in Fig. 1, where the ground layer consists of a protein-protein interaction (PPI) network, in which two proteins are connected if a biological interaction exists between them (e.g. metabolic enyme-coupled interactions), and the upper layer is a drug-drug interaction (DDI) network where a link between two drugs exists if the combined use causes a side-effect. The two layers are interconnected by a DTI (drug-target interaction) network where each drug is linked to its target proteins.

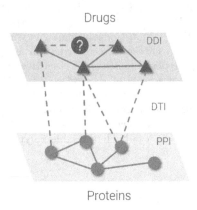

Fig. 1. Structure of a polypharmacy side effect network as a multilayer graph. Triangles represent drugs (layer 1) while circles are proteins (layer 2). Interactions are represented by lines: lines in the upper layer are drug-drug interactions (DDIs), lines in the ground layer are protein-protein interactions (PPIs) and interconnecting lines are drug-target interactions (DTIs).

Before the wide diffusion of deep neural networks (DNN) a common approach to perform machine learning tasks on graphs has been the use of hand crafted features, through graph embedding or graph matching methods. The trend is changing: Graph Neural Networks (GNN), a new class of machine learning methods, emerged, promising to be as disruptive as Convolutional Neural Networks (CNN) have been for image analysis. Gori et al. discussed in [8] the first formulation of a GNN, from that time on, different kind of GNNs have been proposed, as exhaustively described in a recent survey of Wu et al. [18]. Among them,

Graph Convolutional Networks (GCNs) [6,9,11,12] have been proved to be the most suitable to process large and sparse graphs thanks to the definition of a convolution-like operation on graphs that is performed by aggregating the information considering only the local structures around the nodes. An effective way to perform such a local convolution has been discussed by Gilmer et al. [7] with the proposition of a new kind of spatial GCNs named Message Passing Neural Networks (MPNMs). The basic idea is to propagate the local information among the nodes as they are sending messages to each other. This is formalized as a two steps process where the network propagates the messages (message passing step) and then pools the received information (readout step) so as to produce a new representation for the nodes and thus for the whole graph. A Convolution-like operation is then performed during the first phase through a message and an update function. Recently, methods based on the concept of *self-attention* have proved to be very effective in discriminating significant messages coming from adjacent nodes [5,16]. Giving different weights to different connections, self-attention makes the neural network able to deal with large and complex structures, that we usually face in biology, by considering only the connections that are more relevant for the task at hand.

In the last years, different methods have been applied to the problem of polypharmacy side effect prediction, that are mainly based on Knowledge Graph Embedding (KGE) algorithms [13] and Unsupervised Graph Embedding [3]. KGE algorithms work directly over triples (*node1, relation, node2*) and aim to find node embeddings that optimize a margin-like loss function. The distance function is computed in the vector space generated by a relation-specific projection function. Unsupervised methods, likewise, learn latent representation on nodes so that the distance among them reveals existing relations. A recent GCN designed to work on multilayer graphs is Decagon [20]; it is characterized by a two-stage architecture composed of a node embedding stage, having the aim of learning message-sharing functions optimized to predict target relationships, and a prediction stage.

In this paper we propose a relation-wise Graph Attention Network (GAT) obtained by extending the architecture of Decagon with an additional layer realized through a *Relation Attention Module (RAM)* having the purpose of weighting each link on the base of the message exchanged. After a description of the proposed architecture, we will present an experimental validation based on a publicly available dataset, in which this new architecture has been demonstrated to considerably improve the accuracy of the original Decagon method, and to achieve state-of-the-art performance by halving the error rate with respect to the most accurate method available in the literature.

2 Proposed Method

Multilayer graphs extend the expressive power of single layer graphs where all the nodes belongs to same domain such as people, proteins, drugs, and so on; however, they put additional complexity onto GCN working on them. This is

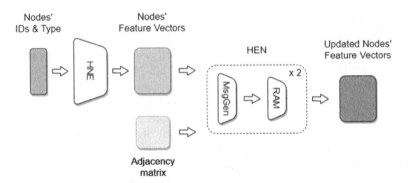

Fig. 2. Embedding Network. The initial embeddings of the node are learned in the *heterogeneous node-embedding (HNE)* layer which assign to each node its own embedding. Then, the initial embedding are updated by the *heterogeneous embedding network (HEN)*. This layer generates the messages shared over the edges by each node in the network and then, it computes the new embedding aggregating them in a weighted manner through the *Relation Attention Module (RAM)*

where an attention mechanism comes into play, giving the GCN the capability to discriminate which relationship is more important when learning the representation of the nodes (Fig. 2).

The proposed GAT is composed of three stages: a *heterogeneous node-embedding*, a *heterogeneous embedding network* based on a *Relation Attention Module* end a *predictor*. Our GAT inherits the advantages of Decagon [20], i.e. end-to-end trained node embeddings for the task we are dealing with and a message passing approach able to share information through edges of different types; but, it increases the expressiveness of the network by weighting the contribution of the messages exchanged over different relations. This is obtained by extending the architecture with the addition of the Relation Attention Module.

In more details, the first stage of the network aims at learning the representation of the graph outputting a d-dimensional node embedding. This stage is composed of a different embedding layer for proteins and drugs respectively, each of them acting as mapping function between a node identifier and a feature vector. The embedding is learned during the end-to-end training of the whole GAT. The representations of the nodes obtained by the heterogeneous node-embedding stage are then provided to the successive stage, the heterogeneous embedding network; this latter generalizes the Message Passing Neural Network paradigm [7] to multilayer graphs where a different message function is used for each type of relation. This stage outputs a new node embedding, where the representation of a node is obtained as the sum of the messages exchanged between the node and its neighborhood. During this process, the additional attention module is used to improve the representativeness of the new node embedding by weighting differently the messages coming from the neighbours on the base of the relationships they have with a considered node. After N message passing iterations, the new node embeddings will contain information depending on both the local properties of the node and its surrounding.

$$h_i^{(k+1)} = \phi \left(\sum_r \alpha_r m_{ir}^{(k)} \right), \quad m_{ir}^{(k)} = \sum_{j \in N_r^i} c_r^{ij} W_r^{(k)} h_j^{(k)} + c_r^i h_i^{(k)} \qquad (1)$$

$$c_r^{ij} = \frac{1}{\sqrt{|N_r^i||N_r^j|}}, \quad c_r^i = \frac{1}{\sqrt{|N_r^i|}} \qquad (2)$$

In Eq. 1 we provide a formalization of the update process realized by the heterogeneous embedding network, where $h_i^{(k+1)}$ represents the features vector of the node i resulting from the k-th iteration of the Message Passing Network ($h_i^{(0)}$ is the input embedding), ϕ is a non-linear activation function, the ReLU in our case, r is the relationship type and $m_{ir}^{(k)}$ are the aggregated messages coming to the node i through an edge of type r. $W_r^{(k)}$ is a projection weights matrix associated to the relationship r, while c_r^{ij} and c_r^i are normalization constant that scale the contribution of the neighbors N_i^r of the node i taking into account only the edges of type r. Finally, $\alpha_r \in [0,1]$ is the attention weight resulting from the Relation Attention Module, that we are going to describe in Sect. 2.1. For the sake of readability, we avoid to explicitly show the dependence with the messages $m_{ir}^{(k)}$ coming from edges of different relations.

Finally, the predictor stage of the proposed method estimates the probability of a polypharmachy side effect taking as input the embeddings of the nodes representing the involved drugs. Similarly to [20], the prediction is performed using a tensor factorization with a shared weights matrix between the different side effects:

$$p(v_i, r, v_j) = \sigma(z_i^T D_r R D_r z_j) \qquad (3)$$

where $p(v_i, r, v_j)$ represents the probability that a side effect r is caused by the combination of drugs v_i and v_j, σ is the sigmoid function, z_i is the node-embedding resulting from the message-passing steps, D_r is a learnable diagonal projection matrix for the side effect r and R is a weights matrix which takes into account all sides effects. Since the predictor function is symmetric the probability of side effects is not influenced by the order of the drugs.

2.1   Relation Attention Module

As discussed above, the heterogeneous embedding network updates the node embedding by aggregating the messages coming from different relationships to the nodes; the contribution of each message is weighted on the base of the relationship r it comes through the attention weight α_r that is computed by the Relation Attention Module. We propose a novel self-attention mechanism for multilayer graphs that works at relation level, inspired to the one proposed in [17], where the attention weights are obtained as a function of the aggregated messages m_{ir} (see Eq. 1) transported by each relation. The proposed module can deal with the absence of edges (i.e. knowledge) since it computes the weights as a function of the effectively shared messages.

To this aim, we build a *relation information matrix* that is obtained by arranging all the aggregated messages of every node in a tensor with dimensions (*features, nodes, relations*). In Fig. 3 we show the way we compute the attention weights starting from the relation information matrix. The first step is to *squeeze* the messages dimension to obtain a feature vector $\in \mathbb{R}^{1 \times relations}$. To this aim, we apply both the average and max pooling, as suggested in [17], in order to carry out more statistics useful for the weights estimation. Then, a Multi-Layer Perceptron (MLP) takes independently the two feature vectors as input and computes the *channel attention maps*. An hidden layer in the MLP acts as bottleneck to extract the useful information for the descriptor and reduce the parameters overhead. To this purpose the hidden layer has dimension C/r_d, where C is the number of relationships and r_d is the reduction ratio. Finally, a sigmoid activation function projects the sum of the two outputs of the MLP in the range $[0, 1]$. Once the attention weights are computed, the node embeddings are weighted and summed along the relation axis to obtain a single embedding for each node (see Fig. 4).

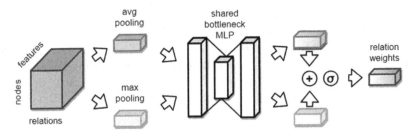

Fig. 3. Relation Attention Module. The module takes as input the relation information matrix and computes both global average and max pooling. The resulting feature vectors are given in input to an bottleneck MLP with shared weights. Finally the attention coefficient are computed by applying the sigmoid function to the sum of the features vectors resulting from the MLP.

3   Experiments

We have compared the proposed relation-wise GAT with other seven state-of-the-art algorithms facing the prediction polypharmacy side effects. The comparison have been performed using the Decagon dataset proposed in [20]. It is a single multilayer graph (see Table 1) composed of a PPI network of 19,085 proteins and 715,612 interactions and a DDI network having 645 drugs and 4,651,131 side effect links. The dataset contains also 18,596 connection among proteins and drugs.

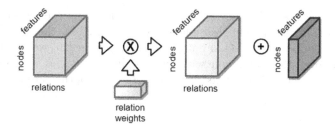

Fig. 4. Relation Attention Module. The embendings resulting from the module are computed as a weighting sum of the aggregated messages transported by each relation.

Table 1. Composition of the dataset

Nodes		Edges	
Drugs	645	DDI	4,651,131
Proteins	19,085	DTI	18,596
		PPI	715,612

The graph contains 964 commonly-occurring polypharmacy side effects which are those ones that occur in at least 500 combinations. As suggested by [20], we remove all polypharmacy side effects that occurs even with just one of the two drugs allowing an unbiased estimation of the model performance. Being built from real data, the dataset is strongly unbalanced with respect to the different kinds of side effects.

As proposed by the authors of the dataset, we have randomly divided it into training, validation and test sets, using 80%, 10% and 10% of the edges of each side effect respectively. For each set and each kind of relation, we equally sampled negative and positive edges.

Regarding the model parameters, we have used embedding layers of size 64 for both the nodes types. The heterogeneous embedding network is composed of 2 layers with sizes respectively equal to 64 and 32. The RAM uses a reduction rate equal to 16. After each message passing step we add a dropout layer with probability 0.056.

The weights of the model have been optimized using the RMSprop algorithm with a learning rate equal to $4.4e-4$. We have used the Binary Cross-Entropy as loss function. Every time that the validation performance does not increase for 5 consecutive epochs (plataue), we have reduced the learning rate using a multiplicative factor equal to 0.1. Finally, we early stopped the model training if the validation performance does not increase for 16 consecutive epochs. An epoch is composed of 100 random batches. To deal with the imbalance of the dataset, we have trained the model with random balanced batches for both the negative/positive edges and the side-effects types. Each batch is composed by 256 positive and 256 negative edges for each type of side effect. The edges considered to evaluate the loss function are excluded from the message passing steps. All and only the training set edges have been used for the evaluation over the validation and the test set. Finally, differently from [20] we applied graph convolutions over the full graph without sampling strategies. Finally, all the hyper-parameters have been tuned using the Bayesian optimization framework.

3.1   Results

The performance have been computed independently for each side-effect in terms of area under the receiver-operating characteristic (AUROC) and area under the precision recall curve (AUPRC). The overall performance of the method have been computed as average of the individual scores.

In Table 2 we have reported the results, in terms of AUROC and AUPRC obtained on the Decagon dataset, the row are ordered by an increasing AUROC. The proposed method outperformed recent approaches proposed in [3, 13, 20] that have achieved accuracies higher than 0.9.

Table 2. Results

Method	Ref.	AUROC	AUPRC
Decagon	[20]	0.872	0.832
KB_{LRN}	[10]	0.899	0.878
DisMult	[19]	0.923	0.898
TransE	[1]	0.949	0.934
ComplEx	[15]	0.965	0.944
TriVec	[13]	0.975	0.966
NBNE	[3]	0.980	–
Our	–	**0.998**	**0.998**

In particular the proposed method achieve an AUROC equal to 0.998 reducing the error of one order of magnitude w.r.t. the most accurate method in the state-of-the-art [3].

It is worth to note that our method significantly outperforms Decagon with a 12% margin. This results may be explained by the following reasons: (i) the embeddings are specialized for the prediction of the side-effects instead of trying to be suitable also for the DTI and PPI prediction; (ii) the choice of using balanced batches allows to avoid the overfitting over more frequent side-effects; (iii) the Relation Attention Module allows to reduce the bias error of the model by weighting the information shared in the different layers of the heterogeneous embedding network. In addition, the proposed method outperforms all KGE models that focus their attention on the DDI network, such as those evaluated in [13]. This result may be justified by the use of a message passing architecture together with the attention mechanism, making our network able to get information from all the layers of the biological network. In fact, our predictor can be seen as a KGE algorithm which, instead of using directly trained embeddings, employs a representation resulting from structured dynamic knowledge.

4   Conclusions

In this paper we have proposed a Graph Attention Network (GAT) to deal with the problem of predicting side effects due to drug interactions. It is an extension of the Decagon GCN proposed by Zitnik et al. in [20] with the addition of an attention module that have considerably improved the accuracy. We have demonstrated that our GAT is very effective on this problem, achieving a state-of-the-art accuracy on a public dataset realized from real data.

For the sake of completeness, it is important to point out that, although the proposed method achieves the highest performance on the Decagon dataset, it depends on the node-type embeddings learned in its first stage. This requires the network to be retrained if new drugs or proteins are added to the graph, while other methods adopt a node embedding that is able to incorporate more easily new kinds of nodes.

In our future works, we will analyze the robustness of the method using new nodes, drugs and side-effects in the test set that were not present in the training phase.

References

1. Bordes, A., Usunier, N., Garcia-Duran, A., Weston, J., Yakhnenko, O.: Translating embeddings for modeling multi-relational data. In: Advances in Neural Information Processing Systems, pp. 2787–2795 (2013)
2. Carletti, V., Foggia, P., Garrison, E., Greco, L., Ritrovato, P., Vento, M.: Graph-based representations for supporting genome data analysis and visualization: opportunities and challenges. In: Conte, D., Ramel, J.-Y., Foggia, P. (eds.) GbRPR 2019. LNCS, vol. 11510, pp. 237–246. Springer, Cham (2019). https://doi.org/10.1007/978-3-030-20081-7_23
3. Chen, D., Jalilifard, A., Veloso, A., Ziviani, N.: Modeling pharmacological effects with multi-relation unsupervised graph embedding. arXiv preprint arXiv:2004.14842 (2020)
4. Consortium, T.C.P.G.: Computational pan-genomics: status, promises and challenges. Brief. Bioinform. **19**, 118–135 (2016). https://doi.org/10.1093/bib/bbw089
5. Fey, M.: Just jump: dynamic neighborhood aggregation in graph neural networks. arXiv preprint arXiv:1904.04849 (2019)
6. Gao, H., Wang, Z., Ji, S.: Large-scale learnable graph convolutional networks. In: Proceedings of the 24th ACM SIGKDD International Conference on Knowledge Discovery and Data Mining. Association for Computing Machinery (2018). https://doi.org/10.1145/3219819.3219947
7. Gilmer, J., Schoenholz, S.S., Riley, P.F., Vinyals, O., Dahl, G.E.: Neural message passing for quantum chemistry. CoRR (2017)

8. Gori, M., Monfardini, G., Scarselli, F.: A new model for learning in graph domains. In: Proceedings of 2005 IEEE International Joint Conference on Neural Networks, vol. 2, pp. 729–734 (2005)

9. Hamilton, W., Ying, Z., Leskovec, J.: Inductive representation learning on large graphs. In: Guyon, I., et al. (eds.) Advances in Neural Information Processing Systems, vol. 30, pp. 1024–1034. Curran Associates, Inc. (2017)

10. Malone, B., García-Durán, A., Niepert, M.: Knowledge graph completion to predict polypharmacy side effects. In: Auer, S., Vidal, M.-E. (eds.) DILS 2018. LNCS, vol. 11371, pp. 144–149. Springer, Cham (2019). https://doi.org/10.1007/978-3-030-06016-9_14

11. Monti, F., Boscaini, D., Masci, J., Rodolà, E., Svoboda, J., Bronstein, M.M.: Geometric deep learning on graphs and manifolds using mixture model CNNs. In: 2017 IEEE Conference on Computer Vision and Pattern Recognition (CVPR), pp. 5425–5434 (2017)

12. Monti, F., Bronstein, M., Bresson, X.: Geometric matrix completion with recurrent multi-graph neural networks. In: Guyon, I., et al. (eds.) Advances in Neural Information Processing Systems, vol. 30, pp. 3697–3707. Curran Associates, Inc. (2017)

13. Nováček, V., Mohamed, S.K.: Predicting polypharmacy side-effects using knowledge graph embeddings. AMIA Summits Transl. Sci. Proc. **2020**, 449 (2020)

14. Pavlopoulos, G.A., et al.: Using graph theory to analyze biological networks. BioData Min. 4(1), 10 (2011). https://doi.org/10.1186/1756-0381-4-10

15. Trouillon, T., Welbl, J., Riedel, S., Gaussier, É., Bouchard, G.: Complex embeddings for simple link prediction. In: Proceedings of the 33rd International Conference on International Conference on Machine Learning (ICML), vol. 48 (2016)

16. Veličković, P., Cucurull, G., Casanova, A., Romero, A., Liò, P., Bengio, Y.: Graph attention networks. In: International Conference on Learning Representations (2018). https://openreview.net/forum?id=rJXMpikCZ

17. Woo, S., Park, J., Lee, J.Y., So Kweon, I.: CBAM: Convolutional block attention module. In: Proceedings of the European Conference on Computer Vision (ECCV), pp. 3–19 (2018)

18. Wu, Z., Pan, S., Chen, F., Long, G., Zhang, C., Yu, P.S.: A comprehensive survey on graph neural networks. IEEE Trans. Neural Netw. Learn. Syst. **32**, 1–21 (2020)

19. Yang, B., Yih, S.W.T., He, X., Gao, J., Deng, L.: Embedding entities and relations for learning and inference in knowledge bases. In: Proceedings of the International Conference on Learning Representations (ICLR), May 2015

20. Zitnik, M., Agrawal, M., Leskovec, J.: Modeling polypharmacy side effects with graph convolutional networks. Bioinformatics **34**(13), i457–i466 (2018). https://doi.org/10.1093/bioinformatics/bty294

21. Zitnik, M., Nguyen, F., Wang, B., Leskovec, J., Goldenberg, A., Hoffman, M.M.: Machine learning for integrating data in biology and medicine: principles, practice, and opportunities. Inf. Fusion **50**, 71–91 (2019)

LGL-GNN: Learning Global and Local Information for Graph Neural Networks

Huan Li[1], Boyuan Wang[1], Lixin Cui[1(✉)], Lu Bai[1], and Edwin R. Hancock[2]

[1] School of Information, Central University of Finance and Economics,
Beijing, China
cuilixin@cufe.edu.cn

[2] Department of Computer Science, University of York, York, UK

Abstract. In this article, we have developed a graph convolutional network model LGL that can learn global and local information at the same time for effective graph classification tasks. Our idea is to concatenate the convolution results of the deep graph convolutional network and the motif-based subgraph convolutional network layer by layer, and give attention weights to global features and local features. We hope that this method can alleviate the over-smoothing problem when the depth of the neural networks increases, and the introduction of motif for local convolution can better learn local neighborhood features with strong connectivity. Finally, our experiments on standard graph classification benchmarks prove the effectiveness of the model.

Keywords: Graph convolutional networks · Graph classification

1 Introduction

In recent years, deep learning has achieved outstanding performance in many fields such as computer vision and natural language processing. The existing deep learning models can handle structured data such as images and speech well, but they are difficult to apply to graph data directly. However, in real life, there are a large number of non-Euclidean data represented in the form of graphs. For example, graphs can be abstracted from social networks, citation networks, protein-interaction networks and other scenarios. Graph is not only ubiquitous, but also can flexibly describe the complex relationships between real things and has a strong structured expression ability. These advantages have inspired researchers to further expand their research horizons to the field of deep learning and graph. However, unlike image data with a regular grid structure, each node in the graph has a different number of neighbor nodes, so basic convolution and pooling operations cannot be used, which poses a huge challenge to the existing convolutional neural network.

B. Wang—Co-first Author: These authors have contributed equally to this work.

A. Torsello et al. (Eds.): S+SSPR 2020, LNCS 12644, pp. 129–138, 2021.
https://doi.org/10.1007/978-3-030-73973-7_13

When extending CNN to the irregular grid structure of graph, two main strategies are adopted, a) Spectral-based [1,2] and b) Spatial-based [3–5] methods. Most existing GCNs are designed under these two strategies. Specifically, Spectral-based GCN defines convolution operations based on spectral graph theory. This method requires graphs to have the same size of structure and is usually used for vertex classification tasks. Spatial-based GCN approximates the spectral convolution operation by defining the layer-by-layer propagation of a node-based one-hop neighborhood. It is not limited to the same size of graph structure, and can be used for graph classification tasks. Although the Spatial-based GCN model can handle graph classification problems, the particularity of the graph structure still brings some difficult problems to GCN. One of the most difficult problems is over-smoothing [6–8]. As the number of network layers increases and the number of iterations increases, the representation of each node tends to converge to the same value, which means that the global information of the entire graph is synchronized to every node, rather than the local structural features we expect.

To overcome the problem of over-smoothing, there are roughly two ways of thinking at present. On the one hand, the SortPooling layer is used to replace SumPooling. The SumPooling layer directly aggregates the learned local vertex features from graph convolution operations into global features. It is difficult to learn rich local vertex topological information from global features, resulting in poor classification results. M. Zhang et al. [9] proposed a novel Deep Graph Convolutional Neural Network (DGCNN), which uses a novel SortPooling layer to sort the extracted multi-scale vertex features instead of summing. DGCNN pays more attention to local vertex features, but only retains the top specified number of vertices when sorting, which may cause a lot of information to be lost. On the other hand, convolve the local subgraph of the node. Z. Zhang et al. [10] designed a local convolution operation based on a subtree. Since the local subgraph only retains the information of the nodes closer to the root node, the design of the graph convolution operation on the subgraph can limit the information interaction with remote nodes, but also loses global information.

The LGL model we proposed is inspired by the simultaneous attention and fusion of global and local information. While using the graph convolutional layer to learn global information, the subgraph convolutional layer is used to learn local node features, and the attention mechanism is introduced to give different weights to them. The framework of the proposed LGL is shown in Fig. 1. Specifically, the main contributions of this paper are summarized as follows:

First, we designed a new local convolution operation based on motif. Motif is a subgraph that appears frequently in graph. Each node in motif has strong connectivity. The using of motif can effectively capture high-quality local neighborhood information.

Second, We have developed a novel hybrid graph convolutional network model for graph classification, which is the LGL model. The LGL model uses the depth graph convolutional network and the subgraph convolutional network to learn global information and local information respectively, and the attention mechanism gives weight to both.

Third, we evaluate the performance of the proposed LGL model on graph classification tasks by means of experience. Experiments on benchmarks demonstrate the effectiveness of the proposed method, when compared to state-of-the-art methods.

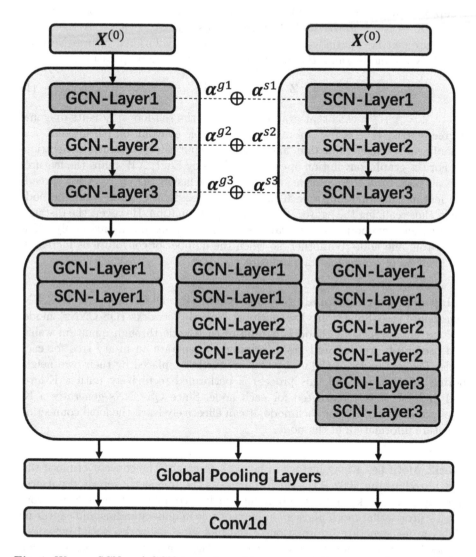

Fig. 1. We use SCN and GCN to perform 3-layer convolution on graph respectively, and splice the convolved features as shown in the figure. Then the spliced the features are sorted into a grid structure, which can be directly subjected to 1-dimensional convolution.

2   Related Works

In this section, we briefly review some important related work of LGL model, including the Deep Graph Convolutional Neural Network (DGCNN) [9], the Subgraph Convolutional Neural Networks (SCN) [10] and motif.

Deep Graph Convolutional Neural Network. Given a graph $\mathbf{G}$ with n nodes, $\mathbf{X} \in \mathbb{R}^{n \times c}$ is the node feature vectors and $\mathbf{A} \in \mathbb{R}^{n \times n}$ is the graph adjacency matrix. Spatially-based Deep Graph Convolutional Neural Network (DGCNN) [9] model takes the following graph convolution operation

$$\mathbf{Z} = f(\tilde{D}^{-1}\tilde{A}XW), \tag{1}$$

where $\tilde{A} = A + I$ means that graph is added to the self-loops, $\tilde{D}$ is its diagonal degree matrix, $\tilde{W} \in \mathbb{R}^{c \times c'}$ is a matrix of trainable convolution parameters, f is a nonlinear activation function, and Z is the output after convolution operation.

For the graph convolution operation defined by Eq. 1, XW maps the features of each node from the c dimension to the c' dimension, $\tilde{A}Y(Y := XW)$ spreads the features information of each node to the neighboring nodes and the node itself, thus realizing the aggregation of nodes information. However, the distance between any two nodes in the graph is relatively close, and it takes only a few steps from one node to another, so when the number of convolutions increases, the problem of over-smoothing appears.

Subgraph Convolutional Neural Network. Different from DGCNN, Quantum-based Subgraph Convolutional Neural Networks (QS-CNNs) model [10] extract the neighborhood subgraph of each node through quantum walks, and use graph grafting and graph pruning to generate an m-ary tree for each node. The leaf nodes of the m-ary tree are further replaced by their own neighboring m-ary trees, and this process is performed recursively until a K-level and m-ary tree is constructed for each node. Since QS-CNNs generates a K-level extended subtree for each node, it can effectively learn the local connection structure information of the node.

Motif. Motif has a long history in network research. The concept of motif was first introduced in 2002 [11], which represents the frequently repeated patterns in complex networks and is the building block of complex networks. Some work [12–14] proves that motif plays an important role in understanding and capturing higher-order structure information of the biological networks, social networks, academic networks, and so on. Capturing the motif structure and its interaction can improve the quality of network embedding. But the current research basically ignores the capture and application of Motif. Several common motifs are shown in Fig. 2.

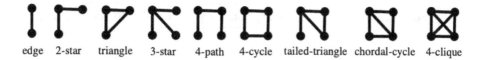

edge   2-star   triangle   3-star   4-path   4-cycle   tailed-triangle   chordal-cycle   4-clique

Fig. 2. Several common motifs

3   Proposed LGL Model

In this section, we first give the overall framework of the proposed model LGL. Subsequently, we introduce the subgraph convolutional layer based on motif, and the feature fusion between subgraph convolutional layer and deep convolutional layer based on attention mechanism.

A. Framework. We develop a novel hybrid convolution structure based on GCN and SCN. Specifically, both GCN and SCN performed the convolution operation three times, and we spliced the features of the convolutional layer as shown in Fig. 1. Each node in SCN only aggregates adjacent motif neighbors, representing the local node information of the graph, and GCN is based on the one-hop neighbor propagation of the node, including the global topological characteristics of the graph. Hybrid graph convolution operation can be described by the following formula:

$$Z^{(l)} = f(\alpha^{g(l)} \tilde{D}^{-1} \tilde{A} X W \oplus \alpha^{s(l)} \tilde{D}_m^{-1} M X W_m), \tag{2}$$

where $\tilde{A} = A + I$ means that graph is added to the self-loops, M is motif adjacency matrix, $\tilde{W} \in \mathbb{R}^{c \times c'}$ and $\tilde{W}_m \in \mathbb{R}^{c \times c'}$ are matrix of trainable convolution parameters, $\alpha^{g(l)}$ and $\alpha^{s(l)}$ are the attention weights of the GCN and SCN layers, f is a nonlinear activation function, and Z is the output after convolution operation.

B. Motif-Based SCN. In order to better learn the strong connection relationship between nodes, we introduced motif and designed SCN based on motif. Specifically, SCN includes four key steps: (1) rank nodes according to their degree; (2) find neighbors based on motif for each node; (3) map the subgraph to the tree: construct a m-ary tree for each node. The leaf nodes of the i-level m-ary tree are replaced by the neighboring m-ary trees, and a K-level m-ary tree is recursively constructed for each node; (4) arrange the tree into a regular grid structure (Fig. 3).

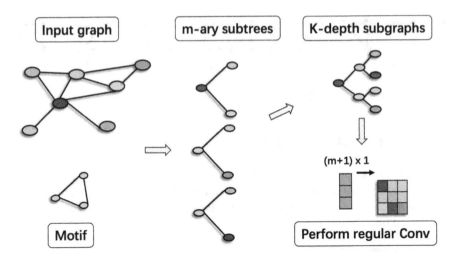

Fig. 3. For the root node (dark blue), find its motif neighbor (light blue, pink) as the leaf node of the root node. Then take the leaf node as the root node, continue to find the leaf node based on the motif pattern, and iterate continuously until it grows into a K-layer m-ary tree. Finally, the trees are arranged into a grid structure. (Color figure online)

C. Attention Layer. Similar to the attention equation mentioned by Vaswan et al. [15], we introduce attention to the splicing process of GCN layer and SCN layer:

$$\text{Attention}^{(l)}(S^{(l)}, G^{(l)}) = \text{softmax}(S^{(l)}W_s^{(l)} \oplus G^{(l)}W_g^{(l)}) \tag{3}$$

where $S^{(l)} \in \mathbb{R}^{N \times d}$ is the last output of SCN, $D^{(l)} \in \mathbb{R}^{N \times d}$ is the last output of GCN, d denotes the dimensions of each vertex and $W_s^{(l)} \in \mathbb{R}^{d \times d}$ and $W_g^{(l)} \in \mathbb{R}^{d \times d_{out}}$ are two learnable matrices.

4   Experimental Results

We set up the experiments on benchmark datasets to evaluate the solid performance of the proposed LGL model against both state-of-the-art graph kernels and other deep learning methods on graph classification problems.

Datasets. We conducted the experiments using three bioinformatics datasets: MUTAG, PROTEINS, PTC-MR and one social networks datasets: IMDB-B. The details of the datasets are shown in Table 1.

Table 1. Information of the benchmark datasets

Datasets	Graphs	Classes	Avg. nodes	Avg. edges	Labels	Description
MUTAG	188	2	17.93	19.79	7	Bioinformatics
PROTEINS	1113	2	39.06	72.82	3	Bioinformatics
PTC-MR	344	2	14.29	14.69	19	Bioinformatics
IMDB-B	1000	2	19.77	96.53	–	Social

Experimental Setting. We compare the performance of the proposed LGL model on graph classification tasks with a) four alternative state-of-the-art graph kernels and b) five alternative state-of-the-art deep learning approaches for graphs. To be specific, the graph kernels include 1) the Weisfeiler-Lehman subtree kernel (WLSK) [16], 2) the shortest path graph kernel (SPGK) [17], 3) the random walk graph kernel (RWGK) [18], and 4) the graphlet count kernel (GK) [19]. The deep learning methods include 1) the deep graph convolutional neural network (DGCNN) [9], 2) the quantum-based subgraph convolutional neural networks (Qs-CNNs) [10], 3) the backtrackless aligned-spatial graph convolutional networks [20], 4) the deep graphlet kernel (DGK) [21], and 5) the diffusion convolutional neural network (DCNN) [22].

Table 2. Classification accuracy (In% ± standard error) for comparisons

Datasets	MUTAG	PROTEINS	PTC-MR	IMDB-B
WLSK	82.88 ± 0.57	73.52 ± 0.43	58.26 ± 0.47	71.88 ± 0.77
SPGK	83.38 ± 0.81	75.10 ± 0.50	55.52 ± 0.46	71.26 ± 1.04
RWGK	80.77 ± 0.72	74.20 ± 0.40	55.91 ± 0.37	67.94 ± 0.77
GK	81.66 ± 2.11	71.67 ± 0.55	52.26 ± 1.41	65.87 ± 0.98
DGCNN	85.83 ± 1.66	75.54 ± 0.94	58.59 ± 2.47	70.03 ± 0.86
Qs-CNNs	93.13 ± 4.67	78.80 ± 4.63	65.99 ± 4.43	–
BASGCN	90.05 ± 0.82	76.05 ± 0.57	61.51 ± 0.77	74.00 ± 0.87
DGK	82.66 ± 1.45	71.68 ± 0.50	57.32 ± 1.13	66.96 ± 0.56
DCNN	66.98	61.29 ± 1.60	58.09 ± 0.53	49.06 ± 1.37
LGL	$\mathbf{90.16 \pm 1.39}$	$\mathbf{78.41 \pm 0.82}$	$\mathbf{65.74 \pm 1.80}$	$\mathbf{66.51 \pm 1.51}$

For the evaluation, we adjust a number of hyperparameters to get the best performance of each dataset, as shown in Table 3. In SCN, we set up two motifs, triangle and 4-cycle, to capture the neighbors of nodes, and construct 2-ary and 3-ary trees respectively. For our model, we perform 10-fold cross-validation to compute the classification accuracy, with nine training folds and one validating fold. For each dataset, we repeat the experiment 10 times and report the average classification accuracy and standard errors in Table 2.

For the alternative graph kernels and deep learning methods except Qs-CNNs, we report the best results collected and experimented by Bai et al. [23]. We report the best results for Qs-CNNs from the original paper [10]. Classification accuracy and standard error of each competing approach are also shown in Table 2.

Experimental Results and Discussions. Table 2 indicates that the proposed LGL model can significantly outperform either the competing graph kernel methods or the deep learning methods for graph classification.

Table 3. Hyperparameters settings for each dataset.

Parameters	K	m	Motif	conv	conv1d	fc	fc_num	batch	lr	L2norm	dropout
MUTAG	5	2	triangle	256	32	64	2	64	0.01	0	0.1
PROTEINS	4	3	4-cycle	32	32	256	3	16	0.0003	0	0
PTC-MR	4	3	4-cycle	256	32	128	2	32	0.01	0	0
IMDB-B	4	2	triangle	256	32	64	2	64	0.01	0	0

Overall, the reasons for the effectiveness of our method are threefold. First of all, as mentioned earlier, most deep learning methods for graph classification cannot well avoid the problems of oversmoothing and retention of rich global and local information. On the contrary, the proposed LGL can alleviate these problems and get better representation learning. Second, the graph kernels with C-SVM classifier are shallow learning methods, while the proposed LGL model can provide an end-to-end deep learning architecture. Thus LGL model can learn better graph characteristics. Third, the use of motif to extract strongly connected neighbor information for nodes simplifies the steps of quantum walk in Qs-CNNs. The splicing of the results of local subgraph convolution and deep graph convolution has achieved a performance exceeding DGCNN. This empirically proves the effectiveness of the proposed LGL model.

5   Conclusions

In this paper, we have shown how to construct motif-based subgraph convolution network for a graph and how to make use of both the global topological arrangement information and local connectivity structures within a graph. Experimental results on graph classification show our LGL model is superior to a number of baseline methods.

It is interesting to notice that different practical problems have different requirements for global and local information. For example, social networks may rely more on the near-end neighbors of nodes, but the properties of chemical molecules may depend on some remote nodes. In addition, the choice of motif also greatly affects the effect of graph classification. Our future plan is to explore the impact of more types of motifs on the experimental results, and a better way to gather global and local information.

Acknowledgments. This work is supported by the National Natural Science Foundation of China (Grant no. 61602535 and 61976235), the Program for Innovation Research in Central University of Finance and Economics, and the Youth Talent Development Support Program by Central University of Finance and Economics, No. QYP1908.

References

1. Kipf, T.N., Welling, M.: Semi-supervised classification with graph convolutional networks. arXiv preprint arXiv:1609.02907 (2016)

2. Bruna, J., Zaremba, W., Szlam, A., et al.: Spectral networks and locally connected networks on graphs. arXiv preprint arXiv:1312.6203 (2013)
3. Defferrard, M., Bresson, X., Vandergheynst, P.: Convolutional neural networks on graphs with fast localized spectral filtering. In: Advances in Neural Information Processing Systems, pp. 3844–3852 (2016)
4. Hamilton, W., Ying, Z., Leskovec, J.: Inductive representation learning on large graphs. In: Advances in Neural Information Processing Systems, pp. 1024–1034 (2017)
5. Wang, Y., Sun, Y., Liu, Z., et al.: Dynamic graph CNN for learning on point clouds. ACM Trans. Graph. (TOG) **38**(5), 1–12 (2019)
6. Li, Q., Han, Z., Wu, X.M.: Deeper insights into graph convolutional networks for semi-supervised learning. arXiv preprint arXiv:1801.07606 (2018)
7. Oono, K., Suzuki, T.: Graph neural networks exponentially lose expressive power for node classification. In: International Conference on Learning Representations (2020)
8. Zhou, J., Cui, G., Zhang, Z., et al.: Graph neural networks: a review of methods and applications. arXiv preprint arXiv:1812.08434 (2018)
9. Zhang, M., Cui, Z., Neumann, M., et al.: An end-to-end deep learning architecture for graph classification. In: Thirty-Second AAAI Conference on Artificial Intelligence (2018)
10. Zhang, Z., Chen, D., Wang, J., et al.: Quantum-based subgraph convolutional neural networks. Pattern Recogn. **88**, 38–49 (2019)
11. Milo, R., Shen-Orr, S., Itzkovitz, S., et al.: Network motifs: simple building blocks of complex networks. Science **298**(5594), 824–827 (2002)
12. Benson, A.R., Gleich, D.F., Leskovec, J.: Higher-order organization of complex networks. Science **353**(6295), 163–166 (2016)
13. Yin, H., Benson, A.R., Leskovec, J., et al.: Local higher-order graph clustering. In: Proceedings of the 23rd ACM SIGKDD International Conference on Knowledge Discovery and Data Mining, pp. 555–564 (2017)
14. Lee, J.B., Rossi, R.A., Kong, X., et al.: Graph convolutional networks with motif-based attention. In: Proceedings of the 28th ACM International Conference on Information and Knowledge Management, pp. 499–508 (2019)
15. Vaswani, A., Shazeer, N., Parmar, N., et al.: Attention is all you need. In: Advances in Neural Information Processing Systems, pp. 5998–6008 (2017)
16. Shervashidze, N., Schweitzer, P., Van Leeuwen, E.J., et al.: Weisfeiler-Lehman graph kernels. J. Mach. Learn. Res. **12**(9), 2539–2561 (2011)
17. Borgwardt, K.M., Kriegel, H.P.: Shortest-path kernels on graphs. In: Fifth IEEE International Conference on Data Mining (ICDM 2005). IEEE (2005). 8 pp
18. Kashima, H., Tsuda, K., Inokuchi, A.: Marginalized kernels between labeled graphs. In: Proceedings of the 20th International Conference on Machine Learning (ICML 2003), pp. 321–328 (2003)
19. Shervashidze, N., Vishwanathan, S.V.N., Petri, T., et al.: Efficient graphlet kernels for large graph comparison. In: Artificial Intelligence and Statistics, pp. 488–495 (2009)
20. Bai, L., Cui, L., Jiao, Y., et al.: Learning backtrackless aligned-spatial graph convolutional networks for graph classification. IEEE Trans. Pattern Anal. Mach. Intell. (2020)
21. Yanardag, P., Vishwanathan, S.V.N.: Deep graph kernels. In: Proceedings of the 21st ACM SIGKDD International Conference on Knowledge Discovery and Data Mining, pp. 1365–1374 (2015)

22. Atwood, J., Towsley, D.: Diffusion-convolutional neural networks. In: Advances in Neural Information Processing Systems, pp. 1993–2001 (2016)
23. Bai, L., Jiao, Y., Cui, L., Hancock, E.R.: Learning aligned-spatial graph convolutional networks for graph classification. In: Brefeld, U., Fromont, E., Hotho, A., Knobbe, A., Maathuis, M., Robardet, C. (eds.) ECML PKDD 2019. LNCS (LNAI), vol. 11906, pp. 464–482. Springer, Cham (2020). https://doi.org/10.1007/978-3-030-46150-8_28

Graph Transformer: Learning Better Representations for Graph Neural Networks

Boyuan Wang[1], Lixin Cui[1(✉)], Lu Bai[1], and Edwin R. Hancock[2]

[1] School of Information, Central University of Finance and Economics, Beijing, China
cuilixin@cufe.edu.cn
[2] Department of Computer Science, University of York, York, UK

Abstract. Graph classifications are significant tasks for many real-world applications. Recently, Graph Neural Networks (GNNs) have achieved excellent performance on many graph classification tasks. However, most state-of-the-art GNNs face the challenge of the over-smoothing problem and cannot learn latent relations between distant vertices well. To overcome this problem, we develop a novel Graph Transformer (GT) unit to learn latent relations timely. In addition, we propose a mixed network to combine different methods of graph learning. We elucidate that the proposed GT unit can both learn distant latent connections well and form better representations for graphs. Moreover, the proposed Graph Transformer with Mixed Network (GTMN) can learn both local and global information simultaneously. Experiments on standard graph classification benchmarks demonstrate that our proposed approach performs better when compared with other competing methods.

Keywords: Graph Convolutional Networks · Graph classification · Graph Transformer

1 Introduction

Graphs are widely used to model complex objects and their dependency relationships in many pattern recognition and machine learning tasks [19]. Along with recent success of deep learning networks, booming interests are focalized on utilizing these methods for analyzing large-scale and high-dimensional regular or Euclidean data [19]. Particularly, Convolutional Neural Networks (CNNs) [20] have become powerful tools to extract useful statistical patterns from large datasets of images, videos, etc. However, because graph structure data is often irregular or non-Euclidean, directly applying CNNs for analyzing such data is difficult. Therefore, great efforts have been devoted to extending CNNs to the graph domain, and a great number of Graph Convolutional Networks (GCNs) [3,19] have been developed for extracting meaningful features for graph classification.

In general, there are two main categories of GCNs, i.e., the spectral methods and the spatial methods [3,19]. Specifically, the former defines convolution operation based on the spectral graph theory [5,6,12] by calculating the eigenvectors

© Springer Nature Switzerland AG 2021
A. Torsello et al. (Eds.): S+SSPR 2020, LNCS 12644, pp. 139–149, 2021.
https://doi.org/10.1007/978-3-030-73973-7_14

of Laplacian matrix. However, due to the heavy computational complexity of calculating eigenvectors, these approaches cannot be expanded to big graphs well. Instead, the latter methods are more flexible by defining operations on neighbor vertices [2,18,19]. For example, Zhang et al. [19] introduced a Subgraph Convolutional Network (SCN) with quantum walk to facilitate regular convolution operations computing on subgraphs. Zhang et al. [18] proposed a novel Deep Graph Convolutional Neural Network (DGCNN) model to consider vertex information both locally and globally. Nevertheless, two major problems arising along with GCNs, i.e., over-smoothing and lack of capturing distant relations. For instance, SCN may lose sight of remote latent connections, DGCNN also cannot learn these potential links well due to the over-smoothing problem. Detailed explanation of these two major issues will be discussed in Sect. 3.1.

To overcome these issues, we propose Graph Transformer with Mixed Network (GTMN) for graph classification problem. The framework of the proposed GTMN is shown in Fig. 1. One important characteristic of the proposed GTMN model is that it can learn latent relations between vertices without using the adjacency matrix. Detailed discussions can be seen in Sect. 3.2 and Sect. 4.

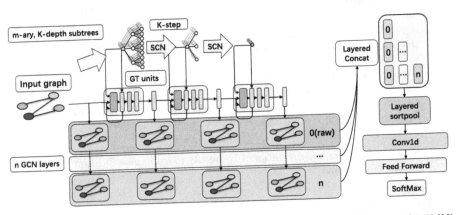

Fig. 1. Overview of the proposed Graph Transformer with Mixed Network (GTMN) architecture. At each step, the proposed Graph Transformer units (GT) take in the input calculated by the Subgraph Convolution Network (SCN) model and the previous output or the input graph. Each GT calculates the latent relationships between substructures with different sizes. Up to K (depth of generated subtree) steps of GT are performed. Outputs then are fed into a set of GCN layers. For n GCN layers, outputs at layer i is concatenated and then $n+1$ big feature matrices are formed. Furthermore, all of them are concatenated and fed into Layered SortPools hierarchically. The final part of the network is to transmute graph features into grid structures and make a prediction for the input graph.

2 Related Works

In this section, we briefly review some important related works of GTMN. Specifically, we first introduce the Subgraph Convolution Network [19] (SCN).

Then we show the operation of the Deep Graph Convolutional Neural Network [18] (DGCNN). Finally, we elucidate the idea of the Transformer.

Subgraph Convolution Network. According to Zhang et al. [19], to operate regular convolution, every vertex is given a QS-Score through quantum walk and each node forms a m-ary subtree by grafting and pruning k-hop neighbor edges. Then a subgraph of K-depth is generated from the subtree for a specific vertex. Thus regular convolution operation is able to process on this subgraph, as shown in Fig. 2.

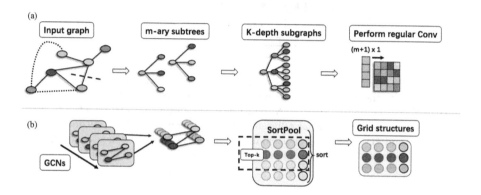

Fig. 2. Detailed procedures of (a) SCN and (b) DGCNN.

Deep Graph Convolutional Neural Network. Zhang et al. [18] proposes an spatially-based operation of GCN, i.e.

$$\mathbf{Z} = f(\tilde{\mathbf{D}}^{-1}\tilde{\mathbf{A}}\mathbf{X}\mathbf{W}), \tag{1}$$

where $\tilde{\mathbf{A}} = \mathbf{A} + \mathbf{I}$ is the adjacency matrix of the graph with added self-loops, $\tilde{\mathbf{D}}$ is its diagonal degree matrix, $\tilde{\mathbf{W}} \in \mathbb{R}^{c \times c'}$ denotes a matrix of trainable parameters, f is a nonlinear function, and $\mathbf{Z}$ is convolution's output. For the defined Eq. 1, it can be explained as each node's aggregation with neighbor features.

Moreover, DGCNN introduces a pooling layer called SortPool based on the Weisfeiler-Lehman (WL) algorithm. Vertices are sorted according to last outputs of GCNs, which represent color labels reflecting topological importance. Detailed procedures are shown in Fig. 2.

Transformer. Vaswan et al. [15] proposed an attention-based architecture Transformer to replace traditional Recurrent Neural Networks (RNNs). The key idea of the Transformer is a Multi-Head attention layer that can learn tokens' connections. Numerous approaches have been developed to apply the transformer to graph domain tasks and have achieved great success [11,17]. Motivated by the idea of transformer, we propose the Graph Transformer unit of our version in Sect. 3.2.

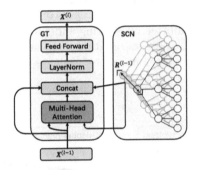

Fig. 3. Detailed structure of Graph Transformer (GT) unit with SCN. Given previous layer's output or the input vertices' features $\mathbf{X}^{(l-1)}$ and preceding roots of all vertices' sub-trees $\mathbf{R}^{(l-1)}$ computed from SCN, multi-head attention can be calculated to learn distant relations beyond original adjacency matrix. Then the output is concatenated with $\mathbf{X}^{(l-1)}$ and $\mathbf{R}^{(l-1)}$ and layer-normalized. Next, the normed concatenation is fed into one linear layer to constrain dimensions. After that, the final output of GT $\mathbf{X}^{(l)}$ is formed.

3   The Proposed Graph Transformer Unit

In this section, we first discuss the major existing problems in GCNs. Moreover, we demonstrate the issue of learning latent relations between substructures. Then we introduce our idea of Graph Transformer (GT) unit associated with the random walk method. Finally, we illustrate how the GT can learn latent relations well from the practical and theoretical perspectives.

3.1   Problems of Existing Spatial GCNs

Over Smoothing. Many spatial-based GCNs are facing the problem of over-smoothing, for example, the typical graph convolution operation proposed by DGCNN [18] in Eq. 1. This equation aims to aggregate each vertex's neighbor information and outputs a new graph representation. However, as the network layer goes deeper, each vertex's features appear to be similar and lose its distinct information. This is called the over-smoothing problem [9]. As analyzed and proved by Liu et al. [9], with more neighbors of larger distances aggregated, each vertex will lose its distinction ultimately.

Latent Relations Between Substructures. To overcome the over-smoothing issue mentioned above, many methods are proposed [8,9]. In particular, it has been shown that methods using random walks [19] are able to overcome this problem through considering sub-graphs or substructures locally. Although more efficient, the proposed methods suffer from the following issue, i.e., two distant substructures' latent relations cannot be learned well. Methods mentioned above either learned these links too late or just ignored these distant links. This leads to significant loss of latent information.

3.2   Graph Transformer Unit

To learn latent relations timely, we propose the Graph Transformer (GT) unit. One main advantage of GT is that it can take distant nodes into consideration. Concretely, the proposed GT unit is composed of a graph attention layer, a layernorm layer and a linear layer. The detailed structure of GT can be seen in Fig. 3. We specifically discuss the graph attention layer in this section.

Attention Layer. Similar to the attention equation mentioned by Vaswan et al. [15], we propose our graph version:

$$\text{Attention}^{(l)}(\mathbf{R}^{(l-1)}, \mathbf{X}^{(l-1)}) = \text{softmax}(\frac{\mathbf{R}^{(l-1)}\mathbf{W}_{qk}^{(l)}\mathbf{X}^{(l-1)T}}{\sqrt{d}})\mathbf{X}^{(l-1)}\mathbf{W}_v^{(l)} \quad (2)$$

where $\mathbf{R}^{(l-1)} \in \mathbb{R}^{N \times d}$ is the root vertices of SCN's last output with shape $N \times d$, $X^{(l-1)} \in \mathbb{R}^{N \times d}$ is the previous output of GT unit or the input graph features, d denotes the dimensions of each vertex and $\mathbf{W}_{qk}^{(l)} \in \mathbb{R}^{d \times d}$ and $\mathbf{W}_v^{(l)} \in \mathbb{R}^{d \times d_{out}}$ are two learnable matrices. Usually, the attention layer requires three inputs including query, key, and value. Here we define the last output of SCN as query, and the previous output of GT unit or input graph features are considered as key and value. Then we develop the graph multi-head attention layer as below.

$$\text{MultiHead}^{(l)}(\mathbf{R}^{(l-1)}, \mathbf{X}^{(l-1)}) = ||_{i=1}^{h} Attention_i^{(l)}(\mathbf{R}^{(l-1)}, \mathbf{X}^{(l-1)})\mathbf{W}^O \quad (3)$$

where $||$ denotes concatenation of all attention heads and $\mathbf{W}^O \in \mathbb{R}^{hd \times out}$ is a projection matrix. Then we concatenate the output and the input altogether and the concatenation is fed to the remaining layers.

3.3   Discussions of the Proposed GT Unit

Practically. Because of the advantages of multi-head attention and the local information learned by SCN, these latent relations can be attained well. According to Eq. 2, similarity between the local info and graph representations are calculated, thus similar substructures are guaranteed to be aggregated no matter how distant they are. Hence, **the GT unit with multi layers can capture meaningful potential relations and learn better representations of multiple scales.**

Theoretically. We show that the core of GT, i.e., a multi-head layer is a more generic type of GCNs. Because each head of multi-head layer can learn a unique adjacency matrix respectively, GT can aggregate vertices or substructures of similar representations and learn both the local and global information well. More specifically, for each $head_i$, the calculation of $softmax$ part in Eq. 2 is $\tilde{\mathbf{A}}_{latent} \in \mathbb{R}^{N \times N}$ and it can be considered as a normalized latent adjacency matrix, where $A_{i,j}$ denotes the hidden relation between $vertex_i$ and $vertex_j$. Thus, Eq. 2 can be transformed to the following equation.

$$\text{Attention}^{(l)}(\mathbf{R}^{(l-1)}, \mathbf{X}^{(l-1)}) = \tilde{\mathbf{A}}_{latent}\mathbf{X}^{(l-1)}\mathbf{W}_v^{(l)} \tag{4}$$

which is quite similar to Eq. 1. **Instead of using existing adjacency matrix, GT is able to learn a better graph representation through a learnable relation matrix.** Hence, each head can learn a distinct latent relation and a multi-head layer is able to learn multiple latent relations. With these better learned representations on hand, GCNs can solve the problems mentioned above better.

4   Mixed Graph Network

This section we propose a novel mixed network structure combined with GT units. As shown in Fig. 1, the detailed procedure can be separated into three main sections: 1. feature extraction (GT with SCN), 2. neighbor aggregation (GCNs), 3. Classification (Layered SortPools with rest parts). Then we discuss the advantages of the proposed mixed graph network.

Feature Extraction. We use the proposed GT units with SCN to learn better graph representations, as GT can extract features without using the adjacency matrix. Specifically, for each vertex of a graph, we generate its m-ary K-depth subgraph for SCN [18]. Instead of using quantum walk, we sorted each node by its degree for simplicity. Then we feed both the root of subgraphs and graph vertices into GT for feature extraction. Note that the channels of GT's output are equal to the input. For K-depth subgraphs, we perform K-step GT extraction and have $K + 1$ outputs (including the raw input).

Neighbor Aggregation. For each output of the first section, we separately feed it to n-layer GCNs. Then we concatenate all outputs of GCN at layer i and denote the concatenation as $\mathbf{X}_i$ (layer 0 contains all raw inputs). The GCN layers help learn structural information that GT cannot well attain, because adjacency matrix are concerned. After this section, n big feature maps are generated.

Classification. For graph classification task, we need to convert the irregular feature maps into grid structures. Inspired by SortPool [18] and the hierarchical approach of Bai et al. [2], we propose a multiple SortPooling layer called Layered SortPool. In general, this is a more generic version of DGCNN's. Detailed procedures can be seen in Fig. 4. Moreover, we expand each output of Layered SortPool and perform 1d-convolution with both step and kernel size equal to the channels of feature maps. This extracts more features for each vertex and outputs a reduction of a big feature map. Then for all conv1d outputs, we concatenate them altogether and feed the concatenation forward to get the final probabilities for each class.

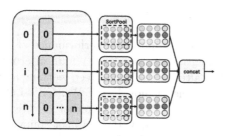

Fig. 4. Detailed procedures for Layered SortPool. Specifically, DGCNN's SortPool uses concatenation of all GCNs' outputs. However, this may lead to some features' loss. Hence, we generalize the SortPool by making it layered, to learn features at different levels.

Discussions of the Proposed Mixed Graph Network. First, because the proposed GT units can learn latent relations between substructures, the proposed GTMN can get better graph representations before feeding into GCN layers. This helps avert the problem of over-smoothing as only a few convolution operations need perform for learning neighboring information with the adjacency matrix. **Second**, we combine SCN with DGCNN through the proposed GT units. More generally, each two graph approaches can be mingled by the proposed GT. This helps the network to take advantages of the two underlying methods and relieve their individual drawbacks. **In conclusion**, GTMN can learn both local and global information simultaneously. With any two methods combined, GTMN is able to well-extract latent relations between substructures and learns better graph representations.

5   Experiments

In this section, we evaluate the performance of the proposed GTMN model, and compare it with the state-of-the-art methods, i.e., traditional graph kernels approaches and deep learning methods for graph classification on five graph benchmarks [10]. Specifically, these benchmarks are abstracted from bioinformatics and social networks. Details of these datasets are summarized in Table 1.

Table 1. Details of the graph benchmarks

Datasets	Graphs	Classes	Avg. nodes	Avg. edges	Labels	Description
MUTAG	188	2	17.93	19.79	7	Bioinformatics
PTC	344	2	14.29	14.69	19	Bioinformatics
PROTEINS	1113	2	39.06	72.82	3	Bioinformatics
IMDB-B	1000	2	19.77	96.53	–	Social

Experimental Setup. We compare the performance of the proposed GTMN model on graph classification tasks with a) four alternative state-of-the-art graph kernels and b) five alternative SOTA deep learning approaches for graphs. Concretely, the graph kernels include 1) the Weisfeiler-Lehman subtree kernel (WLSK) [13], 2) the shortest path graph kernel (SPGK) [4], 3) the random walk graph kernel (RWGK) [7], and 4) the graphlet count kernel (GK) [14]. The deep learning methods include 1) the deep graph convolutional neural network (DGCNN) [18], 2) the quantum-based subgraph convolutional neural networks (Qs-CNN) [19], 3) the backtrackless aligned-spatial graph convolutional networks (BASGCN), 4) the deep graphlet kernel (DGK) [16], and 5) the diffusion convolutional neural network (DCNN) [1].

Table 2. Classification accuracy (In% ± standard error) for comparisons

Datasets	MUTAG	PROTEINS	PTC	IMDB-B
WLSK	82.88 ± 0.57	73.52 ± 0.43	58.26 ± 0.47	71.88 ± 0.77
SPGK	83.38 ± 0.81	75.10 ± 0.50	55.52 ± 0.46	71.26 ± 1.04
RWGK	80.77 ± 0.72	74.20 ± 0.40	55.91 ± 0.37	67.94 ± 0.77
GK	81.66 ± 2.11	71.67 ± 0.55	52.26 ± 1.41	65.87 ± 0.98
DGCNN	85.83 ± 1.66	75.54 ± 0.94	58.59 ± 2.47	70.03 ± 0.86
Qs-CNN	93.13 ± 4.67	78.80 ± 4.63	65.99 ± 4.43	–
BASGCN	90.05 ± 0.82	76.05 ± 0.57	61.51 ± 0.77	74.00 ± 0.87
DGK	82.66 ± 1.45	71.68 ± 0.50	57.32 ± 1.13	66.96 ± 0.56
DCNN	66.98	61.29 ± 1.60	58.09 ± 0.53	49.06 ± 1.37
GTMN	91 ± 3.14	**81.08 ± 0.363**	**68.66 ± 2.49**	69.9 ± 1.15

For the evaluation, we adjust a number of hyperparameters to get the best performance of each dataset, as shown in Table 3. Note that every GT's head number is set equal to its input channel. We use tanh function in SCN and GCNs, leakyReLU in each linear layer and conv1d layer. Also, each linear layer is followed by a dropout rate. To optimize the GTMN model, we use the Adam optimizer with the default parameters. For our model, we perform 10-fold cross-validation to compute the classification accuracy, with nine training folds and one validating fold. For each dataset, we repeat the experiment 10 times and report the average classification accuracy and standard errors in Table 2.

For the alternative graph kernels and deep learning methods except Qs-CNN, we report the best results collected and experimented by Bai et al. [2]. We report the best results for Qs-CNN from the original paper [19]. Classification accuracy and standard error of each competing approach are also shown in Table 2.

Table 3. Hyperparameters settings for each dataset.

Parameters	K	m	gcn_num	conv	conv1d	fc	fc_num	batch	lr	L2norm	dropout
MUTAG	5	6	6	256	32	32	5	16	0.0001	0	0
PROTEINS	4	9	2	32	32	256	3	64	0.0003	0	0.05
PTC	4	5	3	256	64	64	3	128	0.0001	0	0
IMDB-B	4	9	2	64	256	256	2	256	0.001	0	0.5

Experimental Results and Discussions. Table 2 indicates that the proposed GTMN significantly outperforms either the competing graph kernel methods or the deep learning methods for graph classification.

Overall, the reasons for the effectiveness of our method are threefold. First, the graph kernels with C-SVM classifier are shallow learning methods, while the proposed GTMN can provide an end-to-end deep learning architecture. Thus GTMN can learn better graph characteristics. Second, as elucidated earlier, most deep learning approaches of graph classification can not well-avert problems of over-smoothing and learn distant relations. Instead, the proposed GT units can relieve these problems and learn better graph representations. Third, consider the proposed Qs-CNN and DGCNN, GTMN simplify the quantum walk procedure, generalize the SortPool layer and obtain better performance. This empirically demonstrates the effectiveness of the proposed GTMN.

6   Conclusions

In this paper, we introduce a novel spatially-based GCN model, i.e., the Graph Transformer with Mixed Network (GTMN), to learn the latent relations between substructures without using the adjacency matrix and alleviate the problem of over-smoothing. Unlike most existing spatially-based GCN models, we propose an attention-based Graph Transformer with a Mixed Network to learn these potential features and learn better graph representations. Experimental results on graph benchmarks indicate the effectiveness of the proposed GTMN.

For future works, the proposed GTMN only combines SCN with DGCNN, using the most original GCN operations and pooling methods. It would be interesting to combine other existing methods through the proposed GT units. Also, we do not consider the labels of edge and other graph data tasks. This could be another future work.

Acknowledgments. This work is supported by the National Natural Science Foundation of China (Grant no. 61976235, 61602535, 61773415), Program for Innovation Research in Central University of Finance and Economics, and the Youth Talent Development Support Program by Central University of Finance and Economics, No. QYP1908.

References

1. Atwood, J., Towsley, D.: Diffusion-convolutional neural networks. In: Advances in Neural Information Processing Systems, pp. 1993–2001 (2016)
2. Bai, L., Cui, L., Jiao, Y., Rossi, L., Hancock, E.: Learning backtrackless aligned-spatial graph convolutional networks for graph classification. IEEE Trans. Pattern Anal. Mach. Intell. (2020)
3. Bai, L., Jiao, Y., Cui, L., Hancock, E.R.: Learning aligned-spatial graph convolutional networks for graph classification. In: Brefeld, U., Fromont, E., Hotho, A., Knobbe, A., Maathuis, M., Robardet, C. (eds.) ECML PKDD 2019. LNCS (LNAI), vol. 11906, pp. 464–482. Springer, Cham (2020). https://doi.org/10.1007/978-3-030-46150-8_28
4. Borgwardt, K.M., Kriegel, H.P.: Shortest-path kernels on graphs. In: Fifth IEEE International Conference on Data Mining (ICDM 2005), 8-pp. IEEE (2005)
5. Bruna, J., Zaremba, W., Szlam, A., LeCun, Y.: Spectral networks and locally connected networks on graphs. arXiv preprint arXiv:1312.6203 (2013)
6. Henaff, M., Bruna, J., LeCun, Y.: Deep convolutional networks on graph-structured data. arXiv preprint arXiv:1506.05163 (2015)
7. Kashima, H., Tsuda, K., Inokuchi, A.: Marginalized kernels between labeled graphs. In: Proceedings of the 20th International Conference on Machine Learning (ICML 2003), pp. 321–328 (2003)
8. Li, G., Muller, M., Thabet, A., Ghanem, B.: DeepGCNs: can GCNs go as deep as CNNs? In: Proceedings of the IEEE International Conference on Computer Vision, pp. 9267–9276 (2019)
9. Liu, M., Gao, H., Ji, S.: Towards deeper graph neural networks. In: Proceedings of the 26th ACM SIGKDD International Conference on Knowledge Discovery & Data Mining, pp. 338–348 (2020)
10. Morris, C., Kriege, N.M., Bause, F., Kersting, K., Mutzel, P., Neumann, M.: TUDataset: a collection of benchmark datasets for learning with graphs. In: ICML 2020 Workshop on Graph Representation Learning and Beyond (GRL+ 2020) (2020). www.graphlearning.io
11. Nguyen, D.Q., Nguyen, T.D., Phung, D.: Universal self-attention network for graph classification. arXiv preprint arXiv:1909.11855 (2019)
12. Rippel, O., Snoek, J., Adams, R.P.: Spectral representations for convolutional neural networks. In: Advances in Neural Information Processing Systems, pp. 2449–2457 (2015)
13. Shervashidze, N., Schweitzer, P., Van Leeuwen, E.J., Mehlhorn, K., Borgwardt, K.M.: Weisfeiler-Lehman graph kernels. J. Mach. Learn. Res. **12**(9), 2539–2561 (2011)
14. Shervashidze, N., Vishwanathan, S., Petri, T., Mehlhorn, K., Borgwardt, K.: Efficient graphlet kernels for large graph comparison. In: Artificial Intelligence and Statistics, pp. 488–495 (2009)
15. Vaswani, A., et al.: Attention is all you need. In: Advances in Neural Information Processing Systems, pp. 5998–6008 (2017)

16. Yanardag, P., Vishwanathan, S.: Deep graph kernels. In: Proceedings of the 21st ACM SIGKDD International Conference on Knowledge Discovery and Data Mining, pp. 1365–1374 (2015)
17. Yun, S., Jeong, M., Kim, R., Kang, J., Kim, H.J.: Graph transformer networks. In: Advances in Neural Information Processing Systems, pp. 11983–11993 (2019)
18. Zhang, M., Cui, Z., Neumann, M., Chen, Y.: An end-to-end deep learning architecture for graph classification. In: Thirty-Second AAAI Conference on Artificial Intelligence (2018)
19. Zhang, Z., Chen, D., Wang, J., Bai, L., Hancock, E.R.: Quantum-based subgraph convolutional neural networks. Pattern Recogn. **88**, 38–49 (2019)
20. Zhang, Z., Chen, D., Wang, Z., Li, H., Bai, L., Hancock, E.R.: Depth-based subgraph convolutional auto-encoder for network representation learning. Pattern Recognit. **90**, 363–376 (2019)

Graph-Theoretic Methods

Weighted Network Analysis Using the Debye Model

Haoran Zhu[1], Hui Wu[1], Jianjia Wang[1(✉)], and Edwin R. Hancock[2]

[1] School of Computer Engineering and Science, Shanghai University,
Shanghai, China
jianjiawang@shu.edu.cn
[2] Department of Computer Science, University of York, York, UK

Abstract. Statistical mechanics provides effective means for complex network analysis, and in particular the classical Boltzmann partition function has been extensively used to explore network structure. One of the shortcomings of this model is that it is couched in terms of unweighted edges. To overcome this problem and to extend the utility of this type of analysis, in this paper, we explore how the Debye solid model can be used to describe the probability density function for particles in such a system. According to our analogy the distribution of node degree and edge-weight in the network can be derived from the distribution of molecular energy in the Debye model. This allows us to derive a probability density function for nodes, and thus is identical to the degree distribution for the case of uniformly weighted edges. We also consider the case where the edge weights follow a distribution (non-uniformly weighted edges). The corresponding network energy is the cumulative distribution function for the node degree. This distribution reveals a phase transition for the temperature dependence. The Debye model thus provides a new way to describe the node degree distribution in both unweighted and weighted networks.

Keywords: Debye's solid model · Degree distribution · Weighted networks

1 Introduction

The study of complex networks has attracted sustained interest since it allows the otherwise intractable interactions between the different units of complex systems to be represented and analysed [8]. This usually involves the study of the unweighted or weighted "edges" between vertices using methods from graph theory [4]. However, the node degree distribution also plays a critical role, since it describes the topological structure of networks and may determine the evolution characteristics of a network [2]. It is widely confirmed that many different types of real-world network exhibit a power-law degree distribution and this can be induced by a linear preferential evolution mechanism [1]. This property illuminates the statistical nature of structural connections in a network.

© Springer Nature Switzerland AG 2021
A. Torsello et al. (Eds.): S+SSPR 2020, LNCS 12644, pp. 153–163, 2021.
https://doi.org/10.1007/978-3-030-73973-7_15

However, the literature mainly focusses on the analysis of the degree distribution for unweighted networks, and rarely considers the distribution of edge weights. This limits the exploration of the nature of network structure based on information concerning the distribution of edge degree combinations or edge-weights. Recently, sophisticated tools from statistical physics have provided powerful ways to extend this kind of analysis [5,6]. These computationally efficient methods rely on thermodynamic analogies to describe the different structural or topological properties of networks [3]. For example, the Boltzmann distribution provides expressions for the macroscopic thermal characteristics such as temperature, energy and entropy from a microcosmic point of view [7]. This provides a novel framework to analyse and understand the statistical structural properties in weighted networks.

This work aims to establish effective statistical mechanical methods for measuring the probability density function for nodes (and node degree) in weighted networks. We commence from a thermal analogy using Boltzmann statistics, which provides a physical meaning of the temperature and energy states in a network. This allows us to introduce and leverage the Debye solid model to calculate the degree distribution.

The Debye solid model originates a statistical mechanical tool for the analysis of the distribution of phonon energy lattice structures from solid state physics. Specifically, it considers the vibrations (or phonons) of the atomic lattice. This treats the solid as an ensemble of harmonic oscillators. The model exhibits similar connectivity patterns to those found in complex networks. The connected nodes are analogous to the atoms, and the edge weights can be regarded as the phonon energies of the harmonic oscillators. Since, in this more general thermal analogy, the degree in the network has two degrees of freedom, i.e., in-degree and out-degree, the model builds on analogies with two a dimensional crystal.

Using this model, we find that, for a given distribution of edge weights, the node probability in a weighted network not only depends on the node degree but also on the global temperature parameter. Furthermore, the corresponding network energy is just the cumulative distribution function for the node probability. Moreover, this reveals a phase transition for the temperature dependence.

2   Graph Representation

2.1   Preliminaries

Let $G(V, E)$ be an undirected graph with node set V and edge set $E \subseteq V \times V$. The edge-set can be represented by an adjacency A, with elements $A(u, v) = 1$ if $(u, v) \in E$ and $A(u, v) = 0$ otherwise. The diagonal degree matrix D has diagonal elements $D(u, u) = d(u)$, where $d_u = \sum_{v \in V} A_{uv}$ is the degree of node u, and off diagonal elements $D(u, v) = 0$ if $u \neq v$. Then, the Laplacian matrix is given by $L = D - A$.

For a weighted network G_w, the pair of nodes (u, v) has an associated real non-negative weight $w(u, v)$ for each edge, i.e., $u \in V, v \in V$, and $u \neq v$. The adjacency matrix A_w for a weighted network is given by

$$A_w = \begin{cases} w(u,v) & \text{if } (u,v) \in E \\ 0 & \text{otherwise.} \end{cases} \tag{1}$$

where, for the undirected network, the weighted adjacency is symmetric, i.e., $w(u,v) = w(v,u)$ for all pairs of nodes such that $(u,v) \in E, u \neq v$.

2.2 Thermodynamic Representation

Here to model networks using a thermal analogy based on Boltzmann statistics, each network is regarded as an isolated system with a fixed number of both nodes $|V|$ and edges $|E|$. The nodes in the network are mapped onto the particles in the thermal system. Each edge has a unit weight. The corresponding node degrees are analogous to the discrete energy states and the energy associated with each node is proportional to the node degree, that is

$$\omega_u = \varepsilon k \tag{2}$$

where ω_u is the energy per node which is identical to the node weight; and $\varepsilon = 1$ for an unweighted network, k is the degree per node; and $k \in \mathbf{Z}$ which is a positive integer or zero and equal to the number of edges connecting to the node u. Thus, the occupation number of the energy states depends on the degree of the nodes connected by edges.

According to the Boltzmann distribution, the probability for an individual node to be at a particular energy state is given by the exponential function

$$P_u = \frac{1}{Z} e^{-\beta \omega_u} \tag{3}$$

where Z is the partition function subject to the constraint of energy conservation and given by

$$Z = \sum_{u=0}^{|V|} e^{-\beta \omega_u} \tag{4}$$

The average energy then can be derived from the Boltzmann partition function

$$\bar{U} = -\frac{1}{Z}\frac{\partial Z}{\partial \beta} = -\frac{\partial \log Z}{\partial \beta} \tag{5}$$

This allows us to treat a network as a statistical ensemble with associated thermal properties such as a partition function and a total energy.

3 Statistical Ensembles

For a network subject to Boltzmann statistics and in thermal equilibrium with a fixed number of nodes and edges, the entropy can be computed using Boltzmann's law, i.e. $S = \kappa_B \log W(U)$, where $W(U)$ is the multiplicity of states and the total energy in the network is

$$U = \varepsilon |E| \tag{6}$$

which is an integer number equal to the total number of edges when the weight is unity.

The entropy relates to the number of ways for choosing $|E|$ edges among the available $U + |V| - 1$ possibilities. This is given by the combinatorial formula in terms of the factorials

$$W(U) = \frac{(U + |V| - 1)!}{U!(|V| - 1)!} \tag{7}$$

When number of nodes and edges are large, then the expression $\log W(U)$ can be simplified by using Stirling's approximation $\log n! \approx n \log n$ and as a result

$$
\begin{aligned}
S &= \kappa_B \ln W \\
&= \log[(U + |V| - 1)!] - \log(U!) - \log[(|V| - 1)!] \\
&= (U + |V| - 1) \log(U + |V| - 1) - U \log U - (|V| - 1) \log(|V| - 1)
\end{aligned}
\tag{8}
$$

where κ_B is the Boltzmann constant.

For a thermodynamic system of constant volume, the temperature (or equivalently the parameter β, i.e., the inverse temperature) is the rate of change of energy with respect to entropy of the network. That is given by

$$\beta = \left(\frac{\partial S}{\partial U}\right)_{|V|} = \frac{1}{w} \log \frac{U + |V| - 1}{U} \tag{9}$$

Given the temperature the partition function for the equilibrium state of the thermal network system can be represented by the series expansion

$$Z = \sum_{u=0}^{|V|} e^{-\beta \omega_u} = \frac{1 - e^{-|V|\beta\omega}}{1 - e^{-\beta\omega}} \approx \frac{1}{1 - e^{-\beta\omega}} \tag{10}$$

From Eq. (3), the probability for a given node at a particular energy state depends on the node degree

$$P(d_u = k) = \frac{1}{Z} e^{-\beta \omega_u} = \left(1 - e^{-\beta\omega}\right) e^{-\beta \varepsilon k} \tag{11}$$

This leads to definitions of energy and entropy that associated with the network structure.

4   The Debye Model

The above analysis makes the rather limiting assumption that the weight for each edge is uniform and the energy states for each node are discrete. It effectively assumes that the density of states is simply a delta function. It is better to assume a distribution of edge weights to make the nodal energy continuous by replacing a density distribution.

4.1   Node Probability

Hence, we would like to incorporate a function $g(\omega)$ which describes the density of edge weights to allow us to make a more detailed analysis. The number of edge states with weights between ω and $\omega + d\omega$ is given by $g(\omega)d\omega$ and we require that the total edge weights sums to the number of edges, i.e. is given by

$$\int g(\omega)d\omega = |E| \tag{12}$$

Equation (2) is equivalent to assuming that the node energy corresponds to the degree. Here, on the other hand, we allow a more complex vectorial representation which accommodates the more general case of directed networks, which admits both node in-degree and out-degree.

For the space of node in-degree and out-degree, we require two integers to specify each node, i.e. the probability density for each node is bivariate depending on two variables k_{in} and k_{out} and is normalised by the sum order these two integers or equivalently by an integral over the volume element dk_{in}, dk_{out} in the node. The discrete summation can thus be rewritten as the integral, that is

$$\sum_k (\cdots) = \frac{1}{4} \int_0^\infty 2\pi k dk (\cdots) \tag{13}$$

Then, the density of states per node as a function of k is given by

$$g(k)dk = \frac{S}{(2\pi)^2} \cdot 2\pi k dk \cdot 2 = \frac{Sk}{\pi} dk \tag{14}$$

where the nodes in a network are assumed to be the square of area $S = |V|^2$ and the factor 2 corresponds to the two degrees of freedom for edges.

Thus, the corresponding density of weights for each node is given by

$$g(\omega)d\omega = \frac{S}{\pi\varepsilon^2}\omega d\omega \tag{15}$$

To derive the thermal quantities in the Debye model as a function of temperature, we begin by writing down the logarithm of partition function as follows,

$$\log Z = \int_0^{\omega_T} g(\omega)d\omega \log\left[\frac{1}{1-e^{-\beta w}}\right] = -\int_0^{\omega_T} g(\omega)d\omega \log\left[1-e^{-\beta w}\right] \tag{16}$$

Then, from Eq. (5), we can calculate the energy of the network using

$$U = -\frac{\partial \log Z}{\partial \beta} = \int_0^{\omega_T} g(\omega)d\omega \cdot \frac{\omega}{e^{\beta\omega}-1} = \frac{S}{\pi\varepsilon^2} \int_0^{\omega_T} \frac{\omega^2}{e^{\beta\omega}-1} d\omega \tag{17}$$

Substituting Eq. (2) into Eq. (17), the corresponding energy is related to the degree and is given by

$$U = \int_0^{\omega_T} \frac{S\varepsilon}{\pi} \cdot \frac{k^2}{e^{\beta\varepsilon k} - 1} dk = \int_0^{\omega_T} P(\beta, k) dk \qquad (18)$$

As a result the probability of each node given the degree k and temperature β is

$$P(\beta, k) = \frac{S\varepsilon}{\pi} \cdot \frac{k^2}{e^{\beta\varepsilon k} - 1} = \frac{|V|^2}{2\pi|E|} \cdot \frac{k^2}{e^{\beta\varepsilon k} - 1} \qquad (19)$$

where $S = |V|^2, U = 2|E|\varepsilon$. This describes the degree distribution in the weighted network. It not only relates to the node degree, but also depends on the global temperature parameter as well.

4.2 Upper Weight Boundary

Because there is an limit on the total number of edges given the number of nodes in the network, the weight distribution has an upper bound ω_T. This is defined by

$$\int_0^{\omega_T} g(\omega) d\omega = 2|E| \qquad (20)$$

which, using Eq. (15), implies that

$$\omega_T = \left(4\pi \frac{|E|}{|V|^2}\right)^{1/2} \varepsilon \qquad (21)$$

This allows us to rewrite Eq. (15) as

$$g(\omega) d\omega = \frac{4|E|\omega}{\omega_T^2} d\omega \qquad (22)$$

Thus, we now have all the ingredients necessary to apply the Debye model to derive the macroscopic thermal characterisations for the network.

4.3 High- and Low-Temperature Limits

Now we substitute Eq. (22) into Eq. (16) to write the logarithm of the partition function as

$$\log Z = -\frac{4|E|}{\omega_T^2} \int_0^{\omega_T} \omega \log \left[1 - e^{-\beta\omega}\right] d\omega \qquad (23)$$

According to Eq. (7), the average energy is

$$\bar{U} = \frac{4|E|}{\omega_T^2} \int_0^{\omega_T} \frac{\omega^2}{e^{\beta\omega} - 1} d\omega = \frac{4|E|}{\omega_T^2 \beta^3} \int_0^{\frac{x_T}{\beta}} \frac{x^2}{e^x - 1} dx \qquad (24)$$

where $x = \beta\omega = \beta\varepsilon k$. This equation does not lead to a simple temperature dependence of average energy. This is because a) exponential term is both degree and temperature dependent, and b) the integral is degree dependent. However, we can analyse and simplify the low and high temperature limits.

High-Temperature Limits. At high temperature, $\beta \to 0$ and hence $e^x \to 1 + x$. Hence, the average energy $\bar{U}$ behaves as

$$\bar{U} \to \frac{|V|^2}{\pi \varepsilon^2 \beta^3} \int_0^{\varepsilon k} x \, dx = \frac{|V|^2}{2\pi} \cdot \frac{k^2}{\beta} \tag{25}$$

The corresponding node probability in Eq. (19) is

$$P(\beta, k) = \frac{|V|^2}{2\pi |E|} \cdot \frac{k}{\beta} \sim k\beta^{-1} \tag{26}$$

Low-Temperature Limits. At low temperature, $\beta \to \infty$ and hence $e^x \gg 1$. The average energy is given by

$$\bar{U} \to \frac{|V|^2}{\pi \varepsilon^2 \beta^3} \int_0^\infty \frac{x^2}{e^x} dx = \frac{|V|^2}{\pi \varepsilon^2 \beta^3} I_B(2) \tag{27}$$

where $I_B(2) = \zeta(3)\Gamma(3)$ is the Bose integral, where $\zeta(3)$ is a Riemann zeta function and $\Gamma(3)$ a gamma function.

Then, the corresponding node probability in Eq. (19) is

$$P(\beta, k) = \frac{C}{\varepsilon^2 \beta^3} \cdot \frac{1}{k_T^2} \sim k_T^{-2} \beta^{-3} \tag{28}$$

where $C = |V|^2 I_B(2)/\pi$ is a constant, and $k_T = \omega_T/\varepsilon$.

5   Experiments and Evaluations

5.1   Data Set

Data Set 1: Here we use real world complex networks from the KONECT database. This database contains a variety of networks including

- The collaboration graph for authors of scientific papers from the arXiv's High Energy Physics-Theory (hep-th) section. Here an edge between two authors represents a common publication [9]. There are 22,908 vertices and 2,763,133 edges in the network.
- Facebook friendships network is the undirected network containing friendship of users. A node represents a user and an edge represents a friendship between two users [10]. There are 63,731 vertices and 817,035 edges.
- The Orkut network is the social network of Orkut users and their connections. There are 3,072,441 vertices and 117,185,083 edges [11].
- The PPIs dataset extracted from STRING consisting of networks which describe the interaction relationships between histidine kinase and other proteins [13]. There are 216 vertices and 5,389 edges in the network.

Data Set 2: This data comes from the New York Stock Exchange. It consists of the daily closing prices of 3,799 stocks traded continuously on the New York Stock Exchange over 2619 trading days. The stock prices were obtained from the Yahoo! financial database [12]. A total of 415 stock are selected with the historical stock prices from the beginning of January 2010 to the end of June 2020. In the network representation, the nodes correspond to stock and the edges indicate that there is a statistical similarity between the time series associated with the stock closing prices.

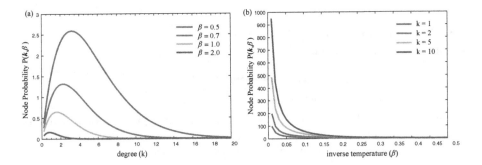

Fig. 1. The node probability varying with the degree k and inverse temperature β in Eq. (19). (a) Node probability with degree; (b) node probability with inverse temperature

5.2   Experimental Results

We first conduct a numerical analysis on the node probability in Eq. (19). Figure 1 plots how the node probability varies with the degree k and inverse

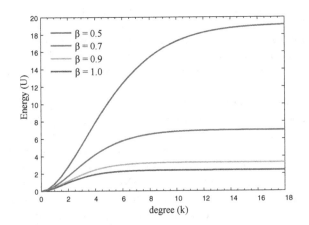

Fig. 2. Network energy varying with degree according to Eq. (18)

temperature β, respectively. In Fig. 1(a), there is a phase transition for the probability varying with the node degree. When the value of inverse temperature β increases, the peak corresponding to the phase transition shifts towards zero. In Fig. 1(b), the node probability exponentially decays with the inverse temperature. The larger value of node degree, the faster the decay.

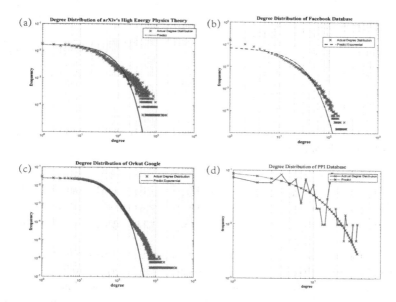

Fig. 3. Degree distributions for real-world networks. The red curves are the actual degree distributions and the blue curves are the result of simulation using Eq. (11). (Color figure online)

Next we analyse the behaviour of the energy given in Eq. (18) with respect to both degree and temperature. The expression in Eq. (18) is quite complicated and it is not obvious by inspection how energy depends on temperature. This is because the exponential term is both degree and temperature dependent and the integral is degree dependent. Figure 2 shows the full degree dependence for the energy. The energy increases with the degree until reaching a plateaux value when the node degree is large. The energy also decreases rapidly when the inverse temperature β increases.

We now turn our attention to the real-world datasets. We examine the predictions of the node probability distribution in Eq. (11) for the complex network dataset. Figure 3 shows four degree distributions for different complex networks. The red curves are the actual degree distributions and the blue curves are the predictions of our model. The four real networks come from the KONECT dataset, and are the arXiv hep-th network, the Facebook network, the Google Orkut user network, and a protein-protein interaction network. It is clear that, instead

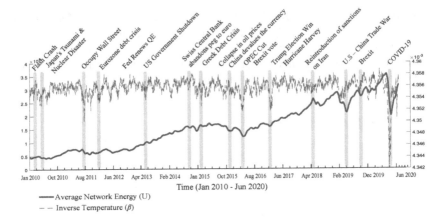

Fig. 4. The average energy and inverse temperature in S&P500 Index Stock Data (2010–2020) for original financial networks.

of following the power law degree distribution, many real world complex networks follow the exponential distribution that we derived in Eq. (11). Actually, our model fits well at the low degree range, and at high-degree the power-law applies.

Finally, we use the time evolving financial networks to evaluate the energy and inverse temperature. Figure 4 plots the derived energy and inverse temperature for the stock exchange networks over the past decade. The stock market networks undergo rapid structural fluctuations during critical financial events. These events are listed in the caption of the figure. Compared with the energy, the temperature is more sensitive to the fluctuations of the financial markets. Sharp peaks in both energy and temperature indicate significant changes in network structure during the different financial events.

In summary, our derived expression for the degree distribution can therefore be used to fit the degree distributions of real complex networks. The corresponding energy and temperature associated with the network structure can also be used to identify abrupt changes of the pattern of edge connectivity.

6   Conclusion

In this paper, we make use of the Debye model to describe the node probability distribution in weighted networks. We commence from a thermal analogy described by the classical Boltzmann distribution. The particles in this thermal system are analogous to the nodes in a network. The energy is determined by the edge weights and node degree, which provides a physical interpretation for temperature. Then, the Debye solid model leads to an exponential expression for the probability density function of node degree. This then depends on the edge weights and the global temperature parameter, both related to the configuration of nodes and edges. The node probability density functions together with the cumulative distribution function for energy reveal a phase transition for both the degree and temperature dependence. Experimental results show that the derived distribution can be used to fit the degree distribution in naturally occurring networks and identify anomalous structure in time evolving networks.

References

1. Petri, G., Scolamiero, M., Donato, I., Vaccarino, F.: Topological strata of weighted complex networks. PLoS One **8**(6) (2013)
2. Anand, K., Krioukov, D., Bianconi, G.: Entropy distribution and condensation in random networks with a given degree distribution. Phys. Rev. E **89**(6), 062807 (2014)
3. Wang, J., Lin, C., Wang, Y.: Thermodynamic entropy in quantum statistics for stock market networks. Complexity **2019**, 1–11 (2019)
4. Cimini, G., Squartini, T., Saracco, F., Garlaschelli, D., Gabrielli, A., Caldarelli, G.: The statistical physics of real-world networks. Nat. Rev. Phys. **1**(1), 58–71 (2019)
5. Wang, J., Richard, W., Edwin, H.: Spin statistics, partition functions and network entropy. J. Complex Netw. **5**(6), 858–883 (2017)
6. Ye, C., et al.: Thermodynamic characterization of networks using graph polynomials. Phys. Rev. E **92**(3), 032810 (2015)
7. Wang, J., Wilson, R.C., Hancock, E.R.: Directed and undirected network evolution from Euler-Lagrange dynamics. Pattern Recognit. Lett. **1**(134), 135–44 (2020)
8. Jean-Charles, D., Anne-Sophie, L.: Centrality measures and thermodynamic formalism for complex networks. Phys. Rev. E **83**, 046117 (2011)
9. Leskovec, J., Kleinberg, J., Faloutsos, C.: Graphs over time: densification laws, shrinking diameters and possible explanations. In: ACM SIGKDD (2005)
10. Viswanath, B., Mislove, A., Cha, M., Gummadi, K.P.: On the evolution of user interaction in Facebook. In: Proceedings of the Workshop on Online Social Networks, pp. 37–42 (2009)
11. Yang, J., Leskovec, J.: Defining and evaluating network communities based on ground-truth. Knowl. Inf. Syst. **42**(1), 181–213 (2015)
12. Silva, F.N., et al.: Modular dynamics of financial market networks. arXiv preprint arXiv:1501.05040. 21 January 2015
13. Szklarczyk, D., Gable, A.L., Lyon, D., et al.: STRING v11: protein-protein association networks with increased coverage, supporting functional discovery in genome-wide experimental datasets. Nucleic Acids Res. **47**(D1), D607–D613 (2019)

Estimating the Manifold Dimension of a Complex Network Using Weyl's Law

Luca Rossi[1]([✉]) and Andrea Torsello[2]

[1] Queen Mary University of London, London, UK
luca.rossi@qmul.ac.uk
[2] Università Ca' Foscari Venezia, Venice, Italy
andrea.torsello@unive.it

Abstract. The dimension of the space underlying real-world networks has been shown to strongly influence the networks structural properties, from the degree distribution to the way the networks respond to diffusion and percolation processes. In this paper we propose a way to estimate the dimension of the manifold underlying a network that is based on Weyl's law, a mathematical result that describes the asymptotic behaviour of the eigenvalues of the graph Laplacian. For the case of manifold graphs, the dimension we estimate is equivalent to the fractal dimension of the network, a measure of structural self-similarity. Through an extensive set of experiments on both synthetic and real-world networks we show that our approach is able to correctly estimate the manifold dimension. We compare this with alternative methods to compute the fractal dimension and we show that our approach yields a better estimate on both synthetic and real-world examples.

Keywords: Manifold dimension · Complex networks · Weyl law

1 Introduction

Graphs have long been used as natural representations for a variety of real-world systems, from biological systems [7] to transportation networks [6] and human interactions [2,10]. These graphs often display non-trivial topological features and are hence referred to as complex networks. The ultimate goal when analysing these networks is that of establishing a link between the structural properties of the networks and their function. To this end, a large number of techniques, from node centralities [9,13] to entropy measures [11], have been introduced to capture the local and global structural properties of a network [14,15,17,18].

Many real-world networks are embedded in either a two-dimensional or a three-dimensional space, such as the network of collaborations between software developers across the world [10] or railway networks [6]. It has been shown that the structural properties of these networks are strongly influenced by the geometry of the underlying space. In the case where the underlying space is hyperbolic, Krioukov et al. [8] have shown that heterogeneous degree distributions and strong

© Springer Nature Switzerland AG 2021
A. Torsello et al. (Eds.): S+SSPR 2020, LNCS 12644, pp. 164–173, 2021.
https://doi.org/10.1007/978-3-030-73973-7_16

clustering naturally emerge as consequences of the negative curvature and metric property of the space. When the underlying space is Euclidean, the network is often referred to as a spatially embedded network and it's been observed that the probability of establishing a connection between two nodes tends to decay exponentially as the distance between them increases [5].

Daqing et al. [5] proposed a way to measure the dimension of spatially embedded networks. This is achieved under the assumption that the Euclidean distance between the nodes is known and it requires measuring the average distance between the nodes of subgraphs of increasing radius centered around randomly chosen seed nodes. What Daqing et al. compute is effectively the fractal dimension of the network [18], a measure of the self-similarity of the network structure. It's easy to show that the fractal dimension of a network is equivalent to the dimension of the embedding space on regular lattices or in general manifold graphs, i.e., graphs that can be seen as discrete representations of the continuous underlying manifold. Interestingly, Daqing et al. [5] showed that the dimension of the network is intimately related to the properties of diffusion and percolation processes on the network.

Song et al. [17] explored instead two alternative methods to estimate the fractal dimension of a network. The first method is very similar to [5] and involves repeatedly sampling a set of random nodes in the network which are used to grow clusters of nodes whose size is used to ultimately estimate the fractal dimension of the network. In practice, this approach is shown to perform poorly in networks with inhomogeneous degree distributions. A second method estimates the fractal dimension of a network based on the box covering algorithm. This is however an NP-hard problem so heuristics are needed to find an approximate solution [17].

In this paper, we propose an alternative way to estimate the manifold dimension of a weighted network, where the weights are not restricted to represent Euclidean distances between the nodes as in [5]. Specifically, we propose to estimate the manifold dimension of a network using Weyl's law [21]. In spectral theory, Weyl's law describes the asymptotic behaviour of the eigenvalues of the Laplacian associated to a bounded domain $\Omega \in \mathbb{R}^d$, establishing a power-law relation between the eigenvalues and their indices. Crucially, the exponent of this power-law relation depends on the dimension of the underlying manifold. As a result, given a network, we are able to estimate the dimension of the underlying manifold from the spectrum of its Laplacian.

The remainder of the paper is organised as follows: Sect. 2 provides a brief overview of the two most commonly used approaches to compute the fractal dimension of a graph. Section 3 reviews Weyl's law and introduces our methodology for estimating the manifold dimension of a network, which is then evaluated on both synthetic and real-world networks in Sect. 4. Finally, Sect. 5 concludes the paper.

2   Background

Similarly to the more general concept of fractal dimension of a set, the fractal dimension of a network tells us something about how the structure of the network changes as we view it under lenses of varying size. In other words, the fractal dimension of a network is a measure of how invariant or self-similar a network is under a length-scale transformation [18].

Existing approaches to estimate the fractal dimension of a network are based either on the box counting method or the cluster growing method. For a given network G and box size l_B, the box counting method (also known as box covering method) defines a box as a set of nodes such that the distance between any two nodes in the set is smaller than l_B. The number of boxes of size l_B required to cover the network is $N_B(l_B)$ and the goal of the box counting method is to find the minimum value of $N_B(l_B)$ for any value of l_B. As shown in [17], this problem can be mapped to the NP-hard graph colouring problem, so it's typically solved using a number of different heuristics. Given the optimal values of $N_B(l_B)$ for a varying number of box sizes, the fractal dimension d_B of the network is given by

$$N_B(l_B) \approx l_b^{-d_B} . \tag{1}$$

Note that, as a consequence of the heuristic nature of the algorithms used to approximate the solution of the box covering problem, the minimum number of boxes for a given size is likely to be overestimated and thus the fractal dimension is instead underestimated.

The cluster growing method instead selects a number of seed nodes at random. For each seed, a cluster is defined as the set of nodes a distance less or equal to l_C from the seed. For each cluster we compute the mass $M_C(l_C)$ as the number of nodes in the cluster. Then the fractional dimension d_C is given by

$$\overline{M_C(l_C)} \approx l_C^{d_C} , \tag{2}$$

where $\overline{M_C(l_C)}$ is the average mass of the clusters for a given value of l_C [18]. The main drawback of this approach is that it performs poorly on networks with inhomogeneous degree distributions. This is because by choosing the seeds at random there is a high probability of including hubs in the clusters, leading to a biased estimate of the fractal dimension [17].

3   Weyl's Law and the Manifold Dimension of a Network

Let $\Omega \in \mathbb{R}^d$ be a bounded domain and λ_j denote the j-th eigenvalue of the Laplacian on this domain. Weyl's law [21] states that

$$\lim_{\lambda \to \infty} \frac{N(\lambda)}{\lambda^{\frac{d}{2}}} = \frac{\omega_d \mathrm{vol}(\Omega)}{(2\pi)^d} \tag{3}$$

where $N(\lambda) = \#\{\lambda_j \le \lambda\}$ is a function that counts the number of eigenvalues less than or equal to λ and ω_d is the volume of the unit ball in $\mathbb{R}^d$. Equation 3 tells us that, for sufficiently large λ,

$$N(\lambda) \approx k\lambda^{\frac{d}{2}} \tag{4}$$

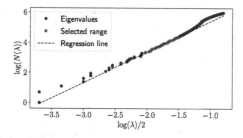

Fig. 1. Estimating the manifold dimension of a two-dimensional 20×20 grid graph. The blue dots correspond to the $(\log N(\lambda_j), \log \lambda_j/2)$ pairs computed on the eigenvalues $\lambda_1, \cdots, \lambda_{400}$ of the graph Laplacian. The manifold dimension is estimated on a selected range of eigenvalues (highlighted in red) to take into account the conditions of Weyl's law and compensate for the boundary effects. (Color figure online)

where we used k to group the constants wrt to λ. Taking the logarithm of both sides of Eq. 4 and ignoring the constant term, we get

$$\log N(\lambda) \approx d\frac{\log \lambda}{2}. \tag{5}$$

3.1 Estimating the Manifold Dimension of a Network

Let G be a manifold graph, or in other words a graph that accurately models the topology of an underlying manifold of dimension d. Then Eq. 3 holds for the eigenvalues of the Laplacian L of G and we can estimate the dimension d from Eq. 5. Specifically, we use the slope of the regression line on the points $(\log N(\lambda_j), \log \lambda_j/2)$ as an estimate of d.

Figure 1 shows the values of the function in Eq. 5 sampled on the Laplacian spectrum of a two-dimensional 20×20 grid graph. The slope of the regression line in Fig. 1 is ~ 2, confirming that in this instance our approach is able to accurately capture the graph manifold dimension.

Note that the linear regression is best computed over a selected range of the spectrum (highlighted in red in the toy example of Fig. 1) which excludes the lowest and highest regions. This is because Eq. 3 doesn't hold for low frequencies and high frequencies end up capturing the local variations of the dimension near the graph boundary.

Dealing with Edge Weights and Node Attributes. Our approach can easily take into account potential information on edge weights and node attributes by incorporating them into the Laplacian. To this end, we turn distances and dissimilarities between the nodes into similarities through a negative exponential transformation.

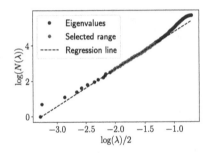 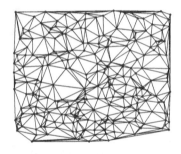

Fig. 2. Sample two-dimensional Delaunay graph (right) over 200 nodes and corresponding log-log plot (left). The slope of the regression line is $d = 2.08$.

4   Experimental Evaluation

We perform an extensive set of experiments on both synthetic and real-world networks to evaluate the proposed approach. We compare our results with those obtained using the Maximum-Excluded-Mass-Burning (MEMB) algorithm of Song et al. [17]. Specifically, we used the implementation made available by Akiba et al. [1] at https://github.com/kenkoooo/graph-sketch-fractality. Unfortunately we were unable to find any implementation of MEMB or alternative algorithms that could take edge weights into account. To our understanding, it should be relatively simple to extend MEMB and similar algorithms to deal with networks where the distance between the nodes is available. Indeed Wei et al. discuss such an extension in [20] but fail to provide an implementation of their algorithm. For this reason, in the following experiments when comparing our method with MEMB we show the results both with and without edge weights.

Finally, as discussed in Sect. 3, our method requires selecting a range of the sorted eigenvalues to sample the values of the function in Eq. 5 and estimate the manifold dimension. Unless otherwise stated, all the experiments in this paper are performed keeping only the eigenvalues in the 7% to 20% range (see Fig. 1).

4.1   Synthetic Networks

Delaunay Graphs. We sampled 200 points uniformly on a two-dimensional plane and we computed their Delaunay triangulation. We repeated this process 100 times and obtained 100 Delaunay graphs. Figure 2 shows one sample Delaunay graph and the eigenvalues plot computed according to Eq. 5. The weights on the edges of these graphs correspond to the Euclidean distance between the corresponding pair of points. Note that these are manifold graphs embedded on a two-dimensional space, so their manifold dimension is 2.

Table 1. Manifold dimension estimated by our method and MEMB [17] for two-dimensional Delaunay graphs and hypercubes of increasing dimension.

Network	Delaunay (d = 2)	Hypercube (d = 2)	Hypercube (d = 3)	Hypercube (d = 4)
Weyl	2.12 ± 0.01	2.00	3.07	4.05
MEMB	1.66 ± 0.01	1.32	1.83	2.16

Table 2. Manifold dimension estimated by our method and MEMB [17] on (u, v)-flower networks with increasing fractal dimension.

Network	(2, 2)-Flower	(2, 3)-Flower	(2, 4)-Flower	(2, 5)-Flower	(2, 6)-Flower
Weyl	2.01	1.91	1.74	1.67	1.62
MEMB	1.37	1.58	1.69	1.86	2.01

Hypercubes. We construct three hypercubes of increasing dimension: 1) one two-dimensional hypercube of side 10, for a total of 100 nodes, 2) one three-dimensional hypercube of side 8, for a total of 512 nodes, and 3) one four-dimensional hypercube of side 6, for a total of 1296 nodes. The manifold dimension of the hypercubes is 2, 3, and 4, respectively.

Table 1 shows the values of the manifold dimension estimated by our method and MEMB [17] on the synthetic datasets. For the Delaunay graphs we report the average value of the dimension $\pm$ standard error. Note that MEMB consistently underestimates the manifold dimension of the graphs. As explained in Sect. 2, this is because computing the optimal box covering is NP-hard and thus the solution found by heuristic approaches like MEMB is likely to overestimate the number of boxes, leading to an underestimation of the manifold dimension. The value estimated with our method, instead, consistently falls very close to the ground truth. This is true even if we drop the edge weights in the Delaunay graphs. In this case, the average manifold dimension is estimated to be 2.31 ± 0.01.

(u, v)-Flowers. We also compare our method and MEMB on an additional set of synthetic graphs where the fractal dimension can be computed analytically [16]. Starting from a cycle graph consisting of $u + v$ nodes, new nodes and edges are iteratively added by replacing each edge by two parallel paths, u and v edges long. When $2 \leq u < v$, it can be shown that the network has a finite fractal dimension equal to $\ln(u + v)/\ln(u)$. By fixing $u = 2$ and letting v grow, we can create networks of increasing fractal dimension. While this trend is correctly captured by MEMB, the dimension estimated with our method decreases as v increases, as Table 2 shows. This isn't surprising, as by fixing u and letting v grow we are effectively creating a network that contains increasingly long unidimensional string-like structures (see for example Fig. 2a in [16]). Indeed, the flower graphs are fractal but not manifold, so our method cannot be applied.

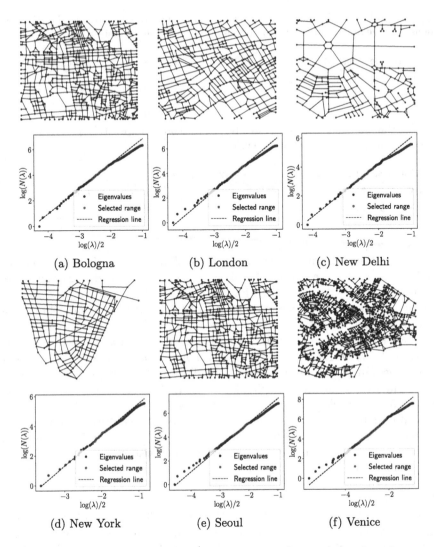

Fig. 3. Graphs (top) and corresponding log-log plots (bottom) for the urban street networks of Bologna ($d = 1.93$), London ($d = 2.11$), New Delhi ($d = 1.80$), New York ($d = 2.09$), Seoul ($d = 2.03$), and Venice ($d = 2.00$).

4.2   Urban Street Networks

We consider 6 urban street networks corresponding to 1-square mile maps of Bologna (541 nodes and 771 edges), London (488 nodes and 729 edges), New Delhi (252 nodes and 328 edges), New York (248 nodes and 418 edges), Seoul (869 nodes and 1,307 edges), and Venice (1,840 nodes and 2,397 edges) [4,9]. The edge weights of these graphs correspond to the length of the road segment connecting the two endpoints. Figure 3 shows the graph of the 6 cities and the corresponding

Table 3. Manifold dimension estimated by our method and MEMB [17] on the urban street networks of 6 cites around the world.

Network	Bologna	London	New Delhi	New York	Seoul	Venice
Weyl	1.93 (*2.05*)	2.11 (*2.11*)	1.80 (*1.78*)	2.09 (*2.12*)	2.03 (*2.07*)	2.00 (*2.00*)
MEMB	1.60	1.55	1.56	1.52	1.59	1.50

log-log plots. Table 3 shows the value of the manifold dimension estimated by our method and MEMB. In all 6 cases our method gives a result that is significantly closer to what we expect to be ground truth for these networks (2), with New Delhi having the lowest dimension. This in turn may be due to the particular structure of the part of New Delhi captured in this dataset, with large areas covered only by a small number of long roads, effectively lowering the estimated dimensionality. As observed for the Delaunay graphs, removing the edge weights has only a minimal effect on the estimate (italic in Table 3).

4.3 Other Real-World Networks

US Power Grid. This network represents the high-voltage power grid of the US (Western states). The nodes (4,941) are transformers, substations, and generators, and the edges (6,594) represent transmission lines [19]. No edge weight information or node coordinates were available for this network. The range of the eigenvalues used to fit the regression line is 1% to 20%.

Dickens. This network represents the most commonly used adjectives and nouns in the novel David Copperfield by Charles Dickens. The network has 112 nodes with 425 edges connecting pairs of adjacent words in the text [12]. The edge weights represent the Levenshtein distance between the words. The range of the eigenvalues used to fit the regression line is 2% to 70%.

C. elegans. This is an unweighted network representing the Caenorhabditis elegans neuronal network, consisting of 279 nodes representing nonpharyngeal neurons and 2,287 edges representing synaptic connections [9,19]. The range of the eigenvalues used to fit the regression line is 7% to 50%.

US Airports. This is the network of flight connections between the 500 US airports with the highest traffic [3,9]. Each node (500) corresponds to an airport and each edge (2,980) has an integer weight corresponding to the total number of seats available on all the direct routes between the two endpoints within a year. The range of the eigenvalues used to fit the regression line is 1% to 30%.

Figure 4 shows the log-log plots for these networks and Table 4 lists the estimated manifold dimensions. While *US power grid* and *Dickens* are clearly manifold, this is less obvious for *C. elegans*, where it is harder to distinguish between boundary effects and non-manifold behaviour. The log-log plot for *Airports*, on the other hand, shows at least two separate linear trends, suggesting that this

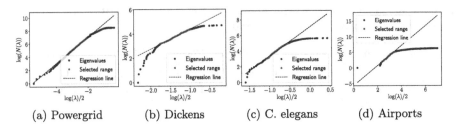

(a) Powergrid   (b) Dickens   (c) C. elegans   (d) Airports

Fig. 4. Log-log plots for the Powergrid ($d = 2.13$), Dickens ($d = 1.65$), C. elegans ($d = 3.75$), and Airports networks (3.19).

Table 4. Manifold dimension estimated by our method and MEMB [17] on the Powergrid, Dickens, C. elegans, and Airports networks.

Network	Powergrid	Dickens	C. elegans	Airports
Weyl	2.13	1.65 (*1.70*)	3.75	3.19 (*3.62*)
MEMB	2.34	2.58	2.98	2.93

is not a manifold network and thus our approach cannot be applied. In general, note that the more manifold the graph is, the more robust the estimation of the dimension wrt the chosen spectral range is.

5   Conclusion

We proposed a way to estimate the manifold dimension of a network using Weyl's law, a mathematical result that describes the asymptotic behaviour of the eigenvalues of the graph Laplacian. We showed through an extensive set of experiments on both synthetic and real-world networks that our approach is able to correctly estimate the manifold dimension, yielding better estimates than an alternative method based on box counting. Future work will investigate the possibility of automatically selecting the spectral range to fit when estimating the manifold dimension. Having access to larger urban networks, it would also be interesting to see if the local manifold dimension of different subgraphs and communities can be related to other quantities of interest.

References

1. Akiba, T., Nakamura, K., Takaguchi, T.: Fractality of massive graphs: scalable analysis with sketch-based box-covering algorithm. In: 2016 IEEE 16th International Conference on Data Mining (ICDM), pp. 769–774. IEEE (2016)
2. Chorley, M., Rossi, L., Tyson, G., Williams, M.: Pub crawling at scale: tapping untappd to explore social drinking. In: Proceedings of the International AAAI Conference on Web and Social Media, vol. 10 (2016)
3. Colizza, V., Pastor-Satorras, R., Vespignani, A.: Reaction-diffusion processes and metapopulation models in heterogeneous networks. Nat. Phys. **3**(4), 276–282 (2007)

4. Crucitti, P., Latora, V., Porta, S.: Centrality measures in spatial networks of urban streets. Phys. Rev. E **73**(3), 036125 (2006)
5. Daqing, L., Kosmidis, K., Bunde, A., Havlin, S.: Dimension of spatially embedded networks. Nat. Phys. **7**(6), 481–484 (2011)
6. Erath, A., Löchl, M., Axhausen, K.W.: Graph-theoretical analysis of the swiss road and railway networks over time. Netw. Spat. Econ. **9**(3), 379–400 (2009)
7. Gursoy, A., Keskin, O., Nussinov, R.: Topological properties of protein interaction networks from a structural perspective. Biochem. Soc. Trans. **36**(Pt 6), 1398–1403 (2008)
8. Krioukov, D., Papadopoulos, F., Kitsak, M., Vahdat, A., Boguná, M.: Hyperbolic geometry of complex networks. Phys. Rev. E **82**(3), 036106 (2010)
9. Latora, V., Nicosia, V., Russo, G.: Complex Networks: Principles, Methods and Applications. Cambridge University Press, Cambridge (2017)
10. Lima, A., Rossi, L., Musolesi, M.: Coding together at scale: GitHub as a collaborative social network. In: Proceedings of 8th AAAI International Conference on Weblogs and Social Media (2014)
11. Minello, G., Rossi, L., Torsello, A.: On the von Neumann entropy of graphs. J. Complex Netw. **7**(4), 491–514 (2019)
12. Newman, M.E.: Finding community structure in networks using the eigenvectors of matrices. Phys. Rev. E **74**(3), 036104 (2006)
13. Rossi, L., Torsello, A., Hancock, E.R.: Node centrality for continuous-time quantum walks. In: Fränti, P., Brown, G., Loog, M., Escolano, F., Pelillo, M. (eds.) S+SSPR 2014. LNCS, vol. 8621, pp. 103–112. Springer, Heidelberg (2014). https://doi.org/10.1007/978-3-662-44415-3_11
14. Rossi, L., Torsello, A., Hancock, E.R.: Measuring graph similarity through continuous-time quantum walks and the quantum Jensen-Shannon divergence. Phys. Rev. E **91**(2), 022815 (2015)
15. Rossi, L., Torsello, A., Hancock, E.R., Wilson, R.C.: Characterizing graph symmetries through quantum Jensen-Shannon divergence. Phys. Rev. E **88**(3), 032806 (2013)
16. Rozenfeld, H.D., Havlin, S., Ben-Avraham, D.: Fractal and transfractal recursive scale-free nets. New J. Phys. **9**(6), 175 (2007)
17. Song, C., Gallos, L.K., Havlin, S., Makse, H.A.: How to calculate the fractal dimension of a complex network: the box covering algorithm. J. Stat. Mech. Theory Exp. **2007**(03), P03006 (2007)
18. Song, C., Havlin, S., Makse, H.A.: Self-similarity of complex networks. Nature **433**(7024), 392–395 (2005)
19. Watts, D.J., Strogatz, S.H.: Collective dynamics of 'small-world' networks. Nature **393**(6684), 440–442 (1998)
20. Wei, D.J., Liu, Q., Zhang, H.X., Hu, Y., Deng, Y., Mahadevan, S.: Box-covering algorithm for fractal dimension of weighted networks. Sci. Rep. **3**(1), 1–8 (2013)
21. Weyl, H.: Über die asymptotische verteilung der eigenwerte. Nachrichten von der Gesellschaft der Wissenschaften zu Göttingen. Mathematisch-Physikalische Klasse **1911**, 110–117 (1911)

Efficient Partitioning of Partial Correlation Networks

Keith Dillon$^{(\boxtimes)}$ (iD)

University of New Haven, New Haven, CT 06516, USA
`kdillon@newhaven.edu`

Abstract. Partial correlation is a popular and principled metric for determining edges between nodes in a graph. However when the goal is to both estimate network connectivity from sample data and subsequently partition the result, methods such as spectral clustering can be applied much more efficiency and at larger scale. We derive a method that can similarly partition partial correlation networks directly from sample data. The method is closely related to spectral clustering, and can be implemented with comparable efficiency. Our results also provide new insight into the success of spectral clustering in many fields, as an approximation to clustering of partial correlation networks.

Keywords: Partial correlation · Spectral clustering · Graphical models · Graph partitioning

1 Introduction

Partial correlation is a natural choice for defining edges in a network. Unlike edges based on affinity or distance, the partial correlation removes the effect of indirect relationships [4] and measures the relationship based on the residual only (Fig. 1). Gaussian graphical models [9] directly relate partial correlation values to the inverse of the sample covariance matrix and to variable prediction via linear regression [6]. Such computations are not efficient for large networks, however; a separate regression problem must be solved for every variable. When the goal is clustering or partitioning of the network, the problem is usually approached, from a different perspective, with graphs formed simply via affinity or univariate correlation. This is used as a starting point for graph-theoretical approaches to partitioning such as spectral clustering [5], a continuous relaxation of the normalized cut algorithm for partitioning graphs.

In simple terms, spectral clustering is typically computed by applying k-means clustering to the rows of the last k eigenvectors of the Laplacian matrix $\mathbf{L}$. In cases where $\mathbf{L}$ must be computed first from a matrix of sample data $\mathbf{A}$, one can instead use the first k singular vectors of $\mathbf{A}$ [3]. As the first k singular vectors can be efficiently calculated for very large data sets, this provides a means to directly estimate the partitioning of a network from sample data. In such partitioning methods the relationships between nodes seems obfuscated

© Springer Nature Switzerland AG 2021
A. Torsello et al. (Eds.): S+SSPR 2020, LNCS 12644, pp. 174–183, 2021.
https://doi.org/10.1007/978-3-030-73973-7_17

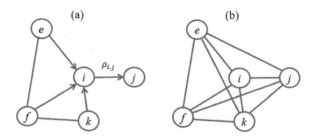

Fig. 1. Depiction of a network, where nodes e, f, and k are strongly interacting with i, but only i is interacting with j; (a) the partial correlation $\rho_{i,j}$ estimates edges by removing contributions from e, f, k, identifying the outlier status of j. (b) edges based on univariate tests such as affinity or distance result in a dense network where all nodes appear connected.

under multiple approximations; first a crude univariate metric is used (Fig. 1); second the partitioning of this graph is approximated with a continuous relaxation. Spectral clustering has empirically shown promise, even in areas where partial correlation networks are preferred due to their sound statistical basis. In [3] it was found that a spectral approach was empirically more generalizable than simpler clustering methods, in terms of its ability to predict node relatedness for held-out data. In [1], it is shown that spectral clustering can approximate the community structure of a partial correlation network.

In this paper, we derive an algorithm for clustering of nodes in partial correlation networks directly from sampled data. The result can be viewed as a variant of spectral clustering with additional correction terms. We show that it can be computed with comparable efficiency to spectral clustering and demonstrate efficient computation for a dense network with roughly 100,000 nodes. We find that the success of spectral clustering noted above can be attributed to the small size of this correction term, particularly in situations where preprocessing steps such as filter are applied to the data.

2   Method

We model the data signal at the ith node as the zero-mean Gaussian random variable a_i. The partial correlation $\rho_{i,j}$ between a_i and a_j is the Pearson correlation between the residuals after regressing out all other nodes except i and j from both. These regression coefficients are defined as the solutions $\beta_{i,j}$ to the linear regression problem

$$a_i = \sum_{k \neq i} \beta_{i,k} a_k + \epsilon_i, \tag{1}$$

where a_i is the ith variable and ϵ_i is the residual. From these $\beta_{i,j}$ we can estimate $\rho_{i,j}$ as [6],

$$\rho_{i,j} = -\beta_{i,j} \sqrt{\frac{\sigma^{i,i}}{\sigma^{j,j}}}, \tag{2}$$

using the residual variances $\sigma^{i,i} = (Var(\epsilon_i))^{-1}$. A common alternative formulation exploits the symmetry of the partial correlation (i.e., that $\rho_{i,j} = \rho_{i,j}$ by definition) and uses the geometric mean to cancel the residual variances [7]

$$\rho_{i,j} = \text{sign}(\beta_{i,j})\sqrt{\beta_{i,j}\beta_{j,i}}. \tag{3}$$

This also has the advantage of enforcing symmetry in sample estimates. If the signs of $\beta_{i,j}$ and $\beta_{j,i}$ differ, $\rho_{i,j}$ is typically set to zero.

To provide sample formulations of the above estimates, we define $\mathbf{A}$ as a matrix containing data, with samples for a_k in the kth column $\mathbf{a}_k$ (which we will assume has been standardized). The vector form of the solution to Eq. (1) can be estimated as $\boldsymbol{\beta}_i = \mathbf{A}_{-i}^{\dagger}\mathbf{a}_i$, where $\mathbf{A}_{-i}^{\dagger}$ is the pseudoinverse of $\mathbf{A}$ after setting its ith column to zeros. Equation (2) in terms of matrices becomes

$$\mathbf{P} = \mathbf{D_d}\mathbf{B}\mathbf{D_d}^{-1}, \tag{4}$$

where $\mathbf{D_d}$ is a diagonal matrix with $D_{i,i} = d_i = \|\mathbf{A}_{-i}\boldsymbol{\beta}_i - \mathbf{a}_i\|$, and $\mathbf{B}$ is a matrix with with $\boldsymbol{\beta}_i$ as columns (where the ith element of $\boldsymbol{\beta}_i$ is zero). $\mathbf{P}$ contains our sample-based estimates of the partial correlations, with $P_{i,j}$ describing the partial correlation between nodes i and j. Again, we can avoid calculating the residual variances using the method of Eq. (3) as follows,

$$\mathbf{P} = \text{sign}(\mathbf{B}) \odot (\mathbf{B} \odot \mathbf{B}^T)^{\circ\frac{1}{2}}, \tag{5}$$

using the Hadamard (element-wise) product $\odot$ and element-wise exponential $\circ\frac{1}{2}$, and where the sign function is taken element-wise.

2.1   Efficient Regularized Estimation of Partial Correlation

Thus far we have merely converted sample estimates of partial correlations into a matrix form. This direct formulation still requires a large amount of computation for each column of $\mathbf{B}$ (i.e., each node), by removing each column of $\mathbf{A}$ in turn, then taking the pseudoinverse of the result. In this subsection we will derive an efficient approach which only requires a single pseudoinverse for the entire network estimate. We will also provide a general form which incorporates regularization, often included as a heuristic data pre-processing step [2].

We incorporate regularization in the derivation by using the regularized pseudoinverse solution, $\boldsymbol{\beta}_i = (\mathbf{A}_{-i})_{\lambda}^{\dagger}\,\mathbf{a}_i$, where λ reflects the regularization parameter used in calculating the pseudoinverse (e.g., the singular value cutoff used). In order to remove the need for individual node pseudoinverses, we first define the matrix $\mathbf{B}^{(0)}$ with $\boldsymbol{\beta}_i^{(0)}$ as its ith column, $\mathbf{B}^{(0)} = \mathbf{A}_{\lambda}^{\dagger}\mathbf{A} = \mathbf{R}$, where $\mathbf{A}_{\lambda}^{\dagger}$ is the regularized pseudoinverse of $\mathbf{A}$, and where we have defined the (regularized) resolution matrix $\mathbf{R}$. For convenience we also define $\bar{\mathbf{A}}_{-i}$ as $\mathbf{A}_{-i}$ with the ith column (which is all zeros) removed. Then without loss of generality, we presume that $\mathbf{A} = (\bar{\mathbf{A}}_{-i}, \mathbf{a}_i)$, in order to simplify the equations (i.e., the order of the variables has been permuted so that the ith variable is last). With this we can write $\mathbf{A}\boldsymbol{\beta}_i^{(0)} = (\bar{\mathbf{A}}_{-i}, \mathbf{a}_i)\,\boldsymbol{\beta}_i^{(0)} = \mathbf{a}_i$. We similarly define $\bar{\boldsymbol{\beta}}_i$ as $\boldsymbol{\beta}_i$ with the ith element removed (recall $\boldsymbol{\beta}_i$ was defined with a zero in the ith element). Then for an ℓ_2-regularized pseudoinverse, the solution for $\bar{\boldsymbol{\beta}}_i$ is

$$\bar{\boldsymbol{\beta}}_i = \bar{\mathbf{A}}_{-i}^T \left(\bar{\mathbf{A}}_{-i} \bar{\mathbf{A}}_{-i}^T + \lambda\mathbf{I} \right)^{-1} \mathbf{a}_i \tag{6}$$

$$= \bar{\mathbf{A}}_{-i}^T \left(\mathbf{A}\mathbf{A}^T + \lambda\mathbf{I} - \mathbf{a}_i\mathbf{a}_i^T \right)^{-1} \mathbf{a}_i \tag{7}$$

We employ the matrix inversion lemma to get

$$\bar{\boldsymbol{\beta}}_i = \bar{\mathbf{A}}_{-i}^T(\mathbf{A}\mathbf{A}^T + \lambda\mathbf{I})^{-1}\mathbf{a}_i - \frac{\bar{\mathbf{A}}_{-i}^T(\mathbf{A}\mathbf{A}^T + \lambda\mathbf{I})^{-1}\mathbf{a}_i\mathbf{a}_i^T(\mathbf{A}\mathbf{A}^T + \lambda\mathbf{I})^{-1}\mathbf{a}_i}{-1 + \mathbf{a}_i^T(\mathbf{A}\mathbf{A}^T + \lambda\mathbf{I})^{-1}\mathbf{a}_i} \tag{8}$$

Meanwhile, the least-squares solution for $\boldsymbol{\beta}_i^{(0)}$ is,

$$\boldsymbol{\beta}_i^{(0)} = \mathbf{A}^T \left(\mathbf{A}\mathbf{A}^T + \lambda\mathbf{I} \right)^{-1} \mathbf{a}_i = \begin{pmatrix} \bar{\mathbf{r}}_i^{(\lambda)} \\ R_{i,i}^{(\lambda)} \end{pmatrix}, \tag{9}$$

where $\bar{\mathbf{r}}_i$ is the ith column of the regularized resolution matrix with the ith element removed, and $R_{i,i}$ is the (i,i)th element. Utilizing these definitions in Eq. (8) gives,

$$\bar{\boldsymbol{\beta}}_i = \left(\frac{1}{1 - R_{i,i}} \right) \bar{\mathbf{r}}_i \tag{10}$$

To write the matrix version of this relation between $\mathbf{B}$ and $\mathbf{R}$, we form the matrix $\mathbf{R}_{-d}$ defined as $\mathbf{R}$ with the values on the diagonal set to zero, and perform the scaling as $\mathbf{B} = \mathbf{R}_{-d}\mathbf{D_s}$ where $\mathbf{D_s}$ is a diagonal matrix with $D_{i,i} = s_i = (1 - R_{i,i})^{-1}$. Inputting this into Eq. (4) gives

$$\mathbf{P}^{(a)} = \mathbf{D_d}\mathbf{R}_{-d}\mathbf{D_s}\mathbf{D_d}^{-1}. \tag{11}$$

We can also write the $\mathbf{d}$ vector (the diagonal of $\mathbf{D_d}$) as

$$d_i = \|\mathbf{A}_{-i}\boldsymbol{\beta}_i - \mathbf{a}_i\| = \left| \frac{1}{1 - R_{i,i}} \right| \|\mathbf{A}\left(\mathbf{A}^\dagger\mathbf{a}_i - \mathbf{e}_i\right)\|, \tag{12}$$

where $\mathbf{e}_i$ is the i column of the identity matrix.

We refer to Eq. (11) as the asymmetric version, as we ignore possible asymmetries $P_{ij} \neq P_{ji}$. Alternatively, we can extend the symmetric version of Eq. (5) by plugging in $\mathbf{B} = \mathbf{R}_{-d}\mathbf{D_s}$, to get,

$$\mathbf{P}^{(s)} = \text{sign}(\mathbf{1}\mathbf{s}^T) \odot (\mathbf{s}\mathbf{s}^T)^{\circ\frac{1}{2}} \odot \mathbf{R}_{-d}. \tag{13}$$

where $\mathbf{s}$ is the vector with elements $s_i = (1 - R_{i,i})^{-1}$. In this version we set $P_{i,j}$ equal to zero when $\text{sign}(s_i) \neq \text{sign}(s_j)$.

1. Choose number of clusters K and initialize cluster centers $\mathbf{c}_k$, $k = 1, ..., K$;
while *Convergence criterion not met* **do**
| 2. Label each column as belonging to nearest cluster center:
| $l_k = \arg\min_i D_{ik}$, using minimum over distance D_{ik} between every column
| $\mathbf{p}_k$ and every cluster center $\mathbf{c}_i$;
| 3. Recalculate cluster centers as mean over data columns with same label:
| $\mathbf{c}_i = \frac{1}{|S_i|} \sum_{j \in S_i} \mathbf{a}_j$, where $S_i = \{k | l_k = i\}$;
end

Algorithm 1: k-means clustering of columns of $\mathbf{P}$.

2.2 Efficient Clustering of P

Next we will show how to partition the partial correlation network directly using the raw data $\mathbf{A}$. A basic k-means clustering algorithm which could be used for clustering the columns of $\mathbf{P}$ is given in Algorithm 1. Of course, for a large network, $\mathbf{P}$ will be extremely large as it has one entry per edge, so N nodes results in a matrix of size $N \times N$. However, as we have eliminated the need for node-specific pseudoinverses for each column, we can compute columns on-the-fly as needed inside the clustering loop, as in

$$
\begin{aligned}
\mathbf{p}_i^{(a)} &= \frac{1}{d_i(1 - R_{i,i})} \mathbf{d} \odot \mathbf{r}_{-i} \\
&= \mathbf{d} \odot \left(\mathbf{A}^\dagger \alpha_i \mathbf{a}_i - R_{i,i} \alpha_i \mathbf{e}_i \right),
\end{aligned}
\tag{14}
$$

where we have defined $\alpha_i = [d_i(1 - R_{i,i})]^{-1}$ and $\mathbf{r}_{-i}$ is the ith column of $\mathbf{R}_{-d}$. We can precompute one pseudoinverse, $\mathbf{A}^\dagger$, the $\mathbf{d}$ vector from Eq. (12), and the diagonal terms $R_{i,i}$ of the resolution matrix. For the symmetric version we compute columns on the fly with a similar form,

$$
\begin{aligned}
\mathbf{p}_i^{(s)} &= \text{sign}(s_i)(s_i \mathbf{s})^{\circ\frac{1}{2}} \odot \mathbf{r}_{-i} \\
&= |\mathbf{s}|^{\circ\frac{1}{2}} \odot (\mathbf{A}^\dagger \sigma_i \mathbf{a}_i - R_{i,i} \sigma_i \mathbf{e}_i),
\end{aligned}
\tag{15}
$$

where $\sigma_i = \text{sign}(s_i)|s_i|^{\frac{1}{2}} = \text{sign}(1 - R_{i,i})|1 - R_{i,i}|^{-\frac{1}{2}}$, which can also be precomputed. Generally we write either Eq. (15) or Eq. (14) as

$$
\mathbf{p}_i = \mathbf{z} \odot (\mathbf{A}^\dagger \zeta_i \mathbf{a}_i - R_{i,i} \zeta_i \mathbf{e}_i),
\tag{16}
$$

for appropriate definitions of $\mathbf{z}$ and ζ.

The squared distances between a given center $\mathbf{c}_i$ and a column $\mathbf{p}_k$ of $\mathbf{P}$ can be calculated as

$$
\begin{aligned}
D_{ik}^2 &= \|\mathbf{c}_i - \mathbf{p}_k\|_2^2 \\
&= \mathbf{c}_i^T \mathbf{c}_i + \mathbf{p}_k^T \mathbf{p}_k - 2\mathbf{c}_i^T \mathbf{p}_k.
\end{aligned}
\tag{17}
$$

Since we are only concerned with the class index i of the cluster with the minimum distance to each column, we can compute the labels as

$$l_k = \arg\min_i D_{ik}^2$$
$$= \arg\min_i \left\{ c_i^T c_i - 2(c_i \odot z)^T (A^\dagger \zeta_k a_k - R_{k,k}\zeta_k e_k) \right\}. \qquad (18)$$

By forming a matrix C_z with weighted cluster centers $c_i \odot z$ as columns, and a weighted data matrix A_ζ with $\zeta_i a_i$ as columns, we can efficiently compute the first part of the cross term for all i and k as $(C_z^T A^\dagger)A_\zeta$, a K by n matrix. The second part of the cross term can be computed by (element-wise) multiplying each row of C_z^T by a vector who's kth element is $R_{k,k}\zeta_k$. Similar tactics can be used to efficiently compute the mean over columns in each cluster, as

$$c_i = \frac{1}{|S|} \sum_{j \in S} p_j$$
$$= \frac{1}{|S|} z \odot \left(A^\dagger \sum_{j \in S} \zeta_j a_j - \sum_{j \in S} R_{j,j}\zeta_j e_j \right) \qquad (19)$$

So in general, we see that clustering of P can be implemented whenever we are able to to implement k-means clustering of the original dataset A, taking roughly double the storage space and computational resources.

In [2] we derived a similar algorithm for clustering columns of the matrix R; in that case $l_k = \arg\min_i \left\{ c_i^T c_i - 2c_i^T A^\dagger a_k \right\}$, and $c_i = \frac{1}{|S|} A^\dagger \sum_{j \in S} a_j$. In [3] we showed that this algorithm could be viewed as a variant of spectral clustering. So the corrections relating partial correlation clustering and spectral clustering are due to the ζ_j and z scaling factors, plus the e_j correction terms.

3 Results

First we produced an artificial dataset with a two-dimensional correlation structure. We generated 3000 samples for each of 300 random variables, with the kth random variable defined as,

$$a_k = x_k s_0 + (1 - x_k)s_1 + y_k t_0 + (1 - y_k)t_1 + \sigma n_k,$$

where s_0, s_1, t_0, and t_1 and n_k are independent unit-variance random variables, and $x_k \in [0,1]$ and $y_k \in [0,1]$ are randomly-chosen points. Effectively, for variables with (x_k, y_k) nearer to each other, the signals a_k are more correlated. We expect a simple partitioning of the x, y domain as a result of clustering this data. Simulated results are given in Fig. 2, using four clusters. By choosing the regularization parameter at a level equivalent to choosing the first two nontrivial eigenvectors, we achieved similar results for spectral and partial correlation clustering even in the presence of high noise, while basic k-means clustering of the noisy signals failed.

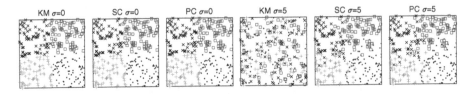

Fig. 2. Simulated results for k-means(KM), spectral clustering (SC), and partial correlation clustering (PC) of data with two-dimensional correlation structure, for noise levels $\sigma = 0$ and 5.

Next we considered a high-dimensional real dataset. Figure 3 demonstrates the algorithm applied to functional Magnetic Resonance Imaging (fMRI) scans for a subject from the Human Connectome Project [8], compared to other clustering approaches. The data was preprocessed by applying spatial smoothing with a 5mm kernel, and regularization was used to achieve a cutoff of 30 percent of singular values. The data contains 96854 time series with 1200 time samples each, resulting in a data matrix $\mathbf{A}$ of size 1200×96854. Each column contains a time-series describing the blood-oxygen-level dependent (BOLD) activity in one voxel of the brain, so a network formed by comparing these signals provides an estimate of the functional connectivity of the brain. A clustering of this network would produce an estimate of the modularity of function in the brain. The network describing the relationships between all pairs of voxels, however, would require a $\mathbf{P}$ matrix of size 96854×96854, which is far too large to fit in RAM. However the limited rank of this matrix means we only need store the 96854×1200 pseudoinverse. In this case the clustering algorithm took 9 s on a desktop computer. We see that clustering of $\mathbf{P}$ produces much more modular segmentation of the regions of the brain, particularly compared to the conceptually-similar approach of clustering the network of univariate correlations of the data instead.

Fig. 3. Clustering functional MRI data for single subject from the HCP project into 100 clusters; k-means clustering of the original time series (left); clustering of univariate correlations between time series (middle); clustering of partial correlations between time-series (right).

We tested the difference between spectral clustering and partial correlation for this dataset. Using identical random initial clusters for both methods, we plot the percentage of nodes which differ in the final clustering results in Fig. 4, as increasing amounts of spatial smoothing are applied. With spatial smoothing, a common preprocessg step in many applications, the removal of the diagonal terms will have less effect, as for example, the same information is increasingly included in the neighboring variables.

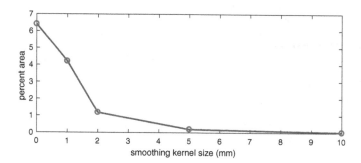

Fig. 4. Plot of difference between clustering partial correlation network versus spectral clustering, showing close agreement between the methods, particularly with spatial smoothing. Spectral clustering in this case uses the covariance matrix of the data as weighted adjacency matrix.

4 Discussion

We derived an efficient approach to partitioning partial correlation networks and demonstrated the approach on neuroimaging data, where we found a close similarity to spectral clustering. This suggests another perspective on the success of spectral clustering methods, as an approximation to clustering of a Gaussian graphical model for the data. A benefit of the proposed approach is its principled basis as a direct estimate for a partial correlation network. As our simulation shows, the benefit of spectral clustering over basic k-means can be viewed, at least in a low-dimensional setting, as resulting from the regularization effect of truncating the eigenvectors. With the experimental data, we also see a close agreement with spectral clustering in higher dimensions, particularly when spatial smoothing is employed. In terms of numerical efficiency, our approach provides a more efficient calculation compared to the brute-force approach of solving a separate regression problem for each node.

The drawbacks of the proposed approach include the moderately increased computational effort, and the potential need to address asymmetric signs in the partial correlation matrix. Though our approach to address the latter (setting them to zero based on a sign test) is the approach commonly used in bioinformatics. There are a number of potential extensions to the approach. Instead of a simple k-means stage, we might apply a more sophisticated clustering algorithm such as fuzzy or hierarchical clustering. Also we could extend the distance calculation to other and more sophisticated statistics or more sophisticated statistical tests.

5   Appendix: Matlab Implementation of Clustering

In this appendix we provide efficient Matlab code for performing partial correlation clustering.

```
A = randn(500,100000); % simulate data matrix
lambda = 1; % regularization parameter
k = 100; % number of clusters
[rows_A,cols_A] = size(A);

% standardize data columns
A = bsxfun(@minus,A,mean(A));
A = bsxfun(@times,A,1./sum(A.^2).^.5);

% compute diagonal of R via sum of squared eigenvectors
[uA,sA,vA] = svd(A,'econ');
r = sum(vA(:,1:rank(A)).^2,2)';
r = r(:);

% compute pseudoinverse efficiently (assuming fewer rows than columns)
iA_lambda = A'*inv(A*A'-lambda*eye(rows_A));

% compute scaling vectors (symmetric version)
s = 1./(1-r(:));
z = abs(s(:)).^.5;
zeta = sign(s).*abs(s(:)).^.5;
Az = bsxfun(@times,A,z(:)'); % precompute scaled version

% randomly assign columns to clusters initially
c = ceil(rand(cols_A,1)*k);
n_change = inf
while (n_change>0) % clustering loop
    M = sparse(1:cols_A,c,1,cols_A,k,cols_A); % cols of M are masks of clusters
    M = bsxfun(@times, M, 1./sum(M,1));   % now M is averaging operator
    P_c_1 = iA_lambda*(Az*M);   % first part of cluster center calc
    P_c_2 = bsxfun(@times,M,r.*zeta); % second park (peak removal)
    P_c = bsxfun(@times,P_c_1-P_c_2,z(:)); % cluster centers
    Pz2_c = sum(P_c.^2,1); % squared term from distance
    Cz = bsxfun(@times,P_c,z(:)); % weighted cluster centers
    D_ct1 = (Cz'*iA_lambda)*Az; % first part of cross-term
    D_ct2 = bsxfun(@times,Cz',r'.*zeta(:)'); % second part of cross term
    D_ct = D_ct1-D_ct2; % cross-term
    Dz = bsxfun(@minus,D_ct,.5*Pz2_c'); % dist metric (sans unnecessary term)
    c_old = c;
    [D_max,c(:)] = max(Dz,[],1); % c is arg of max
    n_change = sum(c~=c_old);
    disp(n_change);
end;
```

References

1. Brownlees, C., Gudmundsson, G.S., Lugosi, G.: Community detection in partial correlation network models. J. Bus. Econ. Stat. 1–11 (2020)
2. Dillon, K., Wang, Y.P.: A regularized clustering approach to brain parcellation from functional MRI data. In: Wavelets and Sparsity XVII, vol. 10394, p. 103940E. International Society for Optics and Photonics, August 2017
3. Dillon, K., Wang, Y.P.: Resolution-based spectral clustering for brain parcellation using functional MRI. J. Neurosci. Methods **335**, 108628 (2020)
4. Ellett, F.S., Ericson, D.P.: Correlation, partial correlation, and causation. Synthese **67**(2), 157–173 (1986)
5. von Luxburg, U.: A tutorial on spectral clustering. Stat. Comput. **17**(4), 395–416 (2007)
6. Pourahmadi, M.: Covariance estimation: the GLM and regularization perspectives. Stat. Sci. **26**(3), 369–387 (2011)
7. Schäfer, J., Strimmer, K.: A shrinkage approach to large-scale covariance matrix estimation and implications for functional genomics. Stat. Appl. Genet. Mol. Biol. **4**(1), 32 (2005)
8. Van Essen, D.C., et al.: WU-Minn HCP consortium: the human connectome project: a data acquisition perspective. NeuroImage **62**(4), 2222–2231 (2012)
9. Whittaker, J.: Graphical Models in Applied Multivariate Statistics. Wiley, New York (2009)

Alzheimer's Brain Network Analysis Using Sparse Learning Feature Selection

Lixin Cui[1], Lichi Zhang[2], Lu Bai[1(✉)], Yue Wang[1], and Edwin R. Hancock[3]

[1] Central University of Finance and Economics, Beijing, China
`bailucs@cufe.edu.cn`
[2] Institute for Medical Imaging Technology, School of Biomedical Engineering,
Shanghai Jiao Tong University, Shanghai, China
[3] University of York, York, UK

Abstract. Accurate identification of Mild Cognitive Impairment (MCI) based on resting-state functional Magnetic Resonance Imaging (RS-fMRI) is crucial for reducing the risk of developing Alzheimer's disease (AD). In the literature, functional connectivity (FC) is often used to extract brain network features. However, it still remains challenging for the estimation of FC because RS-fMRI data are often high-dimensional and small in sample size. Although various Lasso-type sparse learning feature selection methods have been adopted to identify the most discriminative features for brain disease diagnosis, they suffer from two common drawbacks. First, Lasso is instable and not very satisfactory for the high-dimensional and small sample size problem. Second, existing Lasso-type feature selection methods have not simultaneously encapsulate the joint correlations between pairwise features and the target, the correlations between pairwise features, and the joint feature interaction into the feature selection process, thus may lead to suboptimal solutions. To overcome these issues, we propose a novel sparse learning feature selection method for MCI classification in this work. It unifies the above measures into a minimization problem associated with a least square error and an Elastic Net regularizer. Experimental results demonstrate that the diagnosis accuracy for MCI subjects can be significantly improved using our proposed feature selection method.

Keywords: Alzheimer's disease · Feature selection · Elastic net

1 Introduction

Alzheimer's disease (AD) is the most common form of dementia in old people, which severely interferes with their daily life and may eventually cause death [3]. Effective and accurate diagnosis of AD at its early stage may possess crucial significance in preventing progression of detrimental symptoms [3]. Recently, the identification of MCI subjects is important for reducing the risk of developing AD

L. Zhang—Co-First Author.

© Springer Nature Switzerland AG 2021
A. Torsello et al. (Eds.): S+SSPR 2020, LNCS 12644, pp. 184–194, 2021.
https://doi.org/10.1007/978-3-030-73973-7_18

and has attracted much attention recently [11]. However, it is very challenging to identify MCI subjects due to its mild clinical symptoms.

In the literature, MCI is generally believed to be associated with a disconnection syndrome within brain networks. Therefore, constructing brain functional connectivity (FC) networks based on the resting-state fMRI (RS-fMRI) BOLD signals of various brain regions has become promising for MCI classification. In this paper, we use a sliding window approach [9] to partition the RS-fMRI BOLD signal from each voxel into multiple overlapping segments, in order to capture the time-varying interactions between different ROIs and obtain a series of dynamic FC networks. We then extract the corresponding FC features for the subsequent brain network analysis. However, the number of the extracted features is much larger than that of the MCI subjects, and more importantly, many features may be irrelevant to the classification task, thus leading to the overfitting problem.

In pattern recognition and machine learning, feature selection are powerful tools for identifying the most salient features from the original feature space and alleviating the overfitting problem [10]. In this regard, various feature selection methods have been widely applied to detect the most discriminative features for AD prediction. In some early works, Chyzhyk et al. [4] proposed an evolutionary wrapper feature selection using Extreme Learning Machines to determine the most salient features for AD diagnosis. However, wrapper methods are often computational burdensome and the results are biased depending on the classifier [6]. To overcome these issues, many efforts have been devoted to developing LASSO-type feature selection methods for AD diagnosis. For instance, Suk et al. [7] utilized a group sparse representation along with a structural equation model to estimate FC from RS-fMRI. Wee et al. [9] proposed a fused sparse learning algorithm for early MCI identification. Chen et al. [3] developed a two-stage feature selection procedure to select a subset of the original features for MCI classification. However, existing LASSO-type feature selection methods for MCI classification suffer from two common limitations. First, LASSO shows instability and is not very satisfactory for high-dimensional small sample size problem. Second, existing Lasso-type feature selection methods have not simultaneously encapsulate the joint correlations between pairwise features and the target, the correlations between pairwise features, and the joint feature interaction into the feature selection process, thus may lead to suboptimal solutions.

To effectively tackle the issues of existing Lasso-type sparse learning feature selection methods, we propose a new feature selection method, i.e., Tripple-EN for MCI classification. We commence by defining three new information theoretic criteria to measure: 1)the relevancy of pairwise features in relation to the target feature, 2) the redundancy of pairwise features and 3)joint feature interaction. With these measures to hand, we formulate the corresponding feature subset selection problem as a least square regression model associated with an elastic net regularizer to simultaneously maximize relevancy, minimize redundancy, and maximize joint interaction of the selected features. An iterative optimization algorithm based on the alternating direction method of multipliers (ADMM) [1] is proposed to solve the optimization problem.

The advantages of the proposed method are twofold. First, it encapsulates the pairwise feature relevancy, feature redundancy and joint feature interaction into a unified learning model to improve the performance of feature selection. Second, by using the elastic net regularizer, the proposed method can ensure sparsity and also promote a grouping effect of the features. Figure 1 shows an overview of the framework of this paper, which consists of the following steps: (1) constructing brain FC networks using a sliding window strategy, (2) identifying the most discriminative features using a new sparse learning feature selection method, and (3) implementing classification following the C-SVM method. Details of each step are illustrated in the following sections.

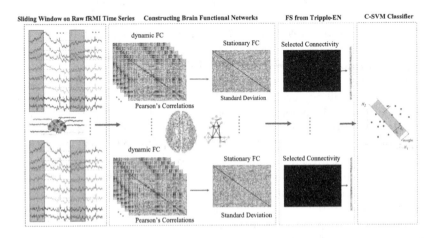

Fig. 1. Framework of this paper.

This paper is organized as follows. Section 2 introduces the construction of the functional connectivity networks from brain networks. Section 3 illustrates the proposed sparse learning feature selection method for MCI classification. Finally, Sect. 5 concludes this paper.

2   Constructing Functional Connectivity Networks

In this section, we will introduce how to construct the functional connectivity networks, which mainly consists of two steps, i.e., generating functional networks using RS-fMRI and feature extraction.

2.1   Generating FC Networks Using RS-fMRI

As in Fig. 1, the preprocessed RS-fMRI data was parcellated using the Automated Anatomical Labeling (AAL) atlas with 116 ROIs [3], which are represented by the time series curves of different colors. We use a sliding window approach to partition the RS-fMRI BOLD signal from each voxel into multiple overlapping segments, in order to capture the time-varying interactions between different ROIs. Specifically, denote the total length of image volumes as M and the length of the sliding windows as N. Then, the total number of segments is $K = \lfloor (M - N)/s \rfloor$. On each segment, within the GM, a regional mean BOLD signal is computed by averaging the BOLD time series over all voxels inside each ROI, which reflects the regional neural activity during a short period of time. We use C_{ij}^k to denote the Pearson's correlation coefficients between ROI i and ROI j on the k-th sliding window. Then we can obtain the interregional dynamic FC (dFC), denoted as $dFC_{ij} = [C_{ij}^1, ..., C_{ij}^k, ..., C_{ij}^K]$, which measures the time-varying interactions of FC between ROI i and ROI j. As shown in Fig. 1, we can obtain a series of dynamic time-varying FC networks. Note that, due to the symmetry of Pearson's correlation, the number of dFC is equal to the total number of ROI pairs.

2.2   Feature Extraction

To extract the features for further analysis, we calculate the standard deviation of a series of dynamic FC networks and obtain one stationary FC network for each subject. Specifically, the corresponding FC network for a series of dynamic time varying networks is obtained by calculating the standard deviation as $\sqrt{\frac{\sum_{k=1}^{K}(C_{ij}^k - \mu)^2}{K}}$, where μ is the mean value of C_{ij}^k. With these FC networks to hand, a total of 6670 features was generated. As shown in Fig. 1, for a series of dynamic time-varying FC networks, we can construct a stationary FC network for each subject, with each node representing a specific ROI and each edge representing the corresponding connection between pairwise ROIs, which incorporates the information from a series of dynamic time-varying FC networks.

3   The Proposed Sparse Learning Feature Selection for MCI Classification

In this section, we focus on the proposed sparse learning feature selection method for identifying the most discriminative FC features. We commence by introducing the proposed information theoretic criteria for measuring the joint relevance (significance) of different pairwise feature combinations with respect to target labels, the redundancy of pairwise features, and the joint feature interaction, respectively. Based on these measures, we develop the corresponding optimization model for feature selection and sparse learning. Finally, an iterative optimization algorithm based on ADMM is proposed to solve the feature selection problem and identify the most discriminative feature subset.

3.1   Proposed Information Theoretic Criteria

Feature Relevancy. For a set of N features $\mathbf{f}_1, ..., \mathbf{f}_i, ..., \mathbf{f}_N$ and the associated target feature $\mathbf{Y}$, the relevancy degree of each feature pair $\{\mathbf{f}_i, \mathbf{f}_j\}$ in relation to the target feature is estimated through Pearson's correlation coefficients, which is defined as

$$W_{(\mathbf{f}_i, \mathbf{f}_j)} = Cor(\mathbf{f}_i, \mathbf{Y}) \times Cor(\mathbf{f}_j, \mathbf{Y}). \tag{1}$$

where Cor is the Pearson's correlation measure. The first term $Cor(\mathbf{f}_i, \mathbf{Y})$ measures the relevance of feature $\mathbf{f}_i$ with respect to the target. Similarly, the second term is the corresponding relevance of feature $\mathbf{f}_j$ with respect to the target. Therefore, $W_{(\mathbf{f}_i, \mathbf{f}_j)}$ is large if and only if both $Cor(\mathbf{f}_i, \mathbf{Y})$ and $Cor(\mathbf{f}_j, \mathbf{Y})$ are large (i.e., both $\mathbf{f}_i$ and $\mathbf{f}_j$ are informative themselves with respect to the target).

Additionally, it is desirable that strongly correlated features should not be in the model together, i.e., the selected features should be less redundant. Therefore, we propose the following criterion to measure the redundancy of pairwise features.

Feature Redundancy. For a set of N features $\mathbf{f}_1, \ldots, \mathbf{f}_i, \ldots, \mathbf{f}_N$, the redundancy of the feature pair $\{\mathbf{f}_i, \mathbf{f}_j\}$ is calculated as

$$U_{(\mathbf{f}_i, \mathbf{f}_j)} = Cor(\mathbf{f}_j, \mathbf{f}_j) \tag{2}$$

where Cor is the Pearson's correlation measure.

Joint Feature Interaction. We propose to use the following criterion to measure the joint interaction of different pairwise feature combinations with respect to target labels. For a set of N features $\mathbf{f}_1, \ldots, \mathbf{f}_i, \ldots, \mathbf{f}_j, \ldots, \mathbf{f}_N$ and the associated continuous target feature $\mathbf{Y}$, the joint interaction degree of the feature pair $\{\mathbf{f}_i, \mathbf{f}_j\}$ is

$$V_{\mathbf{f}_i, \mathbf{f}_j} = \frac{Cor(\mathbf{f}_i, \mathbf{Y}) + Cor(\mathbf{f}_j, \mathbf{Y})}{Cor(\mathbf{f}_j, \mathbf{f}_j)}, \tag{3}$$

where Cor is the Pearson's correlation measure. The above measure consists of three terms. The terms $Cor(\mathbf{f}_i, \mathbf{Y})$ and $Cor(\mathbf{f}_j, \mathbf{Y})$ are the relevance degrees of individual features $\mathbf{f}_i$ and $\mathbf{f}_j$ with respect to the target feature $\mathbf{Y}$, respectively. The term $Cor(\mathbf{f}_j, \mathbf{f}_j)$ measures the relevance between the feature pair $\{\mathbf{f}_i, \mathbf{f}_i\}$. Therefore, $V_{\mathbf{f}_i, \mathbf{f}_j}$ is large if and only if both $Cor(\mathbf{f}_i, \mathbf{Y})$ and $Cor(\mathbf{f}_j, \mathbf{Y})$ are large (i.e., both $\mathbf{f}_i$ and $\mathbf{f}_j$ are informative themselves with respect to the target feature representation $\mathbf{Y}$) and $Cor(\mathbf{f}_j, \mathbf{f}_j)$ is small (i.e., $\mathbf{f}_i$ and $\mathbf{f}_j$ are not correlated).

Furthermore, based on the proposed information theoretic measures, we construct three interacted matrices denoted as $\mathbf{W}$, $\mathbf{U}$, and $\mathbf{V}$ respectively. Specifically, each element $W_{i,j} \in \mathbf{W}$ represents the joint relevancy between a feature pair $\{\mathbf{f}_i, \mathbf{f}_j\}$ based on Eq. (1). Likewise, each element $U_{i,j} \in \mathbf{U}$ represents the redundancy between a feature pair $\{\mathbf{f}_i, \mathbf{f}_j\}$ based on Eq. (2). Moreover, each element $V_{i,j} \in \mathbf{V}$ represents the joint interaction between a feature pair $\{\mathbf{f}_i, \mathbf{f}_j\}$ based on Eq. (3). Given $\mathbf{W}$, $\mathbf{U}$, $\mathbf{V}$ and the N-dimensional feature indicator vector β, where β_i represents the coefficient for the i-th feature, we can identify the informative feature subset by solving the following optimization problem to ensure maximum joint relevancy, minimum redundancy, and maximum joint interaction of the selected features,

$$\max f(\beta) = \max_{\beta \in \Re^N} \nu \beta^T \mathbf{W} \beta - \omega \beta^T \mathbf{U} \beta + \sigma \beta^T \mathbf{V} \beta, \tag{4}$$
$$s.t. \quad \beta \in \Re^N, \beta \geq 0.$$

where ν, ω and σ are the corresponding tuning parameters.

3.2 A Novel Sparse Learning Feature Selection Approach

Our discriminative feature selection approach is motivated by the purpose to ensure maximum joint relevancy, minimum redundancy, and maximum joint interaction of the selected features. In addition, it should simultaneously promote a sparse solution and a grouping effect of the highly correlated features. Therefore, we unify the minimization problem of Eq. (4) with the elastic net regression framework and propose the sparse learning feature selection method as

$$\min_{\beta \in \Re^N} \frac{1}{2}\|\mathbf{y}^T - \beta^T \mathbf{X}\|_2^2 + \lambda_1\|\beta\|_1 + \lambda_2\|\beta\|_2^2 - \lambda_3\beta^T\mathbf{W}\beta + \lambda_4\beta^T\mathbf{U}\beta - \lambda_5\beta^T\mathbf{V}\beta, \quad (5)$$

where λ_1 and λ_2 are the tuning parameters for elastic net, λ_3, λ_4, and λ_5 are the tuning parameters for the relevancy matrix $\mathbf{W}$, the redundancy matrix $\mathbf{U}$, and the joint interaction matrix $\mathbf{V}$, respectively. The first term in Eq. (5) is the least square error term, the second term and the third term encourage sparsity and also promote a grouping effect of the selected feature as in the elastic net model. The fourth term guarantees maximum joint relevancy of selected features. The fifth term ensures minimum redundancy among selected features. Finally, the last term ensures that the selected features are jointly more interacted with the target class.

3.3 An Iterative Optimization Algorithm

To solve the optimization problem (5), we develop an iterative optimization algorithm based on ADMM, which uses a decomposition-coordination procedure. By using ADMM, the solutions to small local subproblems are coordinated to find a solution to a large global problem. This algorithm can be viewed as an attempt to blend the benefits of dual decomposition and augmented Lagrangian methods for constrained optimization.

Firstly, we reformulate the proposed feature selection problem into an equivalent constrained problem in the ADMM form,

$$\min_{\beta \in \Re^N} \frac{1}{2}\|\mathbf{y}^T - \beta^T\mathbf{X}\|_2^2 + \lambda_2\|\beta\|_2^2 - \lambda_3\beta^T\mathbf{W}\beta + \lambda_4\beta^T\mathbf{U}\beta - \lambda_5\beta^T\mathbf{V}\beta + \lambda_1\|\gamma\|_1$$

$$s.t. \quad \beta = \gamma, \tag{6}$$

where γ is an auxiliary variable, which can be regarded as a proxy for vector β. In this way, the objective function can be divided into two separate parts associated with two different variables, i.e., β and γ. This indicates that the hard constrained problem can be solved separately. As in the method of multipliers, we form the augmented Lagrangian function associated with the constrained problem (5) as follows

$$L_\rho(\beta, \gamma, z) = \frac{1}{2}\|\mathbf{y}^T - \beta^T\mathbf{X}\|_2^2 + \lambda_2\|\beta\|_2^2 - \lambda_3\beta^T\mathbf{W}\beta + \lambda_4\beta^T\mathbf{U}\beta - \lambda_5\beta^T\mathbf{V}\beta$$

$$+ \lambda_1\|\gamma\|_1 + <\beta - \gamma, z> + \frac{\rho}{2}\|\beta - \gamma\|_2^2, \tag{7}$$

where $\langle \cdot, \cdot \rangle$ is an Euclidean inner product, z is a dual variable (i.e.,the Lagrange multiplier) associated with the equality constraint $\beta = \gamma$, and ρ is a positive penalty parameter (step size for dual variable update). By introducing an additional variable γ and an additional constraint $\beta - \gamma = 0$, we have simplified the optimization problem (5) by decoupling the objective function into two parts that depend on two different variables. In other words, we can decompose the minimization of $L_\rho(\beta, \gamma, z)$ into two simpler subproblems. Specifically, we solve the original problem (5) by seeking for a saddle point of the augmented Lagrangian by iteratively minimizing $L_\rho(\beta, \gamma, z)$ over β, γ, and z. Then the variables β, γ, and z can be updated in an alternating or sequential fashion, which accounts for the term alternating direction. This updating rule is shown as follows

(1) $\beta^{k+1} = \arg\min_{\beta \in \Re^p} L(\beta, \gamma^k, z^k)$, //$\beta$-minimization
(2) $\gamma^{k+1} = \arg\min_{\beta \in \Re^p} L(\beta^{k+1}, \gamma, z^k)$, //$\gamma$-minimization
(3) $z^{k+1} = z^k + \rho(\beta^{k+1} - \gamma^{k+1})$, //$z$-update

Given the above updating rule, we need to resolve each sub-problem iteratively until the termination criteria is satisfied. We perform the following calculation steps at each iteration.

(a)**Update β**

In the $(k+1)$-th iteration, in order to update β^k, we need to solve the following sub-problem, where the values of γ^k and z^k are fixed

$$\min_{\beta \in \Re^N} \frac{1}{2}\|\mathbf{y}^T - \beta^T\mathbf{X}\|_2^2 + \lambda_2\|\beta\|_2^2 - \lambda_3\beta^T\mathbf{W}\beta + \lambda_4\beta^T\mathbf{U}\beta - \lambda_5\beta^T\mathbf{V}\beta$$
$$+ \lambda_1\|\gamma\|_1 + <\beta - \gamma^k, z^k> + \frac{\rho}{2}\|\beta - \gamma^k\|_2^2. \tag{8}$$

Let the partial derivative with respect to β be equal to zero, we have

$$\frac{\partial[\min_{\beta \in \Re^N} \frac{1}{2}\|\mathbf{y}^T - \beta^T\mathbf{X}\|_2^2 + \lambda_2\|\beta\|_2^2 - \lambda_3\beta^T\mathbf{W}\beta + \lambda_4\beta^T\mathbf{U}\beta - \lambda_5\beta^T\mathbf{V}\beta]}{\partial\beta}$$
$$+ \frac{\partial[\min_{\beta \in \Re^N} \lambda_1\|\gamma\|_1 + <\beta - \gamma^k, z^k> + \frac{\rho}{2}\|\beta - \gamma^k\|_2^2]}{\partial\beta} = 0, \tag{9}$$

because

$$\begin{cases} \frac{\partial(\frac{1}{2}\|\mathbf{y}^T - \beta^T\mathbf{X}\|_2^2)}{\partial\beta} = -\mathbf{X}\mathbf{y} + \mathbf{X}\mathbf{X}^T\beta \\ \frac{\partial(\lambda_2\|\beta\|_2^2)}{\partial\beta} = \lambda_2\beta \\ \frac{\partial(-\lambda_3\beta^T\mathbf{W}\beta)}{\partial\beta} = -2\lambda_3\mathbf{W}\beta \\ \frac{\partial(\lambda_4\beta^T\mathbf{U}\beta)}{\partial\beta} = 2\lambda_4\mathbf{U}\beta \\ \frac{\partial(-\lambda_5\beta^T\mathbf{V}\beta)}{\partial\beta} = -2\lambda_5\mathbf{V}\beta \\ \frac{\partial<\beta - \gamma^k, z^k>}{\partial\beta} = z^k \\ \frac{\partial(\frac{\rho}{2}\|\beta - \gamma^k\|_2^2)}{\partial\beta} = \rho(\beta - \gamma^k), \end{cases} \tag{10}$$

we have

$$-\mathbf{X}\mathbf{y} + \mathbf{X}\mathbf{X}^T\beta + \lambda_2\beta - 2\lambda_3\mathbf{W}\beta + 2\lambda_4\mathbf{U}\beta - 2\lambda_5\mathbf{V}\beta + z^k + \rho(\beta - \gamma^k) = 0, \quad (11)$$

that is,

$$\beta^{k+1} = (\mathbf{X}\mathbf{X}^T + \lambda_2\mathbf{I} - 2\lambda_3\mathbf{W} + 2\lambda_4\mathbf{U} - 2\lambda_5\mathbf{V} + \rho\mathbf{I})^{-1}(\mathbf{X}\mathbf{y} - z^k + \rho\gamma^k). \quad (12)$$

(b)**Update** γ

Based on the results, assume β_i^{k+1} and z_i^k are fixed, for $i = 1, 2, ..., d$, we update γ_i^{k+1} by solving the following sub-optimization problem

$$\min_{\gamma_i} \lambda_1 \sum_{i=1}^{p} \|\gamma_i\|_1 - \sum_{i=1}^{p} <\gamma_i, z_i^k> + \frac{\rho}{2}\sum_{i=1}^{p} \|\beta_i^{k+1} - \gamma_i\|_2^2, \quad (13)$$

$$\frac{\partial[\min_{\gamma_i} \lambda_1 \sum_{i=1}^{p} \|\gamma_i\|_1 - \sum_{i=1}^{p} <\gamma_i, z_i^k> + \frac{\rho}{2}\sum_{i=1}^{p} \|\beta_i^{k+1} - \gamma_i\|_2^2]}{\partial\gamma_i} = 0. \quad (14)$$

We therefore have the following results

$$\gamma_i^{k+1} = \begin{cases} \frac{1}{\rho}(z_i^k + \rho\beta_i^{k+1} - \lambda_1), & \text{if } z_i^k + \rho\beta_i^{k+1} > \lambda_1 \\ \frac{1}{\rho}(z_i^k + \rho\beta_i^{k+1} + \lambda_1), & \text{if } z_i^k + \rho\beta_i^{k+1} < -\lambda_1 \\ 0, & \text{if } z_i^k + \rho\beta_i^{k+1} \in [-\lambda_1, \lambda_1] \end{cases} \quad (15)$$

(c)**Update z**

Then, assume β_i^{k+1} and γ_i^{k+1} are fixed, for $i = 1, 2, ..., d$, we update z_i^{k+1} by using the following equation

$$z_i^{k+1} = z_i^k + \rho(\beta_i^{k+1} - \gamma_i^{k+1}). \quad (16)$$

Based on procedures (a), (b), and (c), we summarize the optimization algorithm below

Input: $\mathbf{X}, \mathbf{y}, \beta^0, z^0, \lambda_1, \lambda_2, \lambda_3, \lambda_4, \lambda_5, \rho$
Step1: While (not converged), **do**
Step2: Update β^{k+1} according to Eq. (12)
Step3: Update $\gamma_i^{k+1}, i = 1, 2, ..., d$ according to Eq. (15)
Step4: Update $\beta_i^{k+1}, i = 1, 2, ..., d$ according to Eq. (16)
End While
Output: β^*.
Algorithm 1: The iterative optimization algorithm for the proposed Tripple-EN method.

4   Experimental Analysis

We evaluate the performance of the proposed feature selection method for MCI classification on the public available Alzheimer's Disease Neuroimaging Initative (ANDI) database. Specifically, 54 MCI patients and 62 NC subjects were selected from ADNI database. The images of each subject were acquired using a 3.0T Philips scanners at centers in different places. The voxel size is $3.13 \times 3.13 \times 3.13\,\mathrm{mm}^3$. SPM8 software package was applied to preprocess the RS-fMRI data. To evaluate the discriminative capabilities of the information captured by our method, we compare the classification results using the selected features from our method (Mu-InElasticNet) with several state-of-the-art feature selection methods, i.e., (a) Lasso [8], (b) ULasso [2], (c) Group Lasso [5], and (d) Elastic Net [12]. For the experiments, due to limited samples, a Leave One Out (LOO) cross-validation associated with C-SVM is applied to benchmark the generalization performance of different methods. Specifically, given N subjects, N-1 subjects are used as training data, and one subject is subsequently evaluated in terms of the classification accuracy. We repeat the procedure L times, and report the averaged classification result. Figure 2(a) exhibits that the C-SVM associated with the proposed method can achieve the best classification accuracy, and the accuracy (y-axis) increases with the increasing number of selected features (x-axis). Moreover, Table 1 shows the best classification accuracy (ACC) for each method associated with the corresponding number of selected features, as well as other four associated indices, i.e., sensitivity (SEN), specificity (SPE), area under the receiver operating characteristic curve (AUC), and F-score. We observe that the proposed method significantly outperforms the remaining methods on all indices. The reason for the effectiveness is that only our method can simultaneously maximize relevancy and minimize redundancy of the selected features. Finally, we also experimentally evaluate the convergence property of the proposed method. Figure 2(b) indicates that the proposed method converges quickly within 50 iterations tend to be stable after 150 iterations.

Table 1. Performance of different methods in MCI classification (NC vs MCI).

Methods	Lasso	ULasso	GroupLasso	ElasticNet	Tripple-EN
ACC	0.6578	0.6842	0.7105	0.7192	**0.7894**
SEN	0.6663	0.6800	0.7143	0.7059	**0.8261**
SPE	0.6783	0.6875	0.7077	0.7143	**0.7647**
AUC	0.6723	0.6821	0.7110	0.7101	**0.7954**
F-score	0.6567	0.6538	0.6796	0.6857	**0.7600**
Feature numbers	60 features	80 features	70 features	80 features	80 features

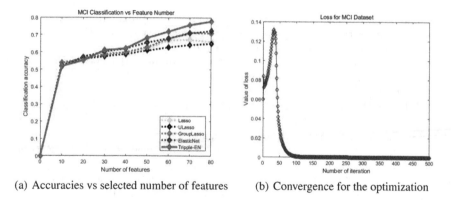

(a) Accuracies vs selected number of features   (b) Convergence for the optimization

Fig. 2. Experiments for the proposed method.

5 Conclusion

In this paper, we have proposed a novel sparse learning feature selection method for MCI classification for AD diagnosis. Specifically, we devised three information theoretic measures to evaluate feature relevancy, feature redundancy and joint feature interaction. These measures are further encapsulated into the least square regression associated with an elastic net regularizer to simultaneously maximize relevancy, minimize redundancy, and maximize joint interaction of the selected features. Experiments demonstrated the effectiveness of our method on MCI classification tasks.

Acknowledgments. This work is supported by the National Natural Science Foundation of China (Grant no. 61602535 and 61976235), the Program for Innovation Research in Central University of Finance and Economics, and the Youth Talent Development Support Program by Central University of Finance and Economics, No. QYP1908.

References

1. Boyd, S.P., Parikh, N., Chu, E., Peleato, B., Eckstein, J.: Distributed optimization and statistical learning via the alternating direction method of multipliers. Found. Trends Mach. Learn. **3**(1), 1–122 (2011)
2. Chen, S., Ding, C.H.Q., Luo, B., Xie, Y.: Uncorrelated lasso. In: Proceedings of the Twenty-Seventh AAAI Conference on Artificial Intelligence, 14–18 July 2013, Bellevue, Washington, USA (2013)
3. Chen, X., Zhang, H., Zhang, L., Shen, C., Lee, S.W., Shen, D.: Extraction of dynamic functional connectivity from brain grey matter and white matter for MCI classification. Human Brain Mapp. **38**, 5019–5034 (2017)
4. Chyzhyk, D., Savio, A., Graña, M.: Evolutionary ELM wrapper feature selection for Alzheimer's disease CAD on anatomical brain MRI. Neurocomputing **128**, 73–80 (2014)
5. Meier, L., Geer, S.V.D., Bhlmann, P.: The group lasso for logistic regression. J. R. Statist. Soc. Ser. B **70**(1), 53–71 (2008)

6. Naghibi, T., Hoffmann, S., Pfister, B.: A semidefinite programming based search strategy for feature selection with mutual information measure. IEEE Trans. Pattern Anal. Mach. Intell. **37**(8), 1529–1541 (2015)

7. Suk, H., Wee, C., Lee, S., Shen, D.: Supervised discriminative group sparse representation for mild cognitive impairment diagnosis. Neuroinformatics **13**(3), 277–295 (2015)

8. Tibshirani, R.: Regression shrinkage and selection via the lasso. J. R. Statist. Soc. Ser. B **58**(1), 267–288 (1996)

9. Wee, C., Yang, S., Yap, P., Shen, D.: Sparse temporally dynamic resting-state functional connectivity networks for early MCI identification. Brain Imag. Behav. **10**(2), 342–356 (2016)

10. Zhang, C., Fu, H., Hu, Q., Zhu, P., Cao, X.: Flexible multi-view dimensionality co-reduction. IEEE Trans. Image Process. **26**(2), 648–659 (2017)

11. Zhang, Y., Zhang, H., Chen, X., Liu, M., Zhu, X., Lee, S., Shen, D.: Strength and similarity guided group-level brain functional network construction for MCI diagnosis. Pattern Recogn. **88**, 421–430 (2019)

12. Zou, H., Hastie, T.: Regularization and variable selection via the elastic net. J. R. Statist. Soc. **67**(5), 301–320 (2005)

The Entropy of Graph Embeddings: A Proxy of Potential Mobility in Covid19 Outbreaks

Francisco Escolano[1,2] (ID), Miguel Angel Lozano[1,2] (ID),
and Edwin R. Hancock[3(✉)] (ID)

[1] University of Alicante, Alicante, Spain
[2] COVID-19 Data Science Task Force, Alicante, Spain
[3] University of York, York, UK
edwin.hancock@york.ac.uk

Abstract. In this paper, we propose a proxy of the R_0 (reproductive number) of COVID-19 by computing the entropy of the mobility graph during the first peak of the pandemic. The study was performed by the *COVID-19 Data Science Task Force* at the Comunidad Valenciana (Spain) during 70 days. Since mobility graphs are naturally attributed, directed and become more and more disconnected as more and more non-pharmaceutical measures are implemented, we discarded spectral complexity measures and classical ones such as network efficiency. Alternatively, we turned our attention to embeddings resulting from random walks and their links with stochastic matrices. In our experiments, we show that this leads to a powerful tool for predicting the spread of the virus and to assess the effectiveness of the political interventions.

Keywords: Graph embeddings · Graph complexity · COVID-19

1 Introduction

Motivation. The outbreak of COVID-19 in Spain activated several research groups addressed to propose non-pharmacologic measures such as: (a) track the impact of global/local lockdowns and (b) model the impact of lockdowns in the progress of the infection. These are part of the objectives of the *COVID-19 Data Science Task Force*. This task force is formed by 20 interdisciplinar scientists of the main universities of the Comunidad Valenciana (CV). Our mission is to interpret, aggregate and make reports to policy makers. It has four areas: *mobile data analysis* (collect and geolocate anonymized cell phone data), *epidemiological models* (formulation and fitting of metapopulation models such as SIR or SEIR and agents-based ones, state of hospitals and ICUs), *predictive models* (hotspot detection, risk-priority maps, etc.) and *citizen's science* (covid impact survey).

BBVA Foundation, Banco de Santander and Spanish Government.

A. Torsello et al. (Eds.): S+SSPR 2020, LNCS 12644, pp. 195–204, 2021.
https://doi.org/10.1007/978-3-030-73973-7_19

Outline of the Paper. The purpose of this paper is to show the link between the R_0 number of the SARS-CoV-2 and the complexity of the mobility graph. The R_0 (basic reproduction number) quantifies how many infectious cases are generated by a single one. Therefore for $R_0 > 1$ we have a spreading infection. The larger it is the more difficult is to control the infection. According to the Imperial College's Report on March 30, 2020[1], $R_0 \approx 5$ in Spain by March 9, 2020, when social distancing was implemented. This number was reduced to $R_0 \approx 1$ after the complete lockdown.

Global or partial lockdowns address the point of stopping the spread of the virus by reducing the mobility of people. This is why in the COVID-19 Data Science Task Force we started to work with anonymized mobility data. One of the objectives was to find a proxy of the spread of the virus by looking at mobility graphs. In principle, we wanted to enrich the prediction power of our stochastic epidemiological model (Sect. 2) where the effects of mobility where shadowed by the big numbers of the metapopulation model.

Then, we turn our attention to look at the topology of the mobility graphs. We followed two complementary strategies. One team studied the degree of fragmentation of the communities as the political measures were implemented. The second team (ours) interpreted the graph in a different way. More precisely, we looked at the stochastic matrices that encode the random walks potentially running on the network. Instead of dealing with a weighted digraph which is difficult to analyze by spectral means, we looked at the powers of the transition matrices. These matrices are the core to several recent embeddings such as *node2vec* [3], *Glove* [5] and *DeepWalk* [8] among others. Their unifying principle is to extract pairs of co-visited nodes and use these statistics (either by deep/shallow learning or SVD factorization) to find vectorial representations of the nodes. With these vectors at hand, one can use a vectorial complexity measure to find a correlation between R_0 and the topology of the graph. In Sect. 3 we show how the factorization of the expected co-occurrence matrix leads to an informative embedding. This information is given by the rank of the co-occurrence matrix and this rank has deep implications in the complexity of several models of graphs. The *key idea* here is to relate the rank with the degrees of freedom of the topology. Summarizing, disconnected mobility graphs lead to low rank (i.e. redundant embeddings) since the random walks running on them are too constrained (they perform a few distinctive hitting patterns). However, more complex graphs are endowed with high-rank embeddings. Herein, the use of vectorial entropy estimator is a computational trick to bypass robust rank estimation.

In Sect. 4, we show our experiments with the SEIR model and the link between R_0 and the square of the vectorial entropy. In Sect. 5, we summarize our conclusions.

---

[1] https://spiral.imperial.ac.uk:8443/handle/10044/1/77731.

2 Mobility in Epidemiological Models

We use cell-phone geolocation data[2] to track the spread of the SARS-CoV-2 within the Comunidad Valenciana (CV) in Spain. We build mobility networks to map 6.8 million people of 324 census block groups (CBGs) between March 15 and May 23, 2020. Each CBG has at least 5,000 people.

Stochastic SEIR. We overlay a metapopulation SEIR model in order to track the infection trajectories, predict the R_0 number and monitor the epidemiological status of the 24 Health Departments of the CV. Each CBG maintains four sub-population: susceptible (S), exposed (E), infectious (I), and removed (R). The differential equations governing these sub-populations are:

$$S_t = \frac{X_t}{n}$$
$$E_t = S_{t-1} - \frac{X_t}{n} + \frac{Y_t}{n}$$
$$I_t = E_{t-1} - \frac{X_t}{n} + \frac{Z_t}{n}$$
$$R_t = R_{t-1} + I_{t-1} - \frac{Z_t}{n} \tag{1}$$

where X, Y and Z are binomial distributions:

$$X_t \sim \mathbb{B}(nS_{t-1}, e^{-\beta I_{t-1}}),\ Y_t \sim \mathbb{B}(nE_{t-1}, e^{-\sigma}),\ Z_t \sim \mathbb{B}(nI_{t-1}, e^{-\gamma}) . \tag{2}$$

and: n is the population, $\sigma = 1/5.1$, $\gamma = 1/12$ and $\beta = R_0\gamma$. The most important parameter is R_0, the reproduction rate (or reproduction number), which indicates the expected number of infectious cases generated by one case. In order to incorporate mobility to the model, we have to consider that the population is divided into N CBGs. As a result we have the conservation rule: $S_t + E_t + I_t + R_t = S_0 + E_0 + I_0 + R_0 = 1$ for all t and

$$S_t = \sum_{i=1}^{N} S_t^{(i)},\ E_t = \sum_{i=1}^{N} E_t^{(i)},\ I_t = \sum_{i=1}^{N} I_t^{(i)},\ R_t = \sum_{i=1}^{N} R_t^{(i)} . \tag{3}$$

The above decomposition applies also for the binomial distributions and we have:

$$X_t = \sum_{i=1}^{N} X_t^{(i \to j)},\ Y_t = \sum_{i=1}^{N} Y_t^{(i \to j)},\ Z_t = \sum_{i=1}^{N} X_t^{(i \to j)}, \tag{4}$$

where the superscript $(i \to j)$ denotes how many movers in the corresponding state for the i-th CBG move to the i-th CBG.

Using this model we predict the infectious cases for the whole CV (see Fig. 1).

---

[2] Provided the INE (National Institute of Statistics) due to an agreement between the Spanish Government and the main phone operators. These data are anonymized and register displacements between INE-CBGs.

Mobility Graphs and Radiation Model. The daily movers between CBGs create a *mobility graph* $G_t = (V, \mathcal{E}_t, \mathcal{W}_t)$ where $|V| = N$, and an edge $\epsilon_{ij} \in \mathcal{E}_t$ exists when $\mathcal{W}_t(i,j) = \frac{M_t(i,j)}{\sum_k M_t(i,k)} > 0$, where $M_t(i,j)$ are the movers from node i to node j at time t.

However, as we have only 324 CGBs we must introduce as much information as possible in order to model mobility fluxes properly. Therefore we use the so called *radiation model* [11] which takes into account the populations of the commuting CGBs as well as the populations of the CGBs in between. In this model $\mathcal{W}_t(i,j)$ is multiplied by

$$T_{ij} = \frac{N_i N_j}{(N_i + S_{ij})(N_i + N_j + S_{ij})} \tag{5}$$

where: N_i and N_j are the populations of CGBs i and j, and S_{ij} is the number of people in a circle centered at i with radius r_{ij}. This model is parameter-free (wrt to others such as the gravitational one) and it predicts better the probability of observing a flux given the known distribution of the populations (origin, destination, in-between).

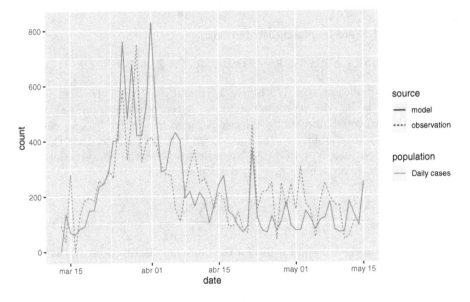

Fig. 1. SEIR model: observed vs predicted cases

3   Entropy of Mobility Graphs

3.1   Embeddings and Random Walks

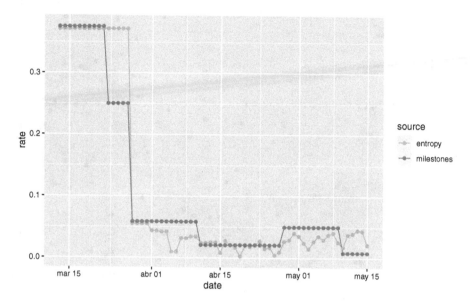

Fig. 2. Correlation between R_0 and graph entropy.

Expected Random Walks. Following the standard factorization approach for network embedding [6], the latent representations for nodes of $G = (V, E)$ are obtained from the SVD of $\hat{\mathbf{M}}_G = \log(\max(\mathbf{M}_G, 1))$, where $\mathbf{M}_G$ is the Pointwise Mutual Information (PMI) matrix. More recently, Qiu et al. [9] show that $\mathbf{M}_G$ can be posed in the following terms

$$\mathbf{M}_G = \frac{vol(G)}{b}\mathbf{S}_G, \ \mathbf{S}_G = \left(\frac{1}{T}\sum_{r=1}^{T}\mathbf{P}_G^r\right)\mathbf{D}_G^{-1}, \tag{6}$$

where $\mathbf{P}_G = \mathbf{D}_G^{-1}\mathbf{W}_G$ is the transition matrix of G. This emphasizes the role of the random walks (RWs) in the resulting embedding. For instance, following [2] $\mathbf{S}_G$ can be seen as the expectation of the *co-ocurrence matrix* $\mathbf{O}_G \in \mathbb{R}^{N \times N}$ where the $\mathbf{O}_{G_{ij}}$ entry contains the number of times nodes i and j are co-visited within a context distance T, i.e. the number of times that a random walk starting an any node hits both i and j at most T steps. The hyperparameter T is called the *window size* and it controls the extent of a nodal context. Thus, for a fixed T Abu-El-Haija et al. define:

$$\mathbb{E}[\mathbf{O}_G; T] = \left(\sum_{r=1}^{T}Pr(c \geq r)\mathbf{P}_G^r\right)\mathbf{P}_G^0, \tag{7}$$

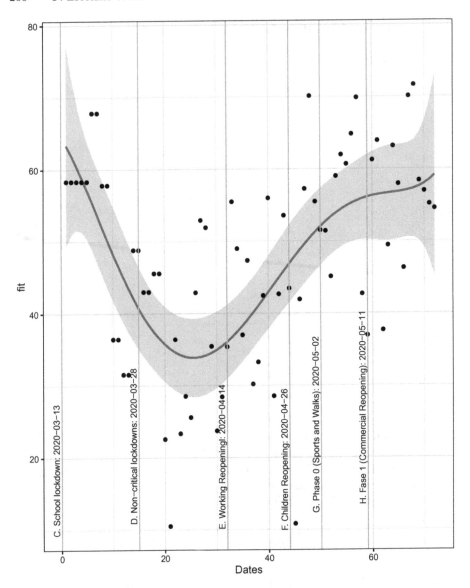

Fig. 3. Fitted (blue line) v real (black dots) graph entropy. (Color figure online)

where $Pr(c \geq r)$ is the probability that node c belonging to the context of any *anchor node* is reached after r steps. Consequently, $\mathbf{S}_G = \mathbb{E}[\mathbf{O}; T]$ results from assuming: (a) $Pr(c \geq r) = \frac{1}{T}$, and (b) $\mathbf{P}_G^0 = \mathbf{D}_G^{-1}$. These simplifying assumptions lead respectively to: (a) Nodes within a context are chosen uniformly and independently of how deep are the random walks, and (b) The probability that a random walk starts at a given node i are inversely proportional to its degree

d_i. Then, looking at $\mathbf{S}_G$ we can interpret $\mathbb{E}[\mathbf{O}_G; T]$ in light of the powers $\mathbf{P}_G^r$ of the transition matrix. More precisely, since the embedding relies on the SVD of $\mathbf{S}_G$, we herein *propose to relate the entropy of G with the rank of* $\mathbf{S}_G$.

Rank and Complexity. We commence by exploring random graphs $G(N, \rho, \mathbf{Q})$ with a community structure. The rank of $\mathbb{E}(\mathbf{P}_G)$ relies on that of $\mathbb{E}(\mathbf{A}_G)$ which in turn is upper bounded by the rank of the $K \times K$ *communicability matrix* $\mathbf{Q}$, where K is the number of communities. The parameter ρ is a discrete probability distribution, i.e. $\sum_{k=1}^{K} \rho_k = 1$, at it induces a node generator τ, where a given node belongs to the $k-$th community with probability ρ_k. In addition, there is an edge between two nodes i and j with probability $\mathbf{Q}_{\tau(i),\tau(j)}$. This is the so called *Stochastic Block Model* (SBM) [1] and its used for community detection. We consider two cases:

(a) If $\mathbf{Q}$ is *symmetric* we have rank$(\mathbb{E}(\mathbf{A}_G)) \leq K$ because the adjacency matrices generated via SBMs have a block structure. One effective way of increasing the latter upper bound is to minimize the entropy of ρ. This leads to (almost) full-connected (complete) dis-assortative graphs with rank$(\mathbb{E}(\mathbf{A}_G)) \approx N$. This includes *ego-nets*, that is networks that code social circles with a large overlap between them such as the Facebook net [7].

(b) If $\mathbf{Q}$ is *not symmetric* it is the degree of asymmetry what determines whether rank$(\mathbb{E}(\mathbf{A}_G)) \approx N$ (the larger the better) or not, independently of the entropy of ρ. This includes *citation networks* such as Cora [10][3], CiteSeer for Document Classification [10][4], and Wiki[5].

Summarizing, dense strictly directed graphs achieve the largest ranks for $\mathbf{A}_G$ and consequently for $\mathbf{P}_G$. This is the case of the mobility networks studied in this paper. This is key, because usually rank$(\mathbf{S}_G) \leq$ rank$(\mathbf{P}_G)$ (matrix powers and matrix addition do not preserve, in general, the rank).

The above facts lead us to interpret rank$(\mathbf{S}_G) \equiv$ rank$(\mathbb{E}[\mathbf{O}_G; T])$ as a proxy of the *complexity of G*:

(a) *Low complexity.* The rank determines the number of independent subspaces of the expected co-ocurrence matrix. Thus, rank$(\mathbb{E}[\mathbf{O}_G; T]) = p$ with $p \ll N$ indicates that the hitting patterns of the RWs are highly redundant, i.e. they collapse in a small number of p clusters, jointly visiting the same nodes in each cluster. Such a redundancy reveals that the random walks are constrained to hit the same subset of nodes independently of how far are them from their anchors. As a result, low rank means also low transport efficiency (and also low graph complexity) without relying on the inverse lengths of the shortest paths.

---

[3] Citation network containing 2708 scientific publications with 5278 links between them. Each publication is classified into one of 7 categories.

[4] Citation network containing 3312 scientific publications with 4676 links between them. Each publication is classified into one of 6 categories.

[5] Contains a network of 2405 web pages with 17981 links between them. Each page is classified into one of 19 categories. https://github.com/thunlp/MMDW/.

(b) *High complexity.* When p is large (ideally $p \approx N$) the co-occurrence patterns are linearly independent because the RWs are less constrained. The absence of bottlenecks favors transport efficiency due to the higher complexity of the graph.

Rank and Entropy. Since rank estimation can be shadowed by numerical errors [12], p typically over-estimates the number of real co-ocurrence clusters. We herein address this problem by estimating the entropy of the embedding. Therefore,

(1) *Embedding.* Let $\mathbf{E}_G = \mathbf{U}_d\sqrt{\Sigma_d}$ the embedding matrix given by the rank$-d$ approximation of $\log(\max(\mathbf{M}_G, 1)) \approx \mathbf{U}_d \Sigma_d \mathbf{V}_d^T$ where $\mathbf{M}_G$ relies on $\mathbf{S}_G$ (see Eq. 6) i.e. on $\mathbb{E}[\mathbf{O}_G; T]$. Then, the i−th row e_i of $\mathbf{E}_G$ is the d−dimensional embedding of the $i - th$ node of G.

(2) *Bypass Entropy.* Given N d−dimensional points, their α−Rényi entropy H_α, with $\alpha > 1$ is consistently estimated by the following functional [4]:

$$\hat{H}_\alpha = \frac{1}{1-\alpha} \log \frac{L_p(F)}{N^{1-p/d}} , \tag{8}$$

where: $p = d(1 - \alpha)$, $F = (\mathcal{V}, \mathcal{E})$ is a k−nn graph whose vertices are the $\mathbf{x}_i = \mathbf{e}_i$ the embedding vectors, the edges $\mathcal{E}_{ij}$ are provided by the k-nn rule. Therefore we have

$$L_p(F) = \sum_{\mathcal{E}_{ij}} ||\mathbf{x}_i - \mathbf{x}_j||^2 \tag{9}$$

and γ is a normalization constant that can be estimated by generating a large sample of points in $[0, 1]^d$ and running the estimator in its k−nn graph. Thus, given the embedding, the bypass entropy returns its entropy. It is desirable to choose $\alpha \approx 1$ (close to the Shannon entropy). In this work we set: $\alpha = 0.99$, $p = 2$ and $k = 4$. The embedding vectors are also normalized before computing the entropy.

4   Experiments

Setting: Mobility Flows. Following the INE protocol[6], for each CBG, a cell phone operator provides the number of terminals that are going to be considered as *living population*: owners of cell phone who spend most of the time at that CBG from 00:00–6:00 am. This is the source CGB, whereas the destination CGB is the most visited CGB from 10:00 am–16:00 pm if the owner is there at least for 2 h. The operators (Telefónica, Orange, Vodafone) report the *number of movements* to the destination CGB if there are at least 10–15 movements (depending on the operator).

---

[6] National Institute of Statistics: https://www.ine.es/covid/covid_movilidad.htm (Technical Project).

Predicting R_0. Our stochastic SEIR model fits well the infection cases (see Fig. 1). However, in absence of any other information the R_0 parameter must be adjusted daily and specially after imposing a non-pharmaceutical measure (social distancing, lockdown). In practice, R_0 can be seen as control parameter that encodes all the measures implemented to slow down the propagation of the virus. However, as most of these measures are somewhat related to mobility we conjectured that there should be a mathematical relationship between the entropy of the embedding (which reflects the degrees-of-freedom of mobility) and R_0. After experimentation, we found that

$$R_0 \propto (H_\alpha)^2 . \tag{10}$$

as we show in Fig. 2, where we plot R_0 and the above estimator at several milestones during the lockdown. We give more details of the milestones and the curve fitting of the entropy in Fig. 3. The quadratic relation $R_0 \propto (H_\alpha)^2$ seems to rely on the L_2 norm used to estimate H_α.

With this mathematical tool at hand we could monitor not only the global evolution of the CV but also the evolution of each of its 24 Health Departments during the first peak of the pandemic.

5   Conclusions

In this paper, we have proposed and successfully tested a proxy of the R_0 number via the complexity of the mobility graph. Such a complexity measure is closely related to the rank of state-of-the-art matrices which encode co-visiting statistics. The key idea is to relate the complexity of a graph with the degrees-of-freedom of the random walks running on it. It these random walks are constrained then the graph is simple (e.g. fragmented as it the COVID-19 mobility graph after political interventions); otherwise the graph is complex.

Future work includes the validation of this model in larger graphs as well as exploring the links between the proposed complexity and other alternatives. The underlying idea is to make this proxy much closer to an early warning system. In addition, the impact of entropy is being tested in a data-driven (Deep Network) wrapping a SIR, feeded with data from 236 regions worldwide and applicable to Comunitat Valenciana.

Acknowledgements. The *COVID-19 Data Science Task Force* is founded by two Spanish Banks: BBVA and Santander, which have opened competitive calls. BBVA funds the project: AYUDAS FUNDACIÓN BBVA A EQUIPOS DE INVESTIGACIÓN CIENTÍFICA SARS-CoV-2 y COVID-19. Santander funds the FONDOS SUPERA. In addition, M.A. Lozano and Francisco Escolano are funded by the project SPRIT (INFORMATION THEORY FOR STRUCTURAL PATTERN RECOGNITION (code RTI2018-096223-B-I00) of the Spanish Government. The COVID-19 Task Force is also grateful to the INE (National Institute of Statistics) who provides the data and the Generalitat Valenciana for its support.

References

1. Abbe, E.: Community detection and stochastic block models: recent developments. J. Mach. Learn. Res. **18**(1), 6446–6531 (2017)
2. Abu-El-Haija, S., Perozzi, B., Al-Rfou, R., Alemi, A.A.: Watch your step: learning node embeddings via graph attention. In: Bengio, S., Wallach, H., Larochelle, H., Grauman, K., Cesa-Bianchi, N., Garnett, R. (eds.) Advances in Neural Information Processing Systems, vol. 31, pp. 9180–9190. Curran Associates, Inc. (2018). https://proceedings.neurips.cc/paper/2018/file/8a94ecfa54dcb88a2fa993bfa6388f9e-Paper.pdf
3. A. Grover, Leskovec, J.: node2vec: scalable feature learning for networks. In: Proceedings of the 22nd ACM SIGKDD International Conference on Knowledge Discovery and Data Mining, pp. 855–864 (2016)
4. Escolano, F., Suau, P., Bonev, B.: Information Theory in Computer Vision and Pattern Recognition, 1st edn. Springer Publishing Company, Inc., Verlag London (2009). https://doi.org/10.1007/978-1-84882-297-9
5. Pennington, J., Socher, R., Manning, C.D.: GloVe: global vectors for word representation. In: Proceedings of the 2014 Conference on Empirical Methods in Natural Language Processing, EMNLP, pp. 1532–1543 (2014)
6. Levy, O., Goldberg, Y.: Neural word embedding as implicit matrix factorization. Adv. Neural Inf. Process. Syst. **27**, 2177–2185 (2014)
7. McAuley, J., Leskovec, J.: Learning to discover social circles in ego networks. In: Bartlett, P.L., Pereira, F., Burges, C.C., Bottou, L., Weinberger, K. (eds.) Advances in Neural Information Processing Systems 25: 26th Annual Conference on Neural Information Processing Systems 2012. Proceedings of a meeting held 3–6 December 2012, Lake Tahoe, Nevada, United States, pp. 548–556 (2012). http://papers.nips.cc/paper/4532-learning-to-discover-social-circles-in-ego-networks
8. Perozzi, B., Al-Rfou, R., Skiena, S.: Deepwalk: online learning of social representations. In: The 20th ACM SIGKDD International Conference on Knowledge Discovery and Data Mining, KDD 2014, New York, NY, USA - August 24–27, 2014, pp. 701–710 (2014). https://doi.org/10.1145/2623330.2623732
9. Qiu, J., Dong, Y., Ma, H., Li, J., Wang, K., Tang, J.: Network embedding as matrix factorization: Unifying deepwalk, LINE, PTE, and node2vec. In: Proceedings of the Eleventh ACM International Conference on Web Search and Data Mining, WSDM 2018, pp. 459–467. ACM, New York, NY, USA (2018). http://doi.acm.org/10.1145/3159652.3159706
10. Sen, P., Namata, G., Bilgic, M., Getoor, L., Gallagher, B., Eliassi-Rad, T.: Collective classification in network data. AI Mag. **29**(3), 93–106 (2008). http://www.aaai.org/ojs/index.php/aimagazine/article/view/2157
11. Simini, F., González, M.C., Maritan, A., Barabási, A.L.: A universal model for mobility and migration patterns. Nature **484**(7392), 96–100 (2012). https://doi.org/10.1038/nature10856
12. Ubaru, S., Saad, Y.: Fast methods for estimating the numerical rank of large matrices. In: Proceedings of the 33rd International Conference on International Conference on Machine Learning, vol. 48, pp. 468–477. ICML 2016 (2016).http://www.jmlr.org

A Novel Data Set for Information Retrieval on the Basis of Subgraph Matching

Kaspar Riesen[1,2]([✉]) [iD], Hans-Friedrich Witschel[2] [iD], and Loris Grether[2] [iD]

[1] Institute of Computer Science, University of Bern, Neubrückstrasse 10,
3012 Bern, Switzerland
`riesen@inf.unibe.ch`
[2] Institute for Informations Systems, University of Applied Sciences Northwestern
Switzerland, Riggenbachstrasse 16, 4600 Olten, Switzerland
{`hansfriedrich.witschel,loris.grether`}`@fhnw.ch`

Abstract. We are facing the challenge of rapidly increasing amounts of data. Moreover, we observe that in many applications the underlying data contains strongly related entities making graphs the most appropriate structure for data modeling. When data is represented by means of a graph, querying corresponds to a graph matching problem. The present paper introduces a novel graph that models information from the medical domain with about 110,000 nodes and 220,000 edges. Additionally we present several basic benchmark queries, i.e. specific subgraphs, from different categories that can be found multiple times in the medical graph. Both the graph and the benchmark can be used to implement, test, and compare novel graph matching algorithms in a real world scenario.

Keywords: Subgraph isomorphism · Graph matching · Graph database

1 Introduction and Related Work

Many of the information repositories available are diverse, large, and often contain strongly related entities. To cope with numerous and arbitrary relations more efficiently, graph based databases are more and more recognized as versatile alternative to relational databases [1]. In fact, in contrast with tabular structures that use foreign keys for relationship modeling, graphs are able to represent not only the values of entities, but can be used to explicitly model structural relations that might exist between different objects by means of edges [2,3]. Moreover, the user's mental model of the data and the actual data structure stored on a device are fully congruent. Hence, visualizations of graphs typically provide an intuitive and clearly understandable overview of the underlying structures and relationships.

Supported by Innosuisse Project Nr. 26281.2 PFES-ES.

A. Torsello et al. (Eds.): S+SSPR 2020, LNCS 12644, pp. 205–215, 2021.
https://doi.org/10.1007/978-3-030-73973-7_20

When a graph is employed for the purpose of data storage, the information retrieved in response to a certain query is typically also a graph, which may be, for instance, a subgraph of the underlying database graph [3]. In the present paper, we employ the concept of *subgraph isomorphism* [4] for information retrieval. Subgraph isomorphism indicates that a smaller graph is contained in a larger graph. Let us assume that we represent a query by means of an attributed graph q, termed query graph. Given q and the database graph G, we can check whether the query graph q is contained in the underlying database G.

About a decade ago one of the authors of the present paper introduced a generalized form of graph isomorphism that is particularly well suited for information retrieval from graphs [5]. This generalized subgraph isomorphism methodology allows to mask out attributes in query graphs that are irrelevant for a particular question. Moreover, the algorithm also allows the definition of certain variables in order to retrieve values of predefined attributes as well as the definition of constraints (for example variables that can assume only certain values). In the meantime these basic concepts for subgraph matching in large graphs have been implemented in many commercial software products (e.g. Neo4j or Amazon Neptune, to name just two prominent examples).

Despite the fact that graph based databases have reached a mature level, graph matching [6], information retrieval on graph-like structures [7], or human interaction with graph models [8] are still active fields of research. The major contribution of the present paper in this particular field is twofold. First, we present a novel large graph that models heterogeneous information from the medical domain by means of about 110,000 nodes and 220,000 edges. Second, we present 21 benchmark queries, i.e. specific subgraphs, from seven different categories that can be found multiple times in the large graph. These categories represent important application scenarios and therefore we need to know how efficiently they can be answered. The present paper is similar in spirit to [9–12] where graphs and benchmark tasks for (sub)graph isomorphism or error-tolerant graph matching are presented.

The remainder of the present paper is organized as follows. In Sect. 2 the basic definitions are introduced. Next, in Sect. 3 the novel medical graph is introduced and thoroughly described. Eventually, in Sect. 4, we define the benchmark tasks in the form of subgraphs that can be found in the medical graph together with the respective matching results and run times. Finally, in Sect. 5, we conclude the paper and discuss some future work ideas.

2   Preliminary Definitions

We employ the *property-graph-model* in our approach. Formally, a graph is a 4-tuple $g = (V, E, \mu, \nu)$, where

- V is the finite set of nodes
- $E \subseteq V \times V$ is the set of edges
- $\mu : V \rightarrow \{(t, \mathbf{x}(t)) | t \in T_{nodes}, \mathbf{x}(t) \in (D_1(t) \times \ldots \times D_{n_t}(t))\}$ is the node property function
- $\nu : V \times V \rightarrow \mathcal{P}(\{t | t \in T_{edges}\}) \setminus \emptyset$ is the edge type function.

Through the node property function μ, each node $u \in V$ in a graph is labeled by a (type, property)-pair $(t, \mathbf{x}(t))$. The first component, the type t, is an element of a finite set of node types T_{nodes}. The node types group nodes together and specify the roles they play within the graph. For example, some nodes could represent objects of type Disease, while others represent nodes of type Symptom or Treatment.

In our scenario, nodes also contain *properties* modeled by means of the second component, i.e. $\mathbf{x}(t) = (x_1, \ldots, x_{n_t})$. In this property vector each attribute x_i belongs to some domain $D_i(t)$. The dimension of vector $\mathbf{x}(t)$, i.e. the number n_t of attributes, as well as each individual attribute domain $D_i(t)$ depends on the actual type t of the node [5]. Possible properties for nodes of type Disease would be, for instance, an ID, the name of the disease, or others.

Formally, edges are pairs of nodes, $(u, v) \in V \times V$ and structure the graph. In our scenario, edges are always directed and always connect exactly one start- with one end-node. In some applications, however, it might be necessary to include more than one edge between the same pair of nodes, because of the existence of multiple relations. In the formal graph model provided above, this can be accomplished by assigning several edge types to the same edge (u, v) by means of the edge type function ν, i.e. $\nu(u, v) = \{t_1, \ldots, t_n\}$. Note that the range of function ν is the power set of all edge types from the finite set T_{edges}. Assigning n types $\{t_1, \ldots, t_n\}$ to edge (u, v) by means of ν is equivalent to providing n individual edges from node u to node v [5]. In our scenario the edges do not contain any further properties[1].

A possible type of an edge between nodes of type Disease and Symptom might be, for instance, causes. The adjacent nodes, the edge's direction and the label of the edge provide semantic clarity to the relationship.

3   Medical Graph

In the present paper we use the data structure $g = (V, E, \mu, \nu)$ defined in the previous section to model diverse information from the medical domain. To this end, we automatically parse data from the following four public domains:

1. Wikidata: wikidata.org
2. SemMED: skr3.nlm.nih.gov/SemMed/
3. Medline: medlineplus.gov
4. DisGeNET: disgenet.org

From Wikidata we parse entities of five different types, viz. diseases, symptoms, treatments, behaviours (such as tobacco addiction or similar), and diagnostic tests. Next, we complement the set of diseases with entities extracted from SemMED. From the same domain as well as from Medline we extract further diagnostic tests (we select the most frequent diagnostic tests by means of a simple heuristic). Also from SemMED we parse research papers and two patient

---

[1] This can be generalized in a straightforward manner.

characteristics (gender and age group) that might be typical for certain diseases. The number of research papers is limited to eight papers per disease and we require the title of the paper to include the name of the respective disease. Finally, From DisGeNET we parse genes and proteins (we select genes with a GeneSymbol-Score greater than, or equal to, 0.3).

All of these entities are modeled by means of nodes of different types and with different sets of properties. In total 110,774 nodes of nine different types are built by means of this procedure. In Table 1 the node types, the number of nodes per type and the properties available in the different node types are summarized.

Table 1. The node types, the number of nodes per type and the properties available in the different node types.

Node type $t \in T_{nodes}$	Count	Properties $\mathbf{x}(t)$
ResearchPaper	60,895	PMID, Abstract
Disease	28,000	CUID, Name, Description, DOID
Protein	9,281	uniprodID
Gene	8,490	Name, GeneID, GeneSymbol Score (level of evidence)
Treatment	1,606	CUID, Name
DiagnosticTest	1,384	CUID, Name
Symptom	588	CUID, Name
Behaviour	489	Name
PatientCharacteristic	41	CUID, Name
Total count	**110,774**	

We use nine different edge types in order to connect the different nodes with each other and build the graph (in total 221,920 edges are inserted). Actually, we build a graph according to the star like scheme with nodes of type `Disease` as most central nodes (see Fig. 1). In Table 2 the edge types, the number of edges per type and the start- and end-node that are connected with the respective edge are summarized. The complete graph is publicly available in a comma separated file at https://github.com/kaspar-riesen/medical-graph.

4   Benchmark Tasks and Results

We divide our benchmark queries on the medical graph into the following seven categories or patterns (see also Fig. 2): *Single, Double, Triple, Triangle, Growing*

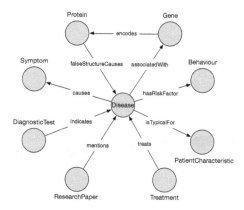

Fig. 1. The nine types of nodes are connected by means of nine different edge types.

Table 2. The edge types, the number of edges per type and the start- and end-node that are connected with the respective edge type.

Edge type $t \in T_{\text{edges}}$	Count	From	To
mentions	62,940	ResearchPaper	Disease
falseStructureCauses	43,088	Protein	Disease
associatedWith	42,610	Disease	Gene
indicates	29,702	DiagnosticTest	Disease
isTypicalFor	26,671	Disease	PatientCharacteristic
encodes	9281	Gene	Protein
treats	5523	Treatment	Disease
causes	1358	Disease	Symptom
hasRiskFactor	747	Disease	Behaviour
Total count	**221,920**		

Star, Top Hub, and *Max Overlap.* Each category represents a different information need ranging from finding information of a single symptom to differentiation between diseases. For each pattern three different specific queries are defined and described in the next paragraphs.

Single (see Fig. 3 (a)). We search for all nodes of ...

Query 1 ... type `DiagnosticTest` where the property `name` contains ...
Query 2 ... type `Disease` where the property `name` contains ...
Query 3 ... any type where any of the available properties contain ...

...the search string `liver`.

Double (see Fig. 3 (b)). We search for all nodes of type...

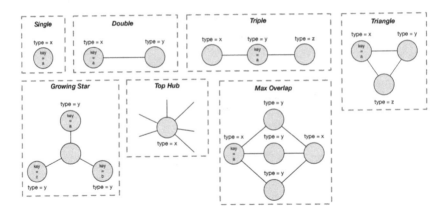

Fig. 2. The seven types of queries.

Query 4 ... Treatment that are connected with an edge treats ...
Query 5 ... DiagnosticTest that are connected with an edge indicates ...
Query 6 ... Gene that are connected with an edge associatedWith ...

...to a node of type Disease with the property name that contains the search string breast cancer.

Triple (see Fig. 3 (c)). We search for all nodes of both types DiagnosticTest and ...

Query 7 ... Symptom that are indirectly connected with two edges indicates and causes via a node of type Disease whose property name contains the search string deficiency.
Query 8 ... Treatment that are indirectly connected with two edges indicates and treats via a node of type Disease whose property name contains the search string periodontitis.
Query 9 ... Behaviour that are indirectly connected with two edges indicates and hasRiskFactor via a node of type Disease whose property name contains the search string arteriosclerosis.

Triangle (see Fig. 4 (a)). We search for all nodes of both types Protein or Gene that are directly connected with an edge encodes with each other and simultaneously connected via edges associatedWith and falseStructureCauses with a node of type Disease whose property name contains the search string ...

Query 10 ... type 1 diabetes.
Query 11 ... hypothermia.
Query 12 ... skin cancer.

Growing Star (see Fig. 4 (b)). We search for all nodes of type Disease that are directly connected via edge causes with at least ...

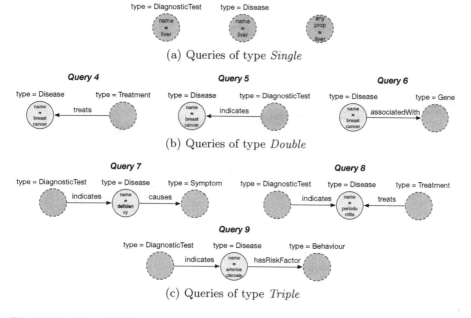

Fig. 3. Queries of different types. Grey colored nodes represent the nodes being searched for.

Query 13 ... two nodes of type Symptom where the property name of the first and second symptom contain the search string fever and fatigue, respectively.

Query 14 ... three nodes of type Symptom where the property name of the first, second, and third symptom contain the search string fever, fatigue, and anorexia, respectively.

Query 15 ... four nodes of type Symptom where the property name of the first, second, third, and fourth symptom contain the search string fever, fatigue, anorexia, and diarrhea, respectively.

Top Hub (see Fig. 4 (c)). We search for the five nodes of type Disease that have the most edges of type ...

Query 16 ... treats (to nodes of type Treatment).
Query 17 ... indicates (to nodes of type DiagnosticTest).
Query 18 ... associatedWith (to nodes of type Gene).

Max Overlap (see Fig. 4 (d)). We search for maximum five nodes of type Disease that share the most ...

Query 19 ... symptoms with a node of type Disease whose property name contains the search string gastroenteritis.

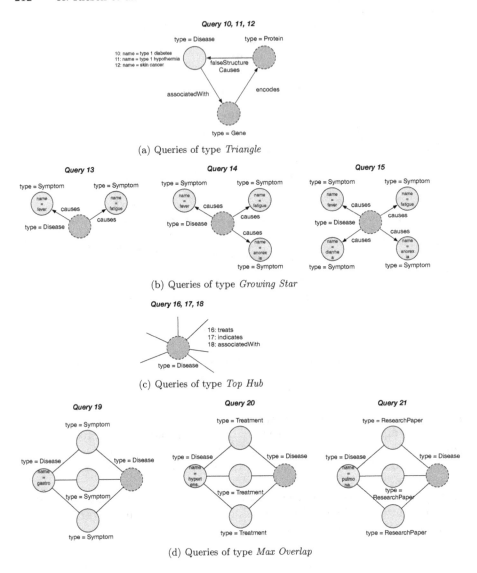

(a) Queries of type *Triangle*

(b) Queries of type *Growing Star*

(c) Queries of type *Top Hub*

(d) Queries of type *Max Overlap*

Fig. 4. Queries of different types. Grey colored nodes represent the nodes being searched for.

Query 20 ... treatments with the node of type `Disease` whose property `name` contains the search string `hypertension`.

Query 21 ... mentions in research papers with the node of type `Disease` whose property `name` contains the search string `pulmonary`.

In Table 3 the size of the result, i.e. the number of nodes that match the given subgraphs, as well as the run time for the actual matching are shown for all queries. We run our experiment on a Intel Core i7 with 16 GB RAM and we have implemented the graph model by means of the graph database implementation `Neo4j`'s `developer edition`[2]. The size of the result sets varies from only one node to more than 1,000 nodes per query. Also for the matching times quite large differences are observable. The ground truth results for these matchings can be found at https://github.com/kaspar-riesen/medical-graph.

Table 3. The result size and matching times of queries 1 to 21.

Query	Size of result	Time [ms]
1	7 nodes	3
2	90 nodes	77
3	1066 nodes	834
4	40 nodes	36
5	30 nodes	82
6	1089 nodes	276
7	19 nodes	7
8	45 nodes	13
9	10 nodes	7
10	6 nodes	155
11	14 nodes	114
12	59 nodes	112
13	18 nodes	12
14	7 nodes	4
15	2 nodes	7
16	5 nodes	84
17	5 nodes	7
18	5 nodes	40
19	5 nodes	3
20	5 nodes	12
21	1 node	41

5 Conclusions and Future Work

Several areas in science and industry are facing the challenge of rapidly increasing amounts of data available, making scalable search methods inevitable. The vast majority of efficient search methods are built for traditional, i.e. tabular data

---

[2] One can define indexes on properties in `Neo4j` – however, we have omitted this possibility in our evaluation.

representations. Yet, we observe that in many modern applications the underlying data is inherently complex, making this limited representation formalism rather inappropriate. Graphs actually allow us to explicitly model relationships between entities. When data is represented by means of a graph, a search for information, or a pattern, exactly corresponds to a graph matching problem. The present paper introduces a novel graph that models information from the medical domain with about 110,000 nodes and 220,000 edges. Additionally we present several basic benchmark queries, i.e. specific subgraphs, from seven different categories that can be found multiple times in the medical graph.

We see several rewarding avenues to be pursued in future work. First, we invite the research community to test their own algorithms for subgraph isomorphism on the publicly available graph. Second, we see great potential to define more complex and more time-consuming benchmark queries. Last but not least, the medical graph could be substantially increased in the number of nodes and edges by accessing and integrating further repositories.

References

1. Robinson, I., Webber, J., Eifrem, E.: Graph Databases. O'Reilly, Springfield (2015)
2. Kandel, A., Bunke, H., Last, M. (eds.): Applied Graph Theory in Computer Vision and Pattern Recognition. Studies in Computational Intelligence, vol. 52. Springer, Heidelberg (2007). https://doi.org/10.1007/978-3-540-68020-8
3. Cook, D., Holder, L.: Mining Graph Data. Wiley-Interscience, Hoboken (2007)
4. Ullmann, J.R.: An algorithm for subgraph isomorphism. J. ACM **23**(1), 31–42 (1976)
5. Brügger, A., Bunke, H., Dickinson, P., Riesen, K.: Generalized graph matching for data mining and information retrieval. In: Perner, P. (ed.) ICDM 2008. LNCS (LNAI), vol. 5077, pp. 298–312. Springer, Heidelberg (2008). https://doi.org/10.1007/978-3-540-70720-2_23
6. Foggia, P., Percannella, G., Vento, M.: Graph matching and learning in pattern recognition in the last 10 years. Int. J. Pattern Recognit. Artif. Intell. **28**(1) (2014)
7. Park, C.-S., Lim, S.: Efficient processing of keyword queries over graph databases for finding effective answers. Inf. Proces. Manag. **51**(1), 42–57 (2015)
8. Witschel, H.F., Riesen, K., Grether, L.: KvGR: a graph-based interface for explorative sequential question answering on heterogeneous information sources. In: Jose, J.M., et al. (eds.) ECIR 2020. LNCS, vol. 12035, pp. 760–773. Springer, Cham (2020). https://doi.org/10.1007/978-3-030-45439-5_50
9. Foggia, P., Sansone, C., Vento, M.: A database of graphs for isomorphism and subgraph isomorphism benchmarking. In: Proceedings of the 3rd International Workshop on Graph Based Representations in Pattern Recognition, pp. 176–187 (2001)
10. Riesen, K., Bunke, H.: IAM graph database repository for graph based pattern recognition and machine learning. In: da Vitoria Lobo, N., et al. (eds.) SSPR /SPR 2008. LNCS, vol. 5342, pp. 287–297. Springer, Heidelberg (2008). https://doi.org/10.1007/978-3-540-89689-0_33

11. Neuen, D., Schweitzer, P.: Benchmark graphs for practical graph isomorphism. CoRR, abs/1705.03686 (2017)
12. Solnon, C., Damiand, G., de la Higuera, C., Janodet, J.-C.: On the complexity of submap isomorphism and maximum common submap problems. Pattern Recogn. **48**(2), 302–316 (2015)

A Graph Pre-image Method Based on Graph Edit Distances

Linlin Jia[1][(✉)], Benoit Gaüzère[1], and Paul Honeine[2]

[1] LITIS, INSA Rouen Normandie, Rouen, France
`linlin.jia@insa-rouen.fr`
[2] LITIS, Université de Rouen Normandie, Rouen, France

Abstract. The pre-image problem for graphs is increasingly attracting attention owing to many promising applications. However, it is a challenging problem due to the complexity of graph structure. In this paper, we propose a novel method to construct graph pre-images as median graphs, by aligning graph edit distances (GEDs) in the graph space with distances in the graph kernel space. The two metrics are aligned by optimizing the edit costs of GEDs according to the distances between the graphs within the space associated with a particular graph kernel. Then, the graph pre-image can be estimated using a median graph method founded on the GED. In particular, a recently introduced method to compute generalized median graphs with iterative alternate minimizations is revisited for this purpose. Conducted experiments show very promising results while opening the computation of graph pre-image to any graph kernel and to graphs with non-symbolic attributes.

Keywords: Pre-image problem · Machine learning · Graph kernels · Graph edit distance.

1 Introduction

Graph structures have been increasingly attracting attention in pattern recognition and machine learning. While they are able to represent a wide range of data, from molecules to social networks, most machine learning methods operate on Euclidean data. Graph kernels allow bridging the gap between the graph structure and machine learning thanks to the kernel trick. This trick consists in implicitly embedding graphs into a Hilbert space, where kernel methods such as Support Vector Machines can be easily operated. The reverse process of the implicit embedding with kernels, namely the so-called pre-image, continues to intrigue researchers. It corresponds to the mapping of elements from the kernel space back to the input space. Many applications require computing the pre-image, such as denoising or feature extraction with kernel principal component

---

This research was supported by CSC (China Scholarship Council) and the French national research agency (ANR) under the grant APi (ANR-18-CE23-0014).

A. Torsello et al. (Eds.): S+SSPR 2020, LNCS 12644, pp. 216–226, 2021.
https://doi.org/10.1007/978-3-030-73973-7_21

analysis [14]. The challenge of finding the pre-image lies in the fact that the reverse mapping does not exist in general and that most elements in the kernel space do not own valid pre-images in the input space. Consequently, various methods have been developed to approximate the solution, namely, to solve the pre-image problem. We refer interested readers to the tutorial [15].

Solving the pre-image problem for graphs opens the door to many interesting applications, such as molecule synthesis and drug design. However, finding the pre-image as a graph inherits the difficulties of traditional pre-image problems. Additionally, unlike inputs considered by traditional pre-image problems, i.e. vectors which are usually lying in continuous spaces, graphs are discrete structures with a variable and non-ordered number of vertices and edges. Furthermore, multiple labels and attributes can be plugged into each vertex and edge in a graph. Given these structure features, the graph pre-image problem is more challenging to address.

Several pioneering works to construct graphs have been proposed. A method based on the random search is proposed in [3]. It is simple to implement, but has a very high computational complexity and is not applicable to continuous real-valued labels, while the quality of the synthesized graph pre-images is not guaranteed. The methods of [2] and [19] infer a graph from path frequency. However, these methods are either restricted to applying a specific sub-structure of graphs or ignoring vertex and edge labels, which are important information for graphs. All these studies do not fully benefit from discrete optimization that needs to be carried out for graph pre-image. In this paper, we propose a novel pre-image method for graphs. To this end, we bridge the gap between graph edit distances (GEDs) and any given graph kernel, which allows uncovering the relationship between graph space and kernel space. GED is a well-known dissimilarity measure between graphs, based on elementary operations that transform one graph to another. By optimizing the edit costs of these operations according to distances between elements in the kernel space, the metrics of the two aforementioned spaces are aligned, thus allowing constructing the graph pre-image by a median graph method based on GEDs. Specifically, a pre-image problem for the median graph of a graph set is addressed, based on the hypothesis that, benefiting from the alignment of the two metrics, the median of the set of graphs corresponds to the mean of their embeddings in the kernel space. We take advantage of recent advances in GED to solve this problem, where an iterative alternate minimization procedure to generate median graphs is adapted [7].

The remainder of the paper is organized as follows. The next section introduces preliminaries for the paper. Section 3 presents the proposed method in two folds, learning edit costs for GEDs by the distances in kernel space (Sect. 3.1) and inferring the graph pre-image (Sect. 3.2). Section 4 gives experiments and analyses. Finally, Sect. 5 concludes the paper.

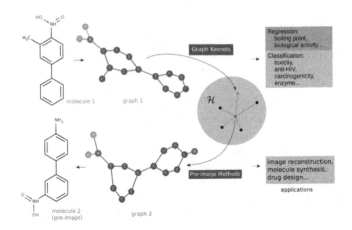

Fig. 1. A graph kernel maps graphs to a kernel space $\mathcal{H}$, while the pre-image provides the reverse procedure, by mapping elements from kernel space back to graphs.

2 Preliminaries

2.1 Graphs, Graph Kernels, and Graph Pre-images

A graph $G = (V, E)$ is an ordered pair of disjoint sets, where V is the vertex set and $E \in V \times V$ is the edge set. A graph can have a label set L from a label space and a labeling function ℓ that assigns a label $l \in L$ to each vertex and/or edge, where l can be symbolic (i.e. discrete values) or non-symbolic (i.e. continuous values). Let φ be the set of vertex labels, Φ the set of edge labels, and n the number of vertices in graph G ($n = |V|$). See [22] for more details.

A positive semi-definite kernel is a symmetric bilinear function that satisfies $\sum_{i=1}^{n} \sum_{j=1}^{n} c_i \, c_j \, k(x_i, x_j) \geq 0$, for all $x_i, \ldots, x_n$ and $c_1, \ldots, c_n \in \mathbb{R}$. These kernels are simply denoted as *kernels* in this paper for conciseness. A kernel corresponds to an inner product between implicit embeddings of input data into an Hilbert space $\mathcal{H}$ (RKHS) thanks to an implicit mapping function $\phi : \mathcal{X} \to \mathcal{H}$.

Graph kernels are kernels defined on graphs. For a given graph kernel, $k(G_i, G_j)$ corresponds to an inner product between the two mapped graphs $\phi(G_i)$ and $\phi(G_j)$ in the kernel space $\mathcal{H}$. More details on graph kernels can be found in [11–13,18]. Given a kernel, the mapping $\phi(\cdot)$ remains implicit and is defined by the kernel itself. It does not have to be explicitly known thanks to the kernel trick. However, the reverse map may be interesting and is difficult to compute in general. Indeed, most combinations $\psi = \sum_i \alpha_i \phi(G_i)$ do not have a valid pre-image, namely a graph $G^\star$ such that $\phi(G^\star) = \psi$; The pre-image problem consists in estimating an approximate solution, namely $\widehat{G}$ such that $\phi(\widehat{G}) \approx \psi$ (Fig. 1).

2.2 Graph Edit Distance

The Graph Edit Distance (GED) between two graphs G_i and G_j is defined as the cost of minimal transformation [21]:

$$d_{GED}(G_i, G_j) = \min_{\pi \in \Pi(G_i, G_j)} C(\pi, G_i, G_j), \tag{1}$$

where $\pi(G_i, G_j)$ is a mapping between $V_i \cup \varepsilon$ and $V_j \cup \varepsilon$ encoding the transformation from G_i to G_j [8]. This transformation consists in a series of six elementary operations: removing or inserting a vertex or an edge, and substituting a label of a vertex or an edge by another label. $C(\pi, G_i, G_j)$ measures the cost associated to π:

$$
\begin{aligned}
C(\pi, G_i, G_j) = &\sum_{\substack{v \in V(G_j) \\ \pi^{-1}(v) \notin V(G_i)}} c_{vfi}(\varepsilon, v) + \sum_{\substack{u \in V(G_i) \\ \pi(u) \notin V(G_j)}} c_{vfr}(u, \varepsilon) + \sum_{\substack{u \in V(G_i) \\ \pi(u) \in V(G_j)}} c_{vfs}(u, \pi(u)) \\
+ &\sum_{\substack{f \in E(G_j) \\ \pi^{-1}(f) \notin E(G_i)}} c_{efi}(\varepsilon, f) + \sum_{\substack{e \in E(G_i) \\ \pi(e) \notin E(G_j)}} c_{efr}(e, \varepsilon) + \sum_{\substack{e \in E(G_i) \\ \pi(e) \in E(G_j)}} c_{efs}(e, \pi(e)),
\end{aligned}
\tag{2}
$$

where $c_{vfr}, c_{vfi}, c_{vfs}, c_{efr}, c_{efi}, c_{efs}$ are the edit cost functions associated to the six edit operations: respectively vertex removal, insertion, substitution and edge removal, insertion and substitution. According to [17], the edit cost functions for graphs with non-symbolic labels can be defined as:

$$
\begin{cases}
c_{vfi}(\varepsilon, v) = c_{vi}, \;\; c_{efi}(\varepsilon, e) = c_{ei}, \;\; c_{vfr}(v, \varepsilon) = c_{vr}, \;\; c_{efr}(e, \varepsilon) = c_{er}, \\
c_{vfs}(u, v) = c_{vs} \| \ell_v(u) - \ell_v(v) \|, \;\; c_{efs}(e, f) = c_{es} \| \ell_e(e) - \ell_e(f) \|,
\end{cases}
\tag{3}
$$

where $c_{vr}, c_{vi}, c_{vs}, c_{er}, c_{ei}, c_{es}$ are the edit costs, namely the coefficients applied to the edit operations. Let $\mathbf{c} = [c_{vr}, c_{vi}, c_{vs}, c_{er}, c_{ei}, c_{es}]^\top$ be the edit cost vector.

By definition, the GED can be regarded as a distance measure between graphs. However, the problem of computing the GED is NP-hard [4]. Many methods have been proposed to approximate GED, such as bipartite [21] and IPFP [8]. For more details on GEDs, we refer interested readers to [4,21].

3 Proposed Graph Pre-image Method

The main motivation of this work is to address the pre-image problem by building connections between graph and kernel spaces. We propose to align the metrics of the two spaces by optimizing the edit costs such that GEDs approximate the distances in kernel space. Then, once GEDs and kernel distances are similar, we propose to recast the pre-image problem as a graph generation problem, based on the assumption that the median of a set of graphs corresponds to the mean of their embeddings in the kernel space. An iterative alternate minimization method is adapted for this purpose, in which the GEDs with the optimized edit cost distances are used. These two steps are detailed next, and the proposed method is summarized in Algorithm 1.

Algorithm 1. Proposed method

Input: Dataset $\mathbb{G}_N$, graph kernel k, thresholds of stopping criteria (r_{max}, i_{max}).
Output: The approximation of the pre-image.
1: Compute $d_{\mathcal{H}}$ as in (7) for $\mathbb{G}_N$.
2: Initialize randomly $\boldsymbol{c}^{(0)} = [c_{vr}^{(0)}, c_{vi}^{(0)}, c_{vs}^{(0)}, c_{er}^{(0)}, c_{ei}^{(0)}, c_{es}^{(0)}]^\top$.
3: Compute kernel distances $d_{\mathcal{H}}$ of all pairs of graphs in $\mathbb{G}_N$ with (4).
4: Let $r = 0$.
5: **while** $r < r_{max}$ **do**
6:     For fixed $\boldsymbol{c}^{(r)}$, estimate $\boldsymbol{W}^{(r)}$ by solving (7) using a GED heuristic (e.g. bipartite or IPFP).
7:     For fixed $\boldsymbol{W}^{(r)}$, estimate $\boldsymbol{c}^{(r+1)}$ by solving (7) using constrained linear least square programming (e.g. CVXPY).
8:     $r = r + 1$.
9: **end while**
10: Find set-median $\widehat{G}^{(0)}$ by (9).
11: Let $i = 0$.
12: **while** $i < i_{max}$ **do**
13:     Compute transformation $\widehat{\pi}_p^{(i+1)}$ by (10) for $\widehat{G}^{(i)}$ with $\boldsymbol{c}^{(r+1)}$.
14:     Generate $\widehat{G}^{(i+1)}$ by (11) with $\widehat{\pi}_p^{(i+1)}$ and $\boldsymbol{c}^{(r+1)}$.
15: **end while**
16: $\widehat{G}^{(i+1)}$ is the graph pre-image.

3.1   Learn Edit Costs by Distances in Kernel Space

When computing GEDs, the choice of edit costs values is essential. In practice, they are determined by domain experts for a given dataset. With our original idea of aligning the GEDs to the kernel metric, we propose to learn the edit costs by the distances of the elements in the kernel space.

On one hand, the distance in $\mathcal{H}$ between two elements $\phi(G_i)$ and $\phi(G_j)$ is:

$$d_{\mathcal{H}}(\phi(G_i), \phi(G_j)) = \sqrt{k(G_i, G_i) + k(G_j, G_j) - 2k(G_i, G_j)}. \tag{4}$$

On the other hand, considering Eq. (1) and the costs defined in Eq. (3), the GED between G_i and G_j is given by:

$$d_{GED}(G_i, G_j) = \boldsymbol{\omega}^\top \boldsymbol{c}, \tag{5}$$

with $\boldsymbol{\omega} = [n_{vr}, n_{vi}, \omega_{vs}, n_{er}, n_{ei}, \omega_{es}]^\top$, where $n_{vr}, n_{vi}, n_{er}, n_{ei}$ are respectively the numbers of vertex removals, insertions, and edge removals, insertions. $\omega_{vs} = \sum_{u \in V(G_i), \pi(u) \in V(G_j)} \|\ell_v(u) - \ell_v(\pi(u))\|$ is the sum of distances of labels between all pairs of vertices; and $\omega_{es} = \sum_{e \in E(G_i), \pi(e) \in E(G_j)} \|\ell_e(e) - \ell_e(\pi(e))\|$ is the sum of distances of labels between all pairs of edges.

A major difficulty, which is not straightforward from (5), is that ω and $\boldsymbol{c}$ are interdependent. For two different edit cost vectors, respective optimal ω may not be equivalent since the costs influence the presence or absence of each edit operation. In addition, ω influences also $\boldsymbol{c}$ since we want to fit GED with kernel distances, i.e., $d_{GED}(G_i, G_j) \approx d_{\mathcal{H}}(\phi(G_i), \phi(G_j))$.

Given a graph space $\mathcal{G}$ of attributed graphs and a kernel space $\mathcal{H}$, we propose to align GEDs in $\mathcal{G}$ with distances in $\mathcal{H}$ between each pair of graphs. In other words, we seek to learn the edit costs of the GED, so that the GED between each pair of graphs in $\mathcal{G}$ is as close as possible to its corresponding distance in

$\mathcal{H}$. To achieve this goal, a least squares optimization on graph dataset $\mathbb{G}_N = \{G_1, G_2, \ldots, G_N\} \subset \mathcal{G}$ is considered, namely

$$\arg\min_{c,\omega} \sum_{i,j=1}^{N} \left(d_{GED}(G_i, G_j) - d_{\mathcal{H}}(\phi(G_i), \phi(G_j)) \right)^2, \tag{6}$$

with d_{GED} depending on c and ω as given in (5), where ω exists for each pair of graphs G_i and G_j in $\mathbb{G}_N$, which will be denoted as $\omega(i,j)$. Moreover, to ensure that the minimum cost edit transformation π in (1) can be found, all edit costs need to be positive, and substituting an element should not be more expensive than removing and inserting it [21]. Thus, the optimization problem becomes:

$$\arg\min_{c,W} \| W^\top c - d_{\mathcal{H}} \|^2 \text{ subject to } c > 0, c_{vr} + c_{vi} \geq c_{vs} \text{ and } c_{er} + c_{ei} \geq c_{es}, \tag{7}$$

where $W^\top \in \mathbb{R}^{N^2 \times 6}$ with rows $\omega(i,j)^\top$ and $d_{\mathcal{H}} \in \mathbb{R}^{N^2}$ encoding the GEDs for each pair of graphs of $\mathbb{G}_N$. To solve this constrained optimization problem, we propose an alternating optimization strategy over c and W. The optimization problem over c, for a fixed W, is a constrained linear least square program problem solved using CVXPY [1,10]. Once the edit costs obtained, the weights W are computed by GED heuristics, such as bipartite and IPFP.

3.2   Generate Graph Pre-image

Given a set of graphs $\mathbb{G}_N \subset \mathcal{G}$, its average point can be easily computed in the kernel space, i.e., $\psi = \sum_{i=1}^{N} \alpha_i \phi(G_i)$ with $\alpha_i = 1/N$. Our objective is to estimate its pre-image, namely the graph $\widehat{G}$ whose image $\phi(\widehat{G})$ is as close as possible to ψ.

With the metric alignment principle (6), $d_{GED}(G_i, G_j) \approx d_{\mathcal{H}}(\phi(G_i), \phi(G_j))$, for all $G_i, G_j \in \mathbb{G}_N$. Therefore, estimating the pre-image is equivalent to estimating the graph median, which can be tackled as the minimization of the sum of distances (SOD) to all the graphs of $\mathbb{G}_N$, namely

$$\widehat{G} = \arg\min_{G \in \mathcal{G}} \sum_{G_{p'} \in \mathbb{G}_N} d_{GED}(G, G_{p'}). \tag{8}$$

A first attempt to solve it is to restrict the solution to the set $\mathbb{G}_N$, namely

$$\widehat{G} = \arg\min_{G_p \in \mathbb{G}_N} \sum_{G_{p'} \in \mathbb{G}_N} d_{GED}(G_p, G_{p'}) = \arg\min_{G_p \in \mathbb{G}_N} \sum_{p'=1}^{N} \min_{\pi_{p'} \in \Pi(G_p, G_{p'})} c(\pi_{p'}, G_p, G_{p'}), \tag{9}$$

where cost $c(\pi_{p'}, G_p, G_{p'})$ consists of two parts, $c_v(\pi_{p'}, \varphi_p, \varphi_{p'})$ and $c_e(\pi_{p'}, A_p, \Phi_p, A_{p'}, \Phi_{p'})$, which are costs of vertex and edge transformation, respectively. This problem can be solved by computing all pairwise GEDs for dataset $\mathbb{G}_N$. The computational complexity is in $\mathcal{O}(aN^2)$, where a is the complexity of computing a GED between two graphs (for instance, by bipartite or IPFP). The resulting pre-image $\widehat{G}$ is also known as the set-median of $\mathbb{G}_N$.

Despite its simplicity, the set-median can only be chosen from the given dataset $\mathbb{G}_N$, which strongly limits the results. To obtain the pre-image from a bigger space, we take advantage of recent advances in [7] where the proposed iterative alternate minimization procedure (IAM) allows generating new graphs. Next, we revisit this method and adapt it for the pre-image problem. The proposed strategy alternates the optimization over all the $\widehat{\pi}_p$ (i.e., transformations from $\widehat{G}$ to G_p) and over the pre-image estimate $\widehat{G}$, namely

$$\widehat{\pi}_p = \arg \min_{\pi_p \in \Pi(\widehat{G}, G_p)} c(\pi_p, \widehat{G}, G_p) \qquad \forall p \in \{1, \ldots, N\}; \tag{10}$$

$$\widehat{G} = \arg \min_{\substack{\varphi \in \mathcal{H}_v^{\widehat{n}} \\ A \in \{0,1\}^{\widehat{n} \times \widehat{n}} \\ \Phi \in \mathcal{H}_e^{\widehat{n} \times \widehat{n}}}} \sum_{p=1}^{N} c_v(\widehat{\pi}_p, \varphi, \varphi_p) + \tfrac{1}{2} c_e(\widehat{\pi}_p, A, \Phi, A_p, \Phi_p). \tag{11}$$

The resolution of (10) is carried out by solving the GED problem N times with time complexity of $\mathcal{O}(aN)$, and the computation of (11) is detailed in [7], where the vertices and edges are updated separately. The new non-symbolic labels assigned for a vertex v (resp. an edge e) are given by the average values of the corresponding labels of the vertices substituted to v (resp. edges substituted to e). The obtained pre-image $\widehat{G}$ is also known as the generalized median of $\mathbb{G}_N$.

4   Experiments

To perform experiments, we implemented[1] Algorithm 1 in Python. The C++ library GEDLIB[2] and its Python interface gedlibpy are used as the core implementation to compute graph edit distances and perform IAM algorithm [5]. We implemented a general edit cost function NonSymbolic[3] for graphs containing only non-symbolic vertex and/or edge labels and an edit cost function Letter2 specifically for dataset *Letter-high* based on NonSymbolic. In these functions, all edit costs can be freely set, which is more convenient for the optimization proposed in Sect. 3.1. We have modified the gedlibpy accordingly[4]. All experiments were carried out on a computer with 8 CPU cores of Intel Core i7-7920HQ @ 3.10 GHz, 32 GB memory, and 64-bit operating system Ubuntu 16.04.3 LTS.

Given the *Letter-high* dataset[5], the goal is to compute, for a given kernel, the pre-image of the average of each class of letters, namely $\psi = \sum_{i=1}^{N} \alpha_i \phi(G_i)$ with $\alpha_i = 1/N$. Two graph kernels are considered, the shortest path (SP) kernel [6] and the structural SP kernel [20], both being able to deal with non-symbolic labels; see [16]. In each class (i.e., a set of distortions of a letter), all 150 graphs are chosen to compose the graph set $\mathbb{G}_N$. To estimate the graph edit distances,

---

[1] https://github.com/jajupmochi/graphkit-learn/tree/master/gklearn/preimage.

[2] GEDLIB: https://github.com/dbblumenthal/gedlib.

[3] https://github.com/jajupmochi/gedlib/tree/master/src/edit_costs.

[4] gedlibpy (modified): https://github.com/jajupmochi/gedlibpy.

[5] http://graphkernels.cs.tu-dortmund.de.

Table 1. Distances in kernel space computed using different methods.

Graph kernels	Algorithms	$d_{\mathcal{H}}$ GM	Running Times (s)		
			Optimization	Generation	Total
Shortest path (SP)	From median set	0.406	-	-	-
	IAM: random costs	0.467	-	142.59	142.59
	IAM: expert costs	0.451	-	30.31	30.31
	IAM: optimized costs	0.460	5968.92	26.55	5995.47
Structural SP (SSP)	From median set	0.413	-	-	-
	IAM: random costs	0.435	-	30.22	30.22
	IAM: expert costs	0.391	-	29.71	29.71
	IAM: optimized costs	0.394	24.79	25.60	50.39

a multi-start counterpart of IPFP (i.e., mIPFP) is applied in both procedures of producing set-median and generalized median, where 40 different solutions to the LSAP are chosen [9]. The maximum number of iterations is set to $r_{max} = 6$.

Table 1 exhibits experimental results. Results of two sets of edit costs are presented. The first set of constants is randomly generated for each class of graphs, while the second set is given by domain experts, where $c_{vi} = c_{vr} = 0.675, c_{ei} = c_{er} = 0.425, c_{vs} = 0.75$ and $c_{es} = 0$ [4]. It is worth noting that these expert values take into account prior knowledge of the data, such as setting c_{es} to 0 as graphs in *Letter-high* do not contain edge labels. Moreover, we also give as a baseline a method to generate median graphs (denoted "From median set"), where the median graph is directly chosen from the median set $\mathbb{G}_N$ whose representation in kernel space is the closest to the true median's (ψ). The average results over all classes are presented for all methods. Column "$d_{\mathcal{H}}$ GM" gives the distances between the embedding of the computed pre-image and the element we want to approximate in the kernel space ($d_{\mathcal{H}}$). The columns "Running Times" give the time to optimize edit costs and generate pre-images.

For the structural SP kernel, when expert and optimized methods are given, applying IAM provides better pre-images than choosing from the median set with respect to $d_{\mathcal{H}}$s of the generalized medians. Compared to $d_{\mathcal{H}}$ of pre-image choosing from median sets, $d_{\mathcal{H}}$ is respectively 5.33% and 4.60% smaller for algorithm with expert and optimized costs. Moreover, $d_{\mathcal{H}}$ of the algorithm with optimized costs is 9.43% smaller than that with random costs and is almost the same as the algorithm with expert costs, which is also the case for the SP kernel. These results show that the algorithm with the optimized costs works better than the one with random costs to generate pre-images as median graphs, and can serve as a method to tune edit costs to help find expert costs, for both median generation problems using IAM and general pre-image problems. Moreover, the running times to optimize edit costs and generate pre-images are acceptable in most cases.

SP, from median set

SP, random costs

SP, expert costs

SP, optimized costs

SSP, from median set

SSP, random costs

SSP, expert costs

SSP, optimized costs

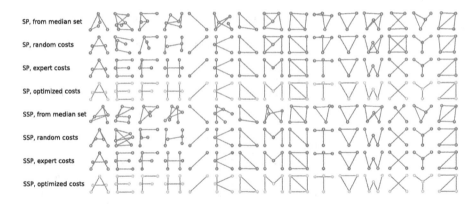

Fig. 2. Pre-images constructed by different algorithms for *Letter-high*, which correspond to the eight rows of Table 1 row by row.

Although these improvements seem trivial, the advantage of our method can be valuated from other aspects. Figure 2 presents the pre-images generated as the median graphs for each letter of the *Letter-high* dataset using aforementioned methods, which correspond to the eight rows of Table 1, row by row. Vertices are drawn according to coordinates determined by their attributes "x" and "y". In this way, plots of graphs are able to display the letters that they represent, which are possible to be recognized by human eyes. When using the SP kernel (the first row to the fourth row), it can be seen that the pre-images chosen directly from the median set (the first row) are illegible in almost all cases, while the IAM with random costs provides more legible results, where letters A, K, Y can be easily recognized (the second row). When the expert and optimized costs are used, almost all letters are readable, despite that the pre-images of letter F are slightly different (the third and fourth rows). The same conclusion can be derived for the structure SP kernel as well (the fifth row to the eighth row).

This analysis indicates that even though the distances $d_{\mathcal{H}}$ are similar, the algorithms applying IAM are able to generate better pre-images, especially when edit costs are optimized. This phenomenon may benefit from the nature of the IAM algorithm. In the update procedure (11), the new non-symbolic labels assigned for a vertex v is given by the average values of the corresponding labels of the vertices substituted to v [7]. It provides a "direction" to construct pre-images with respect to features and structures of graphs . For instance, the "x" and "y" attributes on the vertices of the letter graphs presents the coordinates of the vertices. To this end, it makes sense to compute their average values as the new values of a vertex as the vertex will be re-positioned at the middle of all vertices substituted to it.

5   Conclusion and Future Work

In this paper, we proposed a novel method to estimate graph pre-images. This approach is based on the hypothesis that metrics in both kernel space and graph space can be aligned. We first proposed a method to align GEDs to distances in the kernel space. Within the procedure, the edit costs are optimized. Then the graph pre-image was generated by a new method to construct the graph generalized median, where we revisited the IAM algorithm. Our method can generate better pre-images than other methods, as demonstrated on the *Letter-high* dataset. Future work includes generalizing our method to graphs with symbolic labels and constructing pre-images as arbitrary graphs rather than median graphs. The convergence proof of the iterative procedure will be conducted and the non-constant edit costs will be considered. Using state-of-the-art generative graph neural networks to solve the pre-image problem is also interesting.

References

1. Agrawal, A., Verschueren, R., Diamond, S., Boyd, S.: A rewriting system for convex optimization problems. J. Control Decis. **5**(1), 42–60 (2018)
2. Akutsu, T., Fukagawa, D.: Inferring a graph from path frequency. In: Apostolico, A., Crochemore, M., Park, K. (eds.) CPM 2005. LNCS, vol. 3537, pp. 371–382. Springer, Heidelberg (2005). https://doi.org/10.1007/11496656_32
3. Bakır, G.H., Zien, A., Tsuda, K.: Learning to find graph pre-images. In: Rasmussen, C.E., Bülthoff, H.H., Schölkopf, B., Giese, M.A. (eds.) DAGM 2004. LNCS, vol. 3175, pp. 253–261. Springer, Heidelberg (2004). https://doi.org/10.1007/978-3-540-28649-3_31
4. Blumenthal, D.B., Boria, N., Gamper, J., Bougleux, S., Brun, L.: Comparing heuristics for graph edit distance computation. VLDB J. **29**, 1–40 (2019)
5. Blumenthal, D.B., Bougleux, S., Gamper, J., Brun, L.: GEDLIB: a C++ library for graph edit distance computation. In: Conte, D., Ramel, J.-Y., Foggia, P. (eds.) GbRPR 2019. LNCS, vol. 11510, pp. 14–24. Springer, Cham (2019). https://doi.org/10.1007/978-3-030-20081-7_2
6. Borgwardt, K.M., Kriegel, H.P.: Shortest-path kernels on graphs. In: Fifth IEEE International Conference on Data Mining, pp. 8–pp. IEEE (2005)
7. Boria, N., Bougleux, S., Gaüzère, B., Brun, L.: Generalized median graph via iterative alternate minimizations. In: Conte, D., Ramel, J.-Y., Foggia, P. (eds.) GbRPR 2019. LNCS, vol. 11510, pp. 99–109. Springer, Cham (2019). https://doi.org/10.1007/978-3-030-20081-7_10
8. Bougleux, S., Gaüzère, B., Brun, L.: Graph edit distance as a quadratic program. In: 2016 23rd International Conference on Pattern Recognition (ICPR), pp. 1701–1706 (2016). https://doi.org/10.1109/ICPR.2016.7899881
9. Daller, É., Bougleux, S., Gaüzère, B., Brun, L.: Approximate graph edit distance by several local searches in parallel. In: 7th International Conference on Pattern Recognition Applications and Methods (2018)
10. Diamond, S., Boyd, S.: CVXPY: a python-embedded modeling language for convex optimization. J. Mach. Learn. Res. **17**(83), 1–5 (2016)
11. Gärtner, T.: A survey of kernels for structured data. ACM SIGKDD Explor. Newsl. **5**(1), 49–58 (2003)
12. Gaüzère, B., Brun, L., Villemin, D.: Graph kernels in chemoinformatics. In: Dehmer, M., Emmert-Streib, F. (eds.) Quantitative Graph Theory Mathematical Foundations and Applications, pp. 425–470. CRC Press (2015). https://hal.archives-ouvertes.fr/hal-01201933

13. Ghosh, S., Das, N., Gonçalves, T., Quaresma, P., Kundu, M.: The journey of graph kernels through two decades. Comput. Sci. Rev. **27**, 88–111 (2018)
14. Honeine, P.: Online kernel principal component analysis: a reduced-order model. IEEE Trans. Pattern Anal. Mach. Intell. **34**(9), 1814–1826 (2012). https://doi.org/10.1109/TPAMI.2011.270
15. Honeine, P., Richard, C.: Preimage problem in kernel-based machine learning. IEEE Signal Process. Mag. **28**(2), 77–88 (2011)
16. Jia, L., Gaüzère, B., Honeine, P.: Graph Kernels Based on Linear Patterns: Theoretical and Experimental Comparisons (2019). Working Paper or Preprint. https://hal-normandie-univ.archives-ouvertes.fr/hal-02053946
17. Kaspar, R., Horst, B.: Graph Classification and Clustering Based on Vector Space Embedding, vol. 77. World Scientific, Singapore (2010)
18. Kriege, N.M., Neumann, M., Morris, C., Kersting, K., Mutzel, P.: A unifying view of explicit and implicit feature maps for structured data: systematic studies of graph kernels. arXiv preprint arXiv:1703.00676 (2017)
19. Nagamochi, H.: A detachment algorithm for inferring a graph from path frequency. Algorithmica **53**(2), 207–224 (2009)
20. Ralaivola, L., Swamidass, S.J., Saigo, H., Baldi, P.: Graph kernels for chemical informatics. Neural Netw. **18**(8), 1093–1110 (2005)
21. Riesen, K.: Structural Pattern Recognition with Graph Edit Distance. Approximation Algorithms and Applications. ACVPR. Springer, Cham (2015). https://doi.org/10.1007/978-3-319-27252-8
22. West, D.B., et al.: Introduction to Graph Theory, vol. 2. Prentice Hall, Upper Saddle River (2001)

Multivalent Graph Matching Problem Solved by Max-Min Ant System

Kieu Diem Ho$^{(\boxtimes)}$, Jean Yves Ramel, and Nicolas Monmarché

Laboratoire d'Informatique Fondamentale et Appliquée de Tours,
Université de Tours, Tours, France
{kieu.ho,jean-yves.ramel,nicolas.monmarche}@univ-tours.fr

Abstract. This paper presents a multivalent graph matching problem and proposes a max-min ant system for its resolution. Multivalent graph matching is a very combinatorial problem where a node (edge) in one graph can match with more than one node (edge) in the other graph. We formalize this problem as an extended graph edit distance problem by adding possibilities of splitting and merging operations. Then, we employ an ant colony based optimization algorithm, the max-min ant system, to solve this very combinatorial problem. A local search is also integrated to enhance the solution quality. The efficiency of the proposed approach is verified on a symbol data set in several aspects. The results show that the proposed approach can be very useful in case of noise when the bijective graph matching-based approaches are not usually robust.

Keywords: Multivalent graph matching · Extended graph edit distance · Max-Min Ant System · Symbol recognition · Classification

1 Introduction

In real life, several applications, like medical analysis, symbol recognition, demand determining an explainable similarity measure between objects rather than a numerical value [6,7]. Among the existing approaches, graph matching-based approach is promising since it can provide the matching correspondence between sub-parts of objects besides the similarity measure.

Normally, GM problem is injective, which just allows one-to-one matching of node (edge). However, in some cases, one feature in one object can correspond to multiple features in the other object [1,4]. Furthermore, in pattern recognition problems, distortions of the graph can occur and an error-tolerant graph matching techniques should be used to allow node (edge) association even if they are not exactly similar [5]. All these matching situations are special cases of multivalent matching. Therefore, multivalent GM can be seen as the most general GM problem and we will contribute to bring solution to solve it in this work.

Through our literature review, we found very few works dealing with multivalent GM. In [1], the authors apply GED-based GM technique for recognizing diatoms. The splitting and merging of node operations are added besides the classical operations. However, the edge operations related to splitting/merging

© Springer Nature Switzerland AG 2021
A. Torsello et al. (Eds.): S+SSPR 2020, LNCS 12644, pp. 227–237, 2021.
https://doi.org/10.1007/978-3-030-73973-7_22

on nodes have not been specified. In [3], the authors also propose a GED-based
approach with node merging to compare similarity between images. Likewise the
previous work, the edge relations related to node merging are still unclear. In [2],
the authors have continued the work of [3], but they define more specifically the
edge merging when there are node merging. In our opinion, the proposed method
is not enough robust and lacks of comparisons with other methods. In [4], a randomized construction algorithm combined with local search is employed for a
non-bijective GM. Then, it is applied to search for the correspondence between
the model and its over-segmented images. Because of the context, the authors
suppose that there is only substitution for nodes and edges, and are not clear
about where are splitting and merging. In [10,11], the authors propose a general
similarity measure for multivalent GM and they utilize ant colony optimization
(ACO) and reactive tabu search to tackle the problem. Good results are found
with ACO in terms of solution quality but it is more time-consuming. Nevertheless, the measure is applied to a specific context and just works for symbolic
attributes. This leads to restrict the range of the applications.

From the mentioned works, we see that the GED-based approach is pretty
popular for a multivalent GM problem. This is done by adding extended operations like splitting and merging for nodes or edges. However, these works lack
of formal formulation for the GED-based approach. More precisely, the speci-
fication of edge operations in extended case either is not clear or depends on
the context. Therefore, in this work, we firstly present a formulation of multivalent GM problem based on GED, called extended graph edit distance (ExGED)
(Sect. 2). Secondly, we propose a way to formalize the edit costs including splitting and merging operations on edges induced by the edit operations done on
nodes. All the considered costs are presented in several cost matrices. These
matrices integrate both local structure and semantic information of all nodes in
two compared graphs (Sect. 3). Thirdly, following the works [10,11], we decide
to use the ACO for the multivalent GM problem (Sect. 4). With the formulation
of ExGED, we are able to solve both numeric and symbolic attributes. Moreover, the introduction of cost matrices allows us to accelerate the computational
time for ACO. Specifically, the max-min ant system (MMAS) is applied to solve
the ExGED in this work. The feasibility of the proposed approach is presented
through the numerical experiments in Sect. 5. Finally, Sect. 6 concludes the paper
and prospects for future research directions.

2   From GED to ExGED Problem Formulation

2.1   Graph Edit Distance

Definition 1. *An attributed graph contains 4-tuple $G = (V, E, \mu, \xi)$, where*

V, E are sets of vertices and edges, respectively,
l_V, l_E are sets of vertex and edge labels, respectively
$\mu : V \mapsto l_V$: function that assigns labels to vertices
$\xi : E \mapsto l_E$: function that assigns labels to edges.

Definition 2. *An edit path is a sequence of edit operations (ed_i) to transform one graph to another graph, denoted $\lambda(G_1, G_2) = \{ed_i\}$. A valid edit path should follow these conditions: 1) deleting a vertex implies deleting its related edges; 2) inserting an edge is only permitted if the two vertices already exist; 3) inserting an edge must not create more than one edge between two vertices (selfloops) [5].*

Definition 3. *Given two graphs $G_1 = (V_1, E_1), G_2 = (V_2, E_2)$, the graph edit distance (GED) is a dissimilarity measure between G_1 and G_2 and is defined by:*

$$d_{min}(G_1, G_2) = \min_{\lambda \in \Theta(G_1, G_2)} \sum_{ed_i \in \lambda} c(ed_i), \tag{1}$$

where $\Theta(G_1, G_2)$ is the set containing all valid edit paths λ between G_1 and G_2, $c(ed_i)$ is the cost of each edit operation ed_i [5].

Classical edit operations are given in Table 1. The cost of each operation is defined according to either the node or edge labels.

2.2 Extended Graph Edit Distance Problem Formulation

Extended graph edit distance (ExGED) is also a dissimilarity measure derived from the costs of the edit operations. To consider multivalent GM, we add splitting and merging operations.

Definition 4. *Given two attributed graphs $G_1 = (V_1, E_1, \mu_1, \xi_1)$ and $G_2 = (V_2, E_2, \mu_2, \xi_2)$, we define:*

- *merging the set $S_{mer} = \{u_i \in V_1, i \geq 2\}$ to $v \in V_2$ is noted $S_{mer} \to v$*
- *splitting $u \in V_1$ into the set $S_{spl} = \{v_j \in V_2, j \geq 2\}$ is noted $u \to S_{spl}$*

These two operators are also mentioned in Table 1. Usually, doing node merging and splitting can lead to edge splitting and merging, this will be discussed more precisely when the cost of each operation will be detailed.

Table 1. Availability of edit operations for GED and ExGED (with $u \in V_1, v \in V_2, e_1 \in E_1, e_2 \in E_2, \varepsilon$ the *virtual* vertex or edge). S_{mer} and S_{spl} are subsets of V_1 and V_2 defined in Definition 4.

Operation	Notation	Cost function notation	GED	ExGED
Vertex substitution	$u \to v$	$c(u \to v)$	✓	✓
Vertex deletion	$u \to \varepsilon$	$c(u \to \varepsilon)$	✓	✓
Vertex insertion	$\varepsilon \to v$	$c(\varepsilon \to v)$	✓	✓
Edge substitution	$e_1 \to e_2$	$c(e_1 \to e_2)$	✓	✓
Edge deletion	$e_1 \to \varepsilon$	$c(e_1 \to \varepsilon)$	✓	✓
Edge insertion	$\varepsilon \to e_2$	$c(\varepsilon \to e_2)$	✓	✓
Vertex merging	$S_{mer} \to v$	$c(S_{mer} \to v)$		✓
Vertex splitting	$u \to S_{spl}$	$c(u \to S_{spl})$		✓

3   Cost Matrices for ExGED

3.1   Definition of the Cost Matrix for Node Operations

Following the idea in [9], a cost matrix for ExGED is also constructed. Each block presents each type of node edit operation and its corresponding cost. Formally, given two attributed graphs G_1, G_2 as above, we denote that $n = |V_1|, m = |V_2|$. Let $P_1^k = \{S_{mer}^i\}$ be set of all possibilities for merging of nodes in G_1 and $h = |P_1^k|$, and $P_2^k = \{S_{spl}^j\}$ be set of all possibilities for splitting of nodes in G_2 and $l = |P_2^k|$. k is a parameter that describes the maximum number of nodes that one node can be associated with. Then we can give the cost matrix:

$$
\mathbf{C} =
\begin{array}{c}
\begin{array}{cccccccc}
1 & \cdots & m & \varepsilon & 1 & \cdots & l & \\
\end{array}\\
\left(
\begin{array}{ccc|c|ccc}
c_{1,1} & \cdots & c_{1,m} & c_{1,\varepsilon} & c_{1,S_{spl}^1} & \cdots & c_{1,S_{spl}^l} \\
\vdots & \ddots & \vdots & \vdots & \vdots & \ddots & \vdots \\
c_{n,1} & \cdots & c_{n,m} & c_{n,\varepsilon} & c_{n,S_{spl}^1} & \cdots & c_{n,S_{spl}^l} \\
\hline
c_{\varepsilon,1} & \cdots & c_{\varepsilon,m} & 0 & \infty & \cdots & \infty \\
c_{S_{mer}^1,1} & \cdots & c_{S_{mer}^1,m} & \infty & \infty & \cdots & \infty \\
\vdots & \ddots & \vdots & \vdots & \vdots & \ddots & \vdots \\
c_{S_{mer}^h,1} & \cdots & c_{S_{mer}^h,m} & \infty & \infty & \cdots & \infty \\
\end{array}
\right)
\begin{array}{c}
1 \\ \vdots \\ n \\ \varepsilon \\ 1 \\ \vdots \\ h
\end{array}
\end{array}
\qquad (2)
$$

where $c_{i,j}$ denotes the cost of a node substitution (with $(i,j) \in \{1 \ldots n\} \times \{1 \ldots m\}$), $c_{i,\varepsilon}$ denotes the cost of a node deletion, $c_{\varepsilon,j}$ denotes the cost of a node insertion, $c_{i,S_{spl}}$ denotes the cost of a node splitting and $c_{S_{mer},j}$ denotes the cost of a node merging.

From Eq. (2), the cost matrix $\mathbf{C}$ is not a square matrix like in GED case. A reduction of dimension has been applied on deletion and insertion blocks, from $(n \times n)$ to $(n \times 1)$ for deletion, from $(m \times m)$ to $(1 \times m)$ for insertion, respectively. That is because a square matrix is not necessary for a MMAS solver. By this way, we can decrease the computational time but still preserves the property of a GED cost matrix. Moreover, by introducing splitting and merging operations, the size of matrix $\mathbf{C}$ increases with h rows and l columns. h and l are strongly influenced by k. The bigger k is, the higher values of h, l get. Consequently, the size of $\mathbf{C}$ will grow up significantly. Thus, choosing the number of k would be very important, especially in big graph. Regularly, the parameter k is problem-dependent and based on expert knowledge.

3.2   Definition of the Costs for Edge Operations in Extended Case

In Eq. (2), only the costs of node operations are presented. To enrich $\mathbf{C}$, we integrate the estimated cost of edge operation related to each node operation.

- For node substitution $u_i \rightarrow v_j$, two sets of incident edges of u_i and v_j are computed, called E_{u_i} and E_{v_j}, respectively. Then, a square edge cost matrix $\mathbf{Ce}$ is built from E_{u_i} and E_{v_j} based on the cost functions of edge operations in Table 2. The size of $\mathbf{Ce}$ is $(|E_{u_i}| + |E_{v_j}|) \times (|E_{u_i}| + |E_{v_j}|)$. Finally, the Munkres's algorithm is applied on $\mathbf{Ce}$ to find the minimum sum of edge operation costs [9], or $c_{i,j} \leftarrow c(u_i \to v_j) + \text{Munkres}(\mathbf{Ce})$.

- For node deletion, deleting a node u_i will remove all its adjacent edges, or $c_{i,\varepsilon} \leftarrow c(u_i \to \varepsilon) + \sum_{e \in E_{u_i}} c(e \to \varepsilon)$.
- For node insertion, inserting a node v_j will insert all its adjacent edges, or $c_{\varepsilon,j} \leftarrow c(\varepsilon \to v_j) + \sum_{e \in E_{v_j}} c(\varepsilon \to e)$.
- For node merging $S_{mer} \to v$, two sets of incident edges to nodes in S_{mer} and v are computed first, denoted $E_{S_{mer}}$ and E_v, respectively. Let E_{loop} be the set of edges connecting the nodes $u_i \in S_{mer}$. We have $E'_{S_{mer}} = E_{S_{mer}} \setminus E_{loop}$. Then, an edge cost matrix $\mathbf{Ce}$ for two sets $E'_{S_{mer}}$ and E_v is built similarly as for node substitution. The Munkres's algorithm is also used to find the minimum cost for $E'_{S_{mer}}$ and E_v. Finally, the total cost for node merging is $c_{S_{mer},v} \leftarrow c(S_{mer} \to v) + \text{Munkres}(\mathbf{Ce}) + \sum_{e \in E_{loop}} c(e \to \varepsilon)$.
- For node splitting $u_i \to S_{spl}$, the computational steps of edge cost is similar to node merging: $c_{u,S_{spl}} \leftarrow c(u \to S_{spl}) + \text{Munkres}(\mathbf{Ce}) + \sum_{e \in E_{loop}} c(\varepsilon \to e)$.

3.3 Illustrative Example

Given two unlabelled graphs as in Fig. 1, we have $n = 3$ and $m = 4$. Suppose that $k = 2$, here nodes which have common edges are considered for merging and splitting. So, $P_1^2 = \{12, 23\}$, and $P_2^2 = \{ab, ac, bc, cd\}$ are sets of merging and splitting nodes in G_1 and G_2, respectively. Due to $n < m$, we restrict deleting and merging nodes in G_1 by setting high values for the costs involved with these operations. Suppose that we encourage splitting of nodes in this case, costs for node splitting and involved cost of edge operations will be small. All cost functions are defined below.

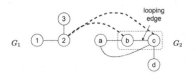

Operation	Node operation cost	Edge operation cost
Substitution	$c(u \to v) = 1$	$c(e_1 \to e_2) = 1$
Insertion	$c(\varepsilon \to v) = 2$	$c(\varepsilon \to e_2) = 2$
Deletion	$c(u \to \varepsilon) = 10$	$c(e_1 \to \varepsilon) = 10$
Merging	$c(S_{mer} \to v) = 10$	$c(e \to \varepsilon) = 10, e \in E_{loop}$
Splitting	$c(u \to S_{spl}) = 1$	$c(\varepsilon \to e) = 0, e \in E_{loop}$

Fig. 1. The example graphs with partial matching $\lambda = \{2 \to b, c\}$ (left) and the cost functions of edit operations (right).

Let have a partial matching $\lambda = \{2 \to b, c\}$ (Fig. 1), then the sets of incident edges to 2 and b, c are $E_2 = \{12, 23\}$ and $E_{spl} = E_{bc} = \{ab, ac, bc, cd\}$. $E_{loop} = \{bc\}$ is set of edges connecting nodes b, c. So, $E'_{bc} = E_{bc} \setminus E_{loop} = \{ab, ac, cd\}$. Then, the edge cost matrix $\mathbf{Ce}$ for E_2 and E'_{bc} is computed in Eq. (3). The total cost for the partial matching $\lambda = \{2 \to b, c\}$ is: $c_{2,bc} = c(2 \to b, c) + \text{Munkres}(\mathbf{Ce}) + c(\varepsilon \to bc) = 1 + 4 + 0 = 5$.

$$
\mathbf{Ce} =
\begin{array}{c|ccc|cc}
 & ab & ac & cd & \varepsilon_{12} & \varepsilon_{23} \\
\hline
12 & c_{12,ab} & c_{12,ac} & c_{12,cd} & c_{12,\varepsilon} & \infty \\
23 & c_{23,ab} & c_{23,ac} & c_{23,cd} & \infty & c_{23,\varepsilon} \\
\varepsilon_{ab} & c_{\varepsilon,ab} & \infty & \infty & 0 & 0 \\
\varepsilon_{ac} & \infty & c_{\varepsilon,ac} & \infty & 0 & 0 \\
\varepsilon_{cd} & \infty & \infty & c_{\varepsilon,cd} & 0 & 0
\end{array}
=
\begin{array}{c|ccc|cc}
 & ab & ac & cd & \varepsilon_{12} & \varepsilon_{23} \\
\hline
12 & 1 & 1 & 1 & 10 & \infty \\
23 & 1 & 1 & 1 & \infty & 10 \\
\varepsilon_{ab} & 2 & \infty & \infty & 0 & 0 \\
\varepsilon_{ac} & \infty & 2 & \infty & 0 & 0 \\
\varepsilon_{cd} & \infty & \infty & 2 & 0 & 0
\end{array}
\tag{3}
$$

4 MMAS for ExGED

4.1 Algorithmic Scheme

The Max-Min Ant System (MMAS) is a variant of Ant Colony Optimization (ACO) [12]. A colony of ants is used to generate solutions of the considered problem in a parallel manner. At each iteration, each ant builds a complete matching based on transition probabilities. A construction graph, denoted G_{ants}, is used to know the possibilities for one ant. G_{ants} is corresponding to the search space, i.e. the possible vertex matching between the 2 graphs. Each node in G_{ants} is corresponding to a possible edit operation done between nodes of G_1 and G_2. Edges in G_{ants} will help to construct incrementally the best edit path. Pheromones of natural ants are here corresponding to probabilities which are shared among ants to build a solution. An initial pheromone value is associated on each vertex of the construction graph. Pheromones are updated according to the performance of the best found matching.

4.2 Construction Graph

The construction graph is complete and undirected. When a solution is built, the ant chooses the next vertex $v_i \in G_{ants}$ according to the previous partial matching. For instance, if the partial matching is $\lambda = \{2 \rightarrow b, c\}$, all candidates related to these nodes will be pruned. That means the current ant could not move on these vertices until the matching is complete.

4.3 Construction of a Complete Matching

At each iteration, every ant starts with an empty matching $\lambda = \{\}$. At each step, it adds a candidate $v_i \in G_{ants}$ to λ until λ is a complete matching. That means λ contains all nodes in G_1 and G_2. The ant chooses the next candidate based on the transition probability. This quantity is derived from two factors: the heuristic and the pheromone. For ExGED problem, the heuristic is built from the cost matrix $\mathbf{C}$ but the pheromone value is laid on the construction graph G_{ants}. Let τ_{v_i}, η_{v_i} be respectively the pheromone and heuristic values of the candidate v_i; $cand = \{v_1, \ldots, v_{|cand|}\}$ be the set of candidates of λ; α, β be respectively the weights of τ and η, the transition probability is:

$$
Pr_{v_i} = \frac{[\tau_{v_i}]^{\alpha} \times [\eta_{v_i}]^{\beta}}{\sum_{j=1}^{|cand|} [\tau_{v_j}]^{\alpha} \times [\eta_{v_j}]^{\beta}}
\tag{4}
$$

4.4 Pheromone Update

Once all ants have built their solutions, pheromones on each vertex $v_i \in G_{ants}$ are updated: they are reinforced and evaporated as in Eq. (5). The reinforcement is done only on the best solution of the current iteration (λ_{itbest}). The evaporation is done for all nodes $v_i \in G_{ant}$ according to the evaporation rate $\rho \in [0, 1]$. After updating, the pheromone value will be adjusted in the interval $[\tau_{min}, \tau_{max}]$ [12].

$$\tau_{v_i} = (1 - \rho) * \tau_{v_i} + \Delta_{v_i}, \quad \Delta_{v_i} = \begin{cases} \frac{1}{1+c(\lambda_{itbest})} & \text{if } \quad v_i \in \lambda_{itbest} \\ 0 & \text{otherwise} \end{cases} \tag{5}$$

5 Experiments

The MMAS algorithm is utilized to find the best matching between two graphs. This approach is used because the search space is huge and the combinatorial underlying problem can not be solved exactly in a reasonable time [2]. But the ACO based algorithms must be tuned according to the problem as they have several parameters. All experiments are implemented in Python 3.7.1 and run on Windows 10 Intel(R) Core(TM) i7-8750 CPU @ 2.20 GHz, RAM of 16.0 Go.

5.1 Data Set and Graph Representation

We use the SESYD data set which contains architectural and electrical symbols. From the original symbols, we add noise to create deformed ones. The noise can break one original line into several lines, rotate or scale the symbol[1].

To transform symbols into graphs, each line is defined as a node and the relations between lines are considered as edges. Each node has its relative length (l). Each edge has its relative angle (θ) and type of relation (rel). The length l is normalized regarding the longest line in each symbol. The rel includes T-Junction (T), Parallel (P), Successive (S), L-Junction (L) and intersection (X) [8]. We give an example in Fig. 2.

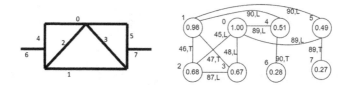

Fig. 2. The symbol (left) and its graph representation (right).

When the original line is deformed, it is possible that the 'S' relation appears (Fig. 6). So, we suppose that the nodes connected by the edges with 'S' relation are merged or split. Also, we set $k = 3$ based on the observation on the data set.

---

[1] Data set link: http://www.rfai.lifat.univ-tours.fr/PublicData/ExGED/home.html.

5.2 Definition of Cost Functions for Edit Operations

From graph representation, costs of node operations depend on its length. Deleting/inserting a long line should cost more than a shorter one. Merging of lines that are split from the initial lines will cost less than the random lines. Because $l \in [0,1], \theta \in [0,180]$, cost of edge operation related to θ should be normalized to be compatible with other costs. Details of the cost functions are in Table 2.

5.3 Parameter Setting for MMAS

The impacts of principal parameters of MMAS and local search on the results of GM problems are examined. Each study of each parameter is run 30 times. The average results are presented. Figure 3 (log scale) presents the effect of the parameters (α, β, ρ) and local search strategies to the final costs. In Fig. 3a, better results are seen when combining both α and β. In Fig. 3b, the lowest cost is obtained when value of ρ is not too high or too low. In Fig. 3c, a significant improvement of the cost is achieved when integrating 3-opt local search with MMAS. So, we choose $\alpha = 2, \beta = 1, \rho = 0.1$, 3-opt local search, $[\tau_{min}, \tau_{max}] = [0.1, 2.0]$, $nb_{ants} = 5, nb_{iters} = 300$ iterations for later experiments.

Table 2. Cost functions of edit operations for the data set (δ is a Dirac function).

Node operation cost	Edge operation cost
$c(u \rightarrow v) = \|l_u - l_v\|$	$c(e_1 \rightarrow e_2) = \frac{\|\theta_{e_1} - \theta_{e_2}\|}{\max(\theta_{e_1}, \theta_{e_2})} + \delta(rel_{e_1} = rel_{e_2})$
$c(\varepsilon \rightarrow v) = l_v$	$c(\varepsilon \rightarrow e_2) = 1$
$c(u \rightarrow \varepsilon) = l_u$	$c(e_1 \rightarrow \varepsilon) = 1$
$c(S_{mer} \rightarrow v) = \|\sum_{u_i \in S_{mer}} l_{u_i} - l_v\|$	$c(e \rightarrow \varepsilon) = 0, e \in E_{loop}$
$c(u \rightarrow S_{spl}) = \|l_u - \sum_{v_i \in S_{spl}} l_{v_i}\|$	$c(\varepsilon \rightarrow e) = 0, e \in E_{loop}$

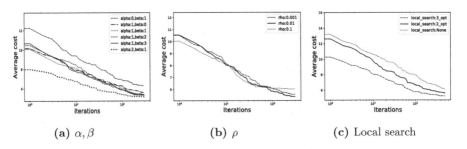

    (a) α, β **(b)** ρ **(c)** Local search

Fig. 3. Influences of parameters and local search to the convergence of MMAS to ExGED (with 5 ants, 300 iterations, $[\tau_{min}, \tau_{max}] = [0.1, 2.0]$).

5.4   Matching Quality Analysis

In this part, matching between nodes and edges of two graphs are considered. The result of MMAS-ExGED is compared to the one of bipartite approach for GED [9], denoted BP-GED. We define a reasonable matching if it maps correspondingly the initial lines in the symbol to their split lines in the distorted one. A group of 6 symbols with 6 levels of distortions for each symbol are used for this evaluation.

Figure 4 shows the average costs given by BP-GED and MMAS-ExGED in terms of distortion levels. In almost cases, MMAS-ExGED obtains lower costs than BP-GED. Go into low-level matching, we see that these biases are really compatible with the obtained mappings. The MMAS-ExGED is able to find the reasonable matching between the symbol and its distortions. An example is given in Fig. 6 and Table 3.

5.5   Symbol Recognition

By Comparison to Perfect Models. We select 20 original symbols from SESYD data set, denoted $M = \{M_i, \ldots, M_{20}\}$. Each symbol has 20 levels of distortions, i.e. $D_i = \{D_j, \ldots, D_{20}\}$, $\forall M_i \in M$. Totally, we have 400 distorted symbols. The distorted symbol is assigned to the model M_i if the distance (cost) from it to this model M_i is minimal. The result shows that MMAS-ExGED recognises well $360/400$ distorted symbols. Also, we compute the average distance from all distorted symbols to all the models. Some results are shown in Fig. 5.

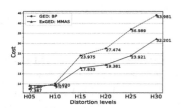

Fig. 4. Average costs of BP-GED and MMAS-GED in terms of distortion levels.

Fig. 5. Dissimilarity between the symbols and their distortions by MMAS-ExGED.

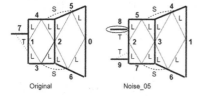

Fig. 6. Drawing of nodes (in bold) and edges (dashed lines) on the symbol 032 and its distortion levels. Ellipse indicates noise position.

Table 3. Difference of node mappings at distorted positions between symbol and its distortions in Fig. 6 given by BP-GED and MMAS-ExGED.

Methods	Level 5	Level 10	Level 15
BP-GED	$\varepsilon \to 9$	$1 \to 5; \varepsilon \to 9$	$2 \to 14; \varepsilon \to 15; 4 \to 16; \varepsilon \to 18$
MMAS-ExGED	$\varepsilon \to 8$	$1 \to \{5, 9\}$	$2 \to \{14, 17\}; 5 \to 15, 4 \to \{16, 18\}$

We see that the average costs from the distorted symbols to their truth symbols are the smallest (in bold).

By Using KNN. For classification problem, 160 difficult symbols of 8 classes are selected. The training and the test sets have 80 symbols in each set. 1NN is used as a classifier. The distances given by the GM algorithms are used to classify. We compare the performance of MMAS for GED and ExGED problems. Table 4 shows that MMAS-ExGED achieves better result than MMAS-GED. However, MMAS is quite time-consuming because the process of building one solution is repeated several times to improve the solution quality.

Table 4. Results of symbol classification given by MMAS for GED and ExGED.

Methods	MMAS-GED	MMAS-ExGED
Classification rate (%)	76.25	**77.50**
Classification time (s)	56762.07	**25563.74**

6   Conclusion

In this paper, an ExGED-approach is proposed to solve a multivalent GM problem. Specifically, the ExGED is introduced by adding splitting and merging operations to classical GED. The problem formulation of ExGED is presented in detail based on concept of cost matrix. Then MMAS with a local search is utilized to tackle the ExGED. The feasibility of the proposed method is shown on a symbol data set. The results are evaluated through several aspects: through an assessment of quality of the provided matching and through different classification strategies. The positive results are obtained with MMAS-ExGED. However, working on reduction of the computational time of MMAS-ExGED will be our upcoming task.

References

1. Ambauen, R., Fischer, S., Bunke, H.: Graph edit distance with node splitting and merging, and its application to diatom identification. In: Hancock, E., Vento, M. (eds.) GbRPR 2003. LNCS, vol. 2726, pp. 95–106. Springer, Heidelberg (2003). https://doi.org/10.1007/3-540-45028-9_9

2. Berretti, S., Bimbo, A.D., Pala, P.: Graph edit distance for active graph matching in content based retrieval applications. Open Artif. Intell. J. **1**, 1–11 (2007)
3. Berretti, S., Del Bimbo, A., Pala, P.: A graph edit distance based on node merging. In: Enser, P., Kompatsiaris, Y., O'Connor, N.E., Smeaton, A.F., Smeulders, A.W.M. (eds.) CIVR 2004. LNCS, vol. 3115, pp. 464–472. Springer, Heidelberg (2004). https://doi.org/10.1007/978-3-540-27814-6_55
4. Boeres, M.C., Ribeiro, C.C., Bloch, I.: A randomized heuristic for scene recognition by graph matching. In: Ribeiro, C.C., Martins, S.L. (eds.) WEA 2004. LNCS, vol. 3059, pp. 100–113. Springer, Heidelberg (2004). https://doi.org/10.1007/978-3-540-24838-5_8
5. Conte, D., Foggia, P., Sansone, C., Vento, M.: Thirty years of graph matching in pattern recognition. Int. J. Pattern Recogn. Artif. Intell. **18**(03), 265–298 (2004)
6. Lladós, J., Martí, E., Villanueva, J.J.: Symbol recognition by error-tolerant subgraph matching between region adjacency graphs. IEEE Trans. Pattern Anal. Mach. Intell. **23**(10), 1137–1143 (2001)
7. Noma, A., Pardo, A., Cesar Jr., R.M.: Structural matching of 2D electrophoresis gels using deformed graphs. Pattern Recogn. Lett. **32**(1), 3–11 (2011)
8. Qureshi, R.J., Ramel, J.Y., Cardot, H., Mukherji, P.: Combination of symbolic and statistical features for symbols recognition. In: 2007 International Conference on Signal Processing, Communications and Networking, pp. 477–482. IEEE (2007)
9. Riesen, K., Bunke, H.: Approximate graph edit distance computation by means of bipartite graph matching. Image Vis. Comput. **27**(7), 950–959 (2009)
10. Sammoud, O., Solnon, C., Ghédira, K.: Ant algorithm for the graph matching problem. In: Raidl, G.R., Gottlieb, J. (eds.) EvoCOP 2005. LNCS, vol. 3448, pp. 213–223. Springer, Heidelberg (2005). https://doi.org/10.1007/978-3-540-31996-2_20
11. Sammoud, O., Sorlin, S., Solnon, C., Ghédira, K.: A comparative study of ant colony optimization and reactive search for graph matching problems. In: Gottlieb, J., Raidl, G.R. (eds.) EvoCOP 2006. LNCS, vol. 3906, pp. 234–246. Springer, Heidelberg (2006). https://doi.org/10.1007/11730095_20
12. Stützle, T., Hoos, H.H.: Max-min ant system. Future Gen. Comput. Syst. **16**(8), 889–914 (2000)

A Metric Learning Approach to Graph Edit Costs for Regression

Linlin Jia[1]($\boxtimes$), Benoit Gaüzère[1], Florian Yger[2], and Paul Honeine[3]

[1] LITIS Lab, INSA Rouen Normandie, Saint-Étienne-du-Rouvray, France
linlin.jia@insa-rouen.fr
[2] LAMSADE, Université Paris Dauphine-PSL, Paris, France
[3] LITIS Lab, Université de Rouen Normandie, Rouen, France

Abstract. Graph edit distance (GED) is a widely used dissimilarity measure between graphs. It is a natural metric for comparing graphs and respects the nature of the underlying space, and provides interpretability for operations on graphs. As a key ingredient of the GED, the choice of edit cost functions has a dramatic effect on the GED and therefore the classification or regression performances. In this paper, in the spirit of metric learning, we propose a strategy to optimize edit costs according to a particular prediction task, which avoids the use of predefined costs. An alternate iterative procedure is proposed to preserve the distances in both the underlying spaces, where the update on edit costs obtained by solving a constrained linear problem and a re-computation of the optimal edit paths according to the newly computed costs are performed alternately. Experiments show that regression using the optimized costs yields better performances compared to random or expert costs.

Keywords: Graph edit distance · Edit costs · Metric learning

1 Introduction

Graphs provide a flexible representation framework to encode relationships between elements. In addition, graphs come with an underlying powerful theory. However, the graph space cannot be endowed with the mathematical tools and properties associated with Euclidean spaces. This issue prevents the use of classical machine learning methods mainly designed to operate on vector representations. To learn models on graphs, several approaches have been designed to leverage this flaw and among these, we can cite graph embeddings strategy [17], graph kernels [3,28] and more recently graph neural networks [8]. Despite their

This research was supported by CSC (China Scholarship Council), the French national research agency (ANR) under the grant APi (ANR-18-CE23-0014), the ANR "Investissements d'avenir" program ANR-19-P3IA-0001 (PRAIRIE 3IA Institute) and grant ESIGMA ANR-17-CE23-0010.

A. Torsello et al. (Eds.): S+SSPR 2020, LNCS 12644, pp. 238–247, 2021.
https://doi.org/10.1007/978-3-030-73973-7_23

state-of-the-art performances, they seldom operate directly in the graph space, hence reducing the interpretability of the underlying operations.

To overcome these issues, one needs to preserve the property of the graph space. For this purpose, one needs to define a dissimilarity measure in the graph space, in order to constitute the minimal requirement to implement simple machine learning algorithms like the k-nearest neighbors. The most used dissimilarity measure between graphs is the graph edit distance (GED) [10,27]. The GED of two graphs G_1 and G_2 can be seen as the minimal amount of distortion required to transform G_1 into G_2. This distortion is encoded by a set of edit operations whose sequence constitutes an edit path. These edit operations include nodes and edges substitutions, removals, and insertions. Depending on the context, each edit operation e included in an edit path γ is associated with a non-negative cost $c(e)$. The sum of all edit operation costs included within the edit path defines the cost $A(\gamma)$ associated with this edit path. The minimal cost[1] among all edit paths $\Gamma(G_1, G_2)$ defines the GED between G_1 and G_2, namely

$$\mathsf{ged}(G_1, G_2) = \min_{\gamma \in \Gamma(G_1,G_2)} A(\gamma). \qquad (1)$$

Evaluating the GED is computationally costly and cannot be done in practice for graphs having more than 20 nodes in general. To avoid this computational burden, strategies to approximate the GED in a limited computational time have been proposed [1] with acceptable classification or regression performances.

An essential ingredient of the GED is the underlying edit cost function $c(e)$, which quantifies the distortion carried by the edit operation e. The values of the edit costs for each edit operation have a major impact on the computation of GED and its performance. Thus, the cost edit function may be different depending on the data encoded by the graph and the one to predict. Generally, they are fixed *a priori* by an expert of the domain, and are provided with the datasets.

However, these predefined costs are not optimal for any prediction tasks, in the same spirit as the no free lunch theorems for machine learning and statistical inference. In addition, these costs may have a great influence on both the prediction performance and the computational time required to compute the graph edit distance. In [9], the authors show that a particular set of edit costs may reduce the problem of computing graph edit distance to well-known problems in graphs like (sub)graph isomorphism or finding the maximum common subgraph of a pair of graphs. This point shows again the importance of the underlying cost function when computing a graph edit distance.

In this paper, we propose a simple strategy to optimize edit costs according to a particular prediction task, and thus avoid the use of predefined costs. The idea is to align the metric in the graph space (namely, the GED) to the prediction space. While this idea has been largely used in machine learning (e.g. with the so-called kernel-target alignment [15]), this is the first time that such a line of attack is investigated to estimate the optimal edit cost. With this distance-preserving principle, we provide a simple linear optimization procedure to optimize a set of

---

[1] Note the GED will be null when comparing two isomorphic graphs.

constant edit costs. The edit costs resulting from the optimization procedure can then be analyzed to understand how the graph space is structured. The relevance of the proposed method is demonstrated on two regression tasks, showing that the optimized costs lead to a lower prediction error.

The remainder of the paper is organized as follows. Section 2 presents related works that aim to compute costs associated with a particular task. Section 3 presents the problem formulation and describes the proposed optimization method. Then, Sect. 4 presents results from conducted experiments. Finally, we conclude and open perspectives on this work.

2   Related Works

As stated in the introduction, the choice of edit costs has a major impact on the computation of graph edit distance, and thus on the performance associated with the prediction task.

The first approach to design these costs is to set them manually, based on the knowledge on a given dataset/task (when such knowledge is available). This strategy leads, for instance, to the classical edit cost functions associated with the IAM dataset [25]. However, it is interesting to challenge these predefined settings and experiment how they can improve the prediction performance.

In order to fit a particular targeted property to predict, tuning the edit costs and thus the GED can be seen as a subproblem of metric learning. Metric learning consists in learning a dissimilarity (or similarity) measure given a training set composed of data instances and associated targeted properties. For the classical metric learning where each data instance is encoded by a real-valued vector, the problem consists in learning a dissimilarity measure, which decreases (resp. increases) where the vectors have similar (resp. different) targeted properties. Many metric learning works focus on Euclidean data, while only a few addresses this problem on structured data [5]. A complete review for general structured data representation is given in [23]. In the following, we will focus on existing studies to learn edit costs for graph edit distance.

A trivial approach to tune the edit costs is to use a grid search strategy among a predefined range. However, the complexity required to compute graph edit distance and the number of different edit costs forbid such an approach.

String edit distance constitutes a particular case of graph edit distance, associated to a lower complexity, where graphs are restricted to be only linear and sequential. In [26], the authors propose to learn edit costs using a stochastic approach. This method shows a performance improvement, hence demonstrating the interest to tune edit costs; it is however restricted to strings.

Another strategy is based on a probabilistic approach [20–22]. By providing a probabilistic formulation for the common edition of two graphs, an Expectation-Maximization algorithm is used to derive weights applied to each edit operation. The tuning is then evaluated in an unsupervised manner. In [21], the strategy consists in modifying the label space associated with nodes and edges such that edit operations occurring more often will be associated to lower edit costs. Conversely, higher values will be associated with edit operations occurring less often.

The learning process was validated on two datasets. However, this approach is computationally too expensive when dealing with general graphs [4].

In [4], the authors propose an interesting way to evaluate whether a distance is a "good" one. This criterion is based on the following concept:

a similarity function is $(\epsilon, \gamma, \tau) - good$ if a $1 - \epsilon$ proportion of examples are on average 2γ more similar to reasonable examples of the same class than to reasonable examples of the opposite class, where a τ proportion of examples must be reasonable.

This principle is then derived to define an objective function to optimize. The matrix encoding the edit costs minimizing this objective function is then used to compute edit distances. However, this approach has only been adapted to strings and trees, but not to general graphs.

Another set of methods that address the problem of learning edit costs for GED is proposed in [13,14]. These methods propose to optimize edit costs to maximize a ground truth mapping between nodes of graphs. This framework requires thus a ground truth mapping, which is not available on many datasets like chemoinformatics.

3   Proposed Method

3.1   Problem Formulation

In this section, we propose an optimization procedure to learn edit costs in the context of regression tasks. Consider a dataset $\mathcal{G}$ of N graphs such that each graph $G_k = (V_k, E_k)$, for $k = 1, 2, \ldots N$, where V_k represents the set of nodes of G_k labeled by a function $f_v : \mathcal{V} \to \mathcal{L}_v$, and E_k encodes the set of edges of G_k, namely $e_{ij} = (v_i, v_j) \in E_k$ iff an edge connects nodes v_i and v_j in G_k.

The graph edit distance between two graphs is defined as the minimal cost associated to an optimal edit path. Given two graphs G_1 and G_2, an edit path between them is defined as a sequence of edit operations transforming G_1 into G_2. An edit operation e can correspond to a node substitution $e = (v_i \to v_j)$, deletion $e = (v_i \to \varepsilon)$ or insertion $e = (\varepsilon \to v_j)$. Similarly, for edges, we have $(e_{ij} \to e_{ab}), (e_{ij} \to \varepsilon)$, and $(\varepsilon \to e_{ab})$. Each edit operation is associated with a cost characterizing the distortion induced by this edit operation on the graph. These costs can be encoded by a cost function c that associates a positive real value to each edit operation, depending on the elements being transformed.

In this paper, we will restrict ourselves to only constant cost functions. Therefore, we can associate each edit operation to a constant value. Let c_{ns}, c_{ni}, c_{nd}, c_{es}, c_{ei}, $c_{ed} \in \mathbb{R}_+$ be the cost values associated with respectively node substitution, insertion, deletion and edge substitution, insertion, deletion.

As shown in [7], any edit path between two graphs G_1 and G_2 can be encoded as two mapping functions. First, $\varphi : V_1 \to V_2 \cup \varepsilon$ encodes the mapping of G_1's nodes to nodes of G_2. If a node v_i is deleted, we have $\varphi(v_i) = \varepsilon$. Similarly, we denote as φ^{-1} the mapping of V_2 to $V_1 \cup \varepsilon$. For the same edit path, we have thus $\varphi(v_i) = v_j \Rightarrow \varphi^{-1}(v_j) = v_i$. Given a mapping and considering constant cost functions, the cost associated to node operations of an edit path represented by φ and φ^{-1} is given by:

$$C_v(\varphi, \varphi^{-1}, G_1, G_2) = \sum_{\substack{v_i \in V_1 \\ \varphi(v_i) \neq \varepsilon}} c_{ns} + \sum_{\substack{v_i \in V_1 \\ \varphi(v_i) = \varepsilon}} c_{ni} + \sum_{\substack{v_i \in V_2 \\ \varphi^{-1}(v_i) = \varepsilon}} c_{nd}. \tag{2}$$

The cost associated with edge operations is defined as:

$$C_e(\varphi, \varphi^{-1}, G_1, G_2) = \sum_{\substack{e=(v_i,v_j) \in E_1 | \\ \varphi(v_i) \neq \varepsilon \wedge \\ \varphi(v_j) \neq \varepsilon \wedge \\ (\varphi(v_i), \varphi(v_j)) \in E_2}} c_{es} + \sum_{\substack{e=(v_i,v_j) \in E_1 | \\ \varphi(v_i) = \varepsilon \vee \\ \varphi(v_j) = \varepsilon \vee \\ (\varphi(v_i), \varphi(v_j)) \notin E_2}} c_{ei} + \sum_{\substack{e=(v_i,v_j) \in E_2 | \\ \varphi^{-1}(v_i) = \varepsilon \vee \\ \varphi^{-1}(v_j) = \varepsilon \vee \\ (\varphi^{-1}(v_i), \varphi^{-1}(v_j)) \notin E_1}} c_{ed}. \tag{3}$$

The final cost is given by:

$$C(\varphi, \varphi^{-1}, G_1, G_2) = C_v(\varphi, \varphi^{-1}, G_1, G_2) + C_e(\varphi, \varphi^{-1}, G_1, G_2). \tag{4}$$

Let $\#ns$ be the number of node substitutions, i.e., the cardinality of the subset of V_1 being mapped onto V_2. This number is given by the number of terms of the first sum in Eq. 2, i.e., $\#ns = |\{v_i \in V_1 \mid \varphi(v_i) \neq \varepsilon\}|$. Similarly:

- The number of node deletions is $\#nd = |\{v_i \in V_1 \mid \varphi(v_i) = \varepsilon\}|$;
- The number of node insertions is $\#ni = |\{v_i \in V_2 \mid \varphi^{-1}(v_i) = \varepsilon\}|$;
- The number of edge substitutions is $\#es = |\{e = (v_i, v_j) \in E_1 \mid \varphi(v_i) \neq \varepsilon \wedge \varphi(v_j) \neq \varepsilon \wedge (\varphi(v_i), \varphi(v_j)) \in E_2\}|$;
- The number of node deletions is $\#ei = |\{e = (v_i, v_j) \in E_1 \mid \varphi(v_i) = \varepsilon \vee \varphi(v_j) = \varepsilon \vee (\varphi(v_i), \varphi(v_j)) \notin E_2\}|$;
- The number of node insertions is $\#ed = |\{e = (v_i, v_j) \in E_2 \mid \varphi^{-1}(v_i) = \varepsilon \vee \varphi^{-1}(v_j) = \varepsilon \vee (\varphi^{-1}(v_i), \varphi^{-1}(v_j)) \notin E_1\}|$.

Then, let $\mathbf{x} \in \mathbb{N}^6$ encode the number of each edit operation as $\mathbf{x} = [\#ns, \#nd, \#ni, \#es, \#ed, \#ei]^\top$. Note that these values depend on both graphs being compared and a given mapping between nodes. Similarly, we define a vector representation of the costs associated with each edit operation by $\mathbf{c} = [c_{ns}, c_{nd}, c_{ni}, c_{es}, c_{ed}, c_{ei}]^\top \in \mathbb{R}_+^6$. Given these representations, the cost associated with an edit path, as defined by Eq. 4, can be rewritten as:

$$C(\varphi, \varphi^{-1}, G_1, G_2, \mathbf{c}) = \mathbf{x}^\top \mathbf{c}. \tag{5}$$

Therefore, the graph edit distance between two graphs is defined as:

$$\mathrm{ged}(G_1, G_2, \mathbf{c}) = \operatorname*{argmin}_{\varphi, \varphi^{-1}} C(\varphi, \varphi^{-1}, G_1, G_2, \mathbf{c}). \tag{6}$$

3.2 Learning the Edit Costs

Consider that each graph $G_k \in \mathcal{G}$ is associated with a particular targeted property $y_k \in \mathcal{Y}$, namely the target in regression tasks (e.g. $\mathcal{Y} \subseteq \mathbb{R}$ for real-valued output regression). Furthermore, a distance $d_{\mathcal{Y}} : \mathcal{Y} \times \mathcal{Y} \rightarrow \mathbb{R}^+$ is defined on this targeted property, such as the Euclidean distance when dealing with a vector space $\mathcal{Y}$, namely $d_{\mathcal{Y}}(y_i, y_j) = \|y_i - y_j\|_2$.

The main idea behind the proposed method is that the best metric in the graph space is the best aligned one to the target distances (i.e., $d_{\mathcal{Y}}$). With this distance-preserving principle, we seek to learn the edit cost vector $\mathbf{c}$ by fitting the distances between graphs to the distances between their targeted properties. Ideally, we seek to preserve the GED between any two graphs G_i and G_j and the distance between their targeted properties. Considering the set of N available graphs $G_1, \ldots, G_N$ and their corresponding targets $y_1, \ldots, y_N$, we seek to have

$$\mathsf{ged}(G_i, G_j, \mathbf{c}) \approx d_{\mathcal{Y}}(y_i, y_j) \qquad \text{for all } i, j = 1, 2, \ldots N. \tag{7}$$

Let $\omega : \mathcal{G} \times \mathcal{G} \times \mathbb{R}_+^6 \rightarrow \mathbb{N}^6$ be the function that computes an optimal edit path between G_i and G_j according to the cost vector $\mathbf{c}$ and returns the vector $\mathbf{x}^\star \in \mathbb{R}_+^6$ of numbers of edit operations associated to this optimal edit path, namely $\mathbf{x}^\star = \omega(G_i, G_j, \mathbf{c})$. This function can be any method computing an exact or sub-optimal graph edit distance [1,6].

For any pair of graphs (G_i, G_j), let $\mathbf{x}_{i,j}$ be a vector encoding the number of each edit operation. Let $\mathbf{X} \in \mathbb{N}^{N^2 \times 6}$ be the matrix of the numbers of edit operations for each pair of graphs, namely its $(iN + j)$-th row is $\mathbf{x}_{i,j}^T$. Then, $\mathbf{Xc}$ is the $N^2 \times 1$ vector composed of edit distances computed according to $\mathbf{c}$ and $\mathbf{X}$ between all pairs of graphs. Let $\mathbf{d} \in \mathbb{R}_+^{N^2}$ be a vector of the differences on targeted properties according to $d_{\mathcal{Y}}$, with $\mathbf{d}(iN + j) = d_{\mathcal{Y}}(G_i, G_j)$. Therefore, the optimization problem can be rewritten as:

$$\underset{\mathbf{c}}{\arg\min} \; \mathcal{L}(\mathbf{Xc}, \mathbf{d}) \qquad \text{subject to } \mathbf{c} > 0, \tag{8}$$

where $\mathcal{L}$ denotes a loss function. Besides the constraint on $\mathbf{c}$ to avoid negative costs, one can also add a constraint to satisfy the triangular inequality, or one to ensure that a deletion cost is equal to an insertion cost [24].

In the case of regression problem, $\mathcal{L}$ can be defined as the sum squares of differences between computed graph edit distances and dissimilarities of the targeted property. Therefore, the final optimization problem is:

$$\underset{\mathbf{c}}{\arg\min} \; \|\mathbf{Xc} - \mathbf{d}\|_2^2 \qquad \text{subject to } \mathbf{c} > 0. \tag{9}$$

Estimating $\mathbf{c}$ by solving this constrained optimization problem allows to linearly fit graph edit distances to a particular targeted property according to the edit paths initially given by ω. However, changing the edit costs may influence the optimal edit path, and thus its description in terms of the numbers of edit operations. There is thus an interdependence between the function ω computing an optimal edit path according to $\mathbf{c}$, and the objective function optimizing $\mathbf{c}$ according to edit paths encoded within $\mathbf{X}$. To solve this interdependence, we propose an alternated optimization strategy, summarized in Algorithm 1 where $\Omega(G, \mathbf{c})$ denotes the computation of $\omega(G_i, G_j, \mathbf{c}), \forall i, j \in 1 \ldots N$. The two main steps of the algorithm are described next:

Algorithm 1. Main algorithm to optimize costs

1: $\mathbf{c} \leftarrow \text{random}(6)$
2: $\mathbf{X} \leftarrow \Omega(G, \mathbf{c})$
3: **while** not converged **do**
4:     $\mathbf{c} \leftarrow \text{argmin}_{\mathbf{c}} \|\mathbf{Xc} - \mathbf{d}\|_2^2, \text{subject to } \mathbf{c} > 0$
5:     $\mathbf{X} \leftarrow \Omega(G, \mathbf{c})$
6: **end while**

- Estimate $\mathbf{c}$ for fixed $\mathbf{X}$ (line 4): This optimization problem is a constrained linear problem that can be resolved using off-the-shelf solvers, such as cvxpy [16] and scipy [29]. This optimization problem can also be viewed as a non-negative least squares problem [19]. For a given set of edit operations between each pair of graphs, this step linearly optimizes the constant costs to be applied such that the difference between graph edit distances and distances between targets is minimized.
- Estimate $\mathbf{X}$ for fixed $\mathbf{c}$ (line 5): The modification performed on costs in the previous step may have an influence on the associated edit path. To address this point, the optimization of costs is followed by a re-computation of the optimal edit paths according to the newly computed $\mathbf{c}$ vector encoding the edit costs. This step can be achieved by any method computing graph edit distance. For the sake of computational time, one can choose an approximated version of GED [6,7].

This alternated optimization is repeated to compute both edit costs and edit operations. Since we do not have theoretical proof of the convergence of this optimization scheme, we limit the number of iterations to 5 in our implementation.

4 Experiments

We conducted experiments[2] on two well-known datasets in chemoinformatics, both composed of molecules and their boiling points. The first dataset is composed of 150 alkanes [11]. An alkane is an acyclic molecule solely composed of carbons and hydrogens. A common representation of such data consists in implicitly encoding hydrogen atoms using the valency of carbon atoms. Such an encoding scheme allows to represent alkanes as acyclic unlabeled graphs. The second dataset is composed of 185 acyclic molecules [12]. In contrast with the previous dataset, these molecules contain several hetero atoms and are thus represented as acyclic labeled graphs.

To evaluate the predictive power of different settings of edit costs, we used a k-nearest-neighbors regression [2] model, where k is the number of the neighbors considered to predict a property. The performances are estimated on ten different random splits. For each split, a test set representing 10% of the graphs in the

[2] Code available at https://gitlab.insa-rouen.fr/bgauzere/fit-distances.

Method	Train errors	Test errors
random	9.60 ± 1.42	11.46 ± 3.50
expert	8.90 ± 0.97	8.28 ± 1.32
fitted	6.24 ± 0.24	6.78 ± 2.11

Method	Train errors	Test errors
random	26.70 ± 5.32	30.43 ± 7.71
expert	29.17 ± 0.66	31.77 ± 2.88
fitted	10.12 ± 0.67	13.69 ± 2.95

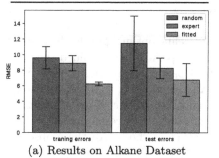

(a) Results on Alkane Dataset

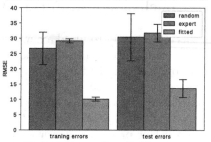

(b) Results on Acyclic Dataset

Fig. 1. Results on each dataset in terms of RMSE for the 10 splits

Table 1. Average and standard deviation of fitted edit costs values

Dataset	c_{ns}	c_{nd}	c_{ni}	c_{es}	c_{ed}	c_{ei}
Acyclic	10.74 ± 0.2	14.26 ± 0.7	14.8 ± 0.6	0.32 ± 0.01	0.23 ± 0.2	0.4 ± 0.2
Alkane	–	26.22 ± 1.0	26.85 ± 0.8	–	0.16 ± 0.1	0.11 ± 0.1

dataset is randomly selected and used to measure the performance of the prediction. The remaining 90% are used to optimize the edit costs and the value of k, where k is optimized through a 5-fold cross-validation (CV) procedure over the candidate values $\{3, 5, 7, 9, 11\}$. The number of iterations for the optimization of the edit costs is fixed to 5.

The proposed optimization procedure is compared to two other edit costs settings: a random set of edit costs and a predefined cost setting as given in [1]; the latter is the so-called expert costs. Tables in Fig. 1 show the average root mean squared errors (RMSE) obtained for each cost settings over the 10 splits, estimated on the training set and on the test set. The $\pm$ sign gives the 95% confidence interval computed over the 10 repetitions. Figures show a different representation of the same results with error bars modeling the 95% confidence interval. As expected, a clear and significant gain in accuracy is obtained when using fitted costs on the two datasets. These promising results confirm the hypothesis that ad-hoc edit costs may help the graph edit distance catch better targeted properties that are associated to a graph, and thus improve the prediction accuracy while still operating in the graph space.

The fitted values of edit costs are summarized in Table 1. From these results, we can observe that insertion and deletion costs are almost similar, hence showing the symmetry of these operations. Also, one can observe that deletion and insertion costs are more important than substitution costs, which shows that the

number of atoms is more important than the atom itself. This is coherent with
the chemistry theory [12]. Finally, we can note that costs associated with nodes
are higher to the ones associated with edges.

5   Conclusion and Future Work

In this paper, we introduced a new principle to define optimal graph edit costs of
a GED for a given regression task. Based on this principle, we defined the optimization problem of fitting the edit costs to a particular metric, measured for
instance on a targeted property to predict. An alternated optimization strategy
was proposed to solve this optimization problem. The conducted experiments on
two well-known datasets showed that the optimization process leads to a GED
with a better predictive power compared to other methods. All these observations confirm that the proposed method helps to fit edit costs and outperforms
other methods. There are still several challenges to address in future work. First,
a clear and complete comparison to other methods cited in the introduction and
related works will be established. Second, we seek to examine other criteria than
the distance-preserving criterion, such as the conformal map for instance [18].
Third, from a theoretical point of view, we are interested in establishing convergence proof on our alternated optimization strategy, and to extend these proofs
to approximate computations of graph edit distances. Fourth, this scheme will
be extended to classification problem and non-constant costs to be applicable
in most application domains. Considering non-constant costs will need to optimize parametric functions rather than scalar values, hence complexifying the
procedure.

References

1. Abu-Aisheh, Z., et al.: Graph edit distance contest: results and future challenges. Pattern Recogn. Lett. **100**, 96–103 (2017)
2. Altman, N.S.: An introduction to kernel and nearest-neighbor nonparametric regression. Am. Stat. **46**(3), 175–185 (1992)
3. Balcilar, M., Renton, G., Héroux, P., Gaüzère, B., Adam, S., Honeine, P.: When spectral domain meets spatial domain in graph neural networks. In: Proceedings of ICML 2020 - Workshop on Graph Representation Learning and Beyond (GRL+ 2020), Vienna, Austria, 12–18 July 2020
4. Bellet, A., Habrard, A., Sebban, M.: Good edit similarity learning by loss mini-mization. Mach. Learn. **89**(1–2), 5–35 (2012)
5. Bellet, A., Habrard, A., Sebban, M.: A survey on metric learning for feature vectors and structured data. arXiv preprint arXiv:1306.6709 (2013)
6. Blumenthal, D.B., Boria, N., Gamper, J., Bougleux, S., Brun, L.: Comparing heuristics for graph edit distance computation. VLDB J. **29**(1), 419–458 (2020)
7. Bougleux, S., Brun, L., Carletti, V., Foggia, P., Gaüzère, B., Vento, M.: Graph edit distance as a quadratic assignment problem. Pattern Recogn. Lett. **87**, 38–46 (2015)

8. Bronstein, M.M., Bruna, J., LeCun, Y., Szlam, A., Vandergheynst, P.: Geometric deep learning: going beyond Euclidean data. IEEE Signal Process. Mag. **34**(4), 18–42 (2017)
9. Bunke, H.: Error correcting graph matching: on the influence of the underlying cost function. IEEE Trans. Pattern Anal. Mach. Intell. **21**(9), 917–922 (1999). https://doi.org/10.1109/34.790431
10. Bunke, H., Allermann, G.: Inexact graph matching for structural pattern recognition. Pattern Recogn. Lett. **1**(4), 245–253 (1983)
11. Cherqaoui, D., Villemin, D.: Use of a neural network to determine the boiling point of alkanes. J. Chem. Soc. Faraday Trans. **90**, 97–102 (1994)
12. Cherqaoui, D., Villemin, D., Mesbah, A., Cense, J.M., Kvasnicka, V.: Use of a neural network to determine the normal boiling points of acyclic ethers, peroxides, acetals and their sulfur analogues. J. Chem. Soc. Faraday Trans. **90**, 2015–2019 (1994)
13. Cortés, X., Conte, D., Cardot, H.: Learning edit cost estimation models for graph edit distance. Pattern Recogn. Lett. **125**, 256–263 (2019). https://doi.org/10.1016/j.patrec.2019.05.001
14. Cortés, X., Serratosa, F.: Learning graph-matching edit-costs based on the optimality of the oracle's node correspondences. Pattern Recogn. Lett. **56**, 22–29 (2015)
15. Cristianini, N., Shawe-Taylor, J., Elisseeff, A., Kandola, J.: On kernel-target alignment. In: Advances in Neural Information Processing Systems, pp. 367–373 (2002)
16. Diamond, S., Boyd, S.: CVXPY: a python-embedded modeling language for convex optimization. J. Mach. Learn. Res. **17**(1), 2909–2913 (2016)
17. Gibert, J., Valveny, E., Bunke, H.: Graph embedding in vector spaces by node attribute statistics. Pattern Recogn. **45**(9), 3072–3083 (2012)
18. Honeine, P., Richard, C.: Preimage problem in kernel-based machine learning. IEEE Signal Process. Mag. **28**(2), 77–88 (2011)
19. Lawson, C.L., Hanson, R.J.: Solving Least Squares Problems. SIAM, Philadelphia (1995)
20. Neuhaus, M., Bunke, H.: A probabilistic approach to learning costs for graph edit distance. Proc. ICPR **3**(C), 389–393 (2004)
21. Neuhaus, M., Bunke, H.: Self-organizing maps for learning the edit costs in graph matching. IEEE Trans. Syst. Man Cybern. **35**(3), 503–514 (2005)
22. Neuhaus, M., Bunke, H.: Automatic learning of cost functions for graph edit distance. Inf. Sci. **177**(1), 239–247 (2007). https://doi.org/10.1016/j.ins.2006.02.013
23. Ontañón, S.: An overview of distance and similarity functions for structured data. Artif. Intell. Rev. (2020). https://doi.org/10.1007/s10462-020-09821-w
24. Riesen, K.: Structural Pattern Recognition with Graph Edit Distance. ACVPR. Springer, Cham (2015). https://doi.org/10.1007/978-3-319-27252-8
25. Riesen, K., Bunke, H.: IAM graph database repository for graph based pattern recognition and machine learning. In: da Vitoria Lobo, N., et al. (eds.) SSPR /SPR 2008. LNCS, vol. 5342, pp. 287–297. Springer, Heidelberg (2008). https://doi.org/10.1007/978-3-540-89689-0_33
26. Ristad, E.S., N.yianilos, P.: Learning string-edit distance. IEEE Trans. Pattern Anal. Mach. Intell. **20**(5), 522–532 (1998). https://doi.org/10.1109/34.682181
27. Sanfeliu, A., Fu, K.S.: A distance measure between attributed relational graphs for pattern recognition. IEEE Trans. Systems, Man Cybern. **13**(3), 353–362 (1983)
28. Shervashidze, N., Schweitzer, P., Van Leeuwen, E.J., Mehlhorn, K., Borgwardt, K.M.: Weisfeiler-lehman graph kernels. J. Mach. Learn. Res. **12**(9) (2011)
29. Virtanen, P., et al.: Scipy 1.0: fundamental algorithms for scientific computing in python. Nat. Method **17**(3), 261–272 (2020)

Parallel Subgraph Isomorphism on Multi-core Architectures: A Comparison of Four Strategies Based on Tree Search

Vincenzo Carletti[✉], Pasquale Foggia, Antonio Greco, and Mario Vento

Department of Information Engineering, Electrical Engineering and Applied
Mathematics, University of Salerno, Fisciano, Italy
{vcarletti,pfoggia,mvento}@unisa.it
https://mivia.unisa.it

Abstract. Subgraph isomorphism is one of the most challenging problems on graph-based representations. Despite many efficient sequential algorithms have been proposed over the last decades, solving this problem on large graphs is still a time demanding task. For this reason, there is a recently growing interest in realizing effective parallel algorithms able to exploit at their best the modern multi-core architectures commonly available on servers and workstations. We propose a comparison of four parallel algorithms derived from the state-of-the-art sequential algorithm VF3-Light; two of them were presented in previous works, while the other two are introduced in this paper. In order to evaluate strong points and weaknesses of each algorithm, we performed a benchmark over six datasets of random large and dense graphs, both labelled and unlabelled, measuring memory usage, speed-up and efficiency. We also add a comparison with a different parallel algorithm, named Glasgow, that is not derived from VF3-Light.

Keywords: Exact graph matching · Subgraph isomorphism · Parallel algorithms · VF3

1 Introduction

Graphs are mathematical structures aimed at representing sets of objects together with their relationships formalized, as nodes and edges respectively. Many real world domains cannot be entirely represented using vectors only; we need graphs indeed, to have a proper understanding of connections, interactions and relationships among the entities under analysis. Examples are social and biological networks [3,5,8,9] where graphs are the most natural representation for protein structures, gene interactions or social interactions. Unfortunately, the great expressive power of structural representations is paid with the high computational complexity required to perform even simple operations, such as the comparison of two objects, as discussed in [10,12,22]. An important but expensive task is the search for occurrences of *pattern graphs* as subgraphs of

© Springer Nature Switzerland AG 2021
A. Torsello et al. (Eds.): S+SSPR 2020, LNCS 12644, pp. 248–258, 2021.
https://doi.org/10.1007/978-3-030-73973-7_24

a larger *target graph*. This task is useful in many practical applications, from motif search in biology to graph database querying, and is usually formalized as *subgraph isomorphism* [10], a NP-Complete problem in the general case. Due to its complexity, several sequential algorithms have been proposed over the last twenty years [3,4,11,13,17,18,21]. However, because of the increasing quantity of data collected, the graphs to be processed are becoming larger and larger, making the problem extremely time consuming (days or weeks can be required) even for the most effective algorithms, like VF3 [4] or RI [3].

Therefore, moving to parallel solutions is the way to speed up the algorithms on large graphs and here is where the research is taking the next step. The goal is exploiting at the best the modern multi-processor, multi-core and GPU-based architectures. Although the latter are the most interesting in terms of parallel computing, the modern GPUs are SIMD (Single Instruction Multiple Data) architectures designed to be very effective on matrix operations, but less general-purpose than a common multi-core CPU that provides a thread-level parallelism. This makes very difficult to realize graph algorithms on GPUs that are really efficient in using the underlying hardware, as discussed also in [14,23]. GPU-based algorithms have been proposed in [20,24]; they all approach the problem using the *filtering-and-joining* [19], but differ from each other on the representation adopted and the way they realize preprocessing phases, filtering and join heuristics. The result is a significant speed-up that is, nevertheless, far from being proportional to the number of processing units used, thus yielding a low efficiency.

We decided to focus on the most diffused and affordable parallel architecture, the multi-core CPUs. Although the level of hardware parallelism is not comparable with GPUs, multi-core CPUs allow to design parallel algorithms that achieve a very high efficiency and speed-up. In the recent literature, state-of-the-art algorithms designed for multi-core architectures are Glasgow [1,16] a parallel constraint programming approach based on the LAD (Local All Different) algorithm by [18], MPMatch [15] a parallel formulation of CFL-Match [2], and more recently VF3P [7] a parallel algorithm derived from VF3-Light [6].

The aim of this paper is to extend the analysis on the algorithms proposed in [7] by adding two new algorithms based on VF3-Light [6] and moving to a larger number of threads per algorithms, 8 and 16 respectively. To be fair in our analysis we have also compared these algorithms with Glasgow, that approaches the problem using constraint programming. The analysis has been performed on six graph datasets, evaluating the speed-up, the efficiency and the memory usage, both when enumerating all the solutions, and when stopping the algorithm after the first solution is found.

2   Subgraph Isomorphism

A graph is defined as a pair $G = (V, E)$, where V is the set of nodes and $E \subset V \times V$ is the set of the edges. In addition to its structure, a graphs can have semantic information through attributes (or labels) on edge and nodes. *Graph Matching* is the problem of finding a mapping between the nodes of two graphs G_1 and G_2 that satisfies a given set of constraints. If all the nodes of the first graph are involved, the mapping can be expressed as a *mapping function*, $M : V_1 \rightarrow V_2$. The subgraph isomorphism [10] is the problem of searching instances of a *pattern* graph inside a larger *target* one. In this case the mapping function M must be injective and the corresponding nodes must have the same structure, i.e. given two nodes $u, u' \in G_1$ and two node $v = M(u), v' = M(u') \in G_2$, the edge (u, u') is in G_1 *if and only if* (v, v') is in G_2.

For over ten years, VF2 [11], VF3 [4] and recently VF3L [6] have demonstrated to be among the most effective methods to deal with exact graph matching problems. The overall approach is based on a depth-first search (DFS) with backtracking over a tree-structured state space where each state represents a partial matching between two graphs. More formally, each state s represents a partial solution $M(s)$ that maps a subset of nodes $V_1(s) \subseteq V_1$ to a subset of nodes $V_2(s) \subseteq V_2$. The root state s_0 has $M(s_0) = \emptyset$. As shown in Fig. 1, the algorithm iteratively adds a new couple of nodes (u, v) to the matching $M(s)$ of the currently visited state (see ExtendState) so as to obtain a new state s' such that $M(s') = M(s) \cup (u, v)$. The algorithm has to search only those states verifying the constraints imposed by the subgraph isomorphism (*consistent states*). This process ends on the goal states where all the nodes of G_1 have been mapped. The consistency of the visited states is ensured by the rules introduced in VF2 [11], namely the *feasibility rule*, used to verify the necessary and sufficient conditions for the subgraph isomorphism. When a consistent state cannot be further extended to generate a new consistent one the algorithm backtracks (see RestoreState). Thus, the exploration stops when no more consistent states have to be extended.

```
procedure Match(s_n, G_1, G_2, N_G_1, out Results)
    if IsGoal(s_n) then append M(s) to Results
    else
        for (u_n, v_n) ∈ NextCandidates(s_n, N_G_1, G_1, G_2)
            if IsFeasible(s_n, u_n, v_n) then
                s_n+1 := ExtendState(s_n, u_n, v_n)
                Match(s_n+1, G_1, G_2, N_G_1, Results)
                RestoreState(s_n+1, u_n, v_n)
```

Fig. 1. Outline of the sequential matching procedure

3 Parallel Subgraph Isomorphism Solvers

Solving the subgraph isomorphism as a DFS over a state space allows to effectively exploit a domain decomposition instead of a functional one. Indeed, the search process can be performed by exploiting a state-grained parallelism where each thread is responsible to visit a single state. Under the assumptions that the subset of the state space explored by the threads are not overlapped and the number of state is equally distributed over the threads, n threads will, ideally, explore n states in parallel. While the first assumption can be guaranteed using an appropriate communication strategy among the threads, having a uniform workload among the threads is hard to achieve in the case of subgraph isomorphism because it requires to know the distribution of the states over the search space before starting the matching.

Realistically, the threads have to self-balance their workload and share with each other the states to be explored; therefore, the communication among them is a crucial point to realize an efficient algorithm. To this aim a stack, namely the *Global State Stack* (GSS), is used by thread to share states with the others and pull a state to be explored. Of course, this design choice entails a communication overhead that grows together with the number of the threads. Hence, to reduce this overhead, an additional private stack, namely the *Local State Stack* (LSS), is used by each thread to explore autonomously a subset of the state space without requiring to access the GSS (Fig. 2).

```
procedure Explore(G₁, G₂, N_{G₁}, GSS, out Results)
    while StateCounter > 0 do
        PullFromStack(GSS, sₙ)
        if sₙ is not null then
            StateCounter = StateCounter − 1
            ExploreState(sₙ, G₁, G₂, N_{G₁}, GSS, out Results)

procedure ExploreState(sₙ, G₁, G₂, N_{G₁}, GSS, out Results)
    if IsGoal(sₙ) then
        append M(sₙ) to Results
    else
        for (uₙ, vₙ) ∈ NextCandidates (sₙ, N_{G₁}, G₁, G₂)
            if IsFeasible(sₙ, uₙ, vₙ) then
                s_{n+1} := ExtendState(sₙ, uₙ, vₙ)
                StateCounter = StateCounter + 1
                PushInStack(s_{n+1}, GSS)
```

Fig. 2. Outline of the procedures executed by each thread.

The overall structure of the parallel algorithm is composed of a static thread-pool so as to control the exact degree of parallelism, also considering the underlying hardware architecture, and avoid the overhead required to create threads on-demand. Each thread executes the procedure Explore (see Algorithm 2), where it checks if there are states to be explored, through a shared state counter (*StateCounter*), then pulls a state from the GSS. It worths pointing out that the use of such a shared counter is needed to keep track of all the states, both those that have to be explored (stored in the GSS or in the LSS of each thread) and the ones currently extracted from the one of the stacks that are under exploration. Since the communication among the threads is performed only by sharing states and there are no threads responsible to orchestrate the work, the information carried by the shared state counter is essential for the threads to understand when no more work is required. Indeed, when the state counter is zero a thread can safely stop: there are no more state to be explored and the other threads are not exploring a state so they are not going to push new ones. The consistency of this information is guaranteed by updating the counter both in the `Explore` procedure and while pushing and pulling new states. The `ExploreState` procedure is used to visit a state just retrieved from one of the stacks; it has the same structure of the `Match` procedure used by the sequential algorithm, except for the backtracking that is not used in the parallel exploration.

Table 1. Summary of all the strategies considered in the analysis.

Algorithm	Policy	LSS	Description
$VF3P$	GSS-only	No	The algorithm only uses the GSS
$VF3P_{LS}$	LSS with Limited Depth	Yes	The algorithm limits the maximum number of state stored in the LSS. If the LSS is full, new states are pushed into the GSS
$VF3P_{DL}$	LSS with Descendant Limit	Yes	The algorithm limits the maximum number of states explored by a thread without using the GSS. Each thread keeps a count of how many states were explored since the last pull from the GSS; if this count is over a threshold, new states are pushed into the GSS
$VF3P_{AWS}$	LSS with Active Work Sharing	Yes	The algorithm ensures that a thread periodically shares part of its work using the GSS. Each thread checks periodically if there are inactive threads; if so, it moves some states from its LSS to the GSS

As discussed in [7], the policy adopted to balance the joint use of LSS and GSS, can significantly affect the efficiency of the algorithm. In the same paper, two policies have been proposed; the first is the very basic one that uses only the GSS, while the other adopt the LSS by limiting its depth so as to set a maximum number of states stored privately by each thread and push them to share the surplus. Hereinafter we name these policies *GSS-only* ($VF3P$) and *LSS with Limited Depth* ($VF3P_{LS}$) respectively. In addition to them, we propose to new strategies to extend the analysis proposed in [7], both of them using the LSS. The first approach, named *LSS with Descendants Limit* ($VF3P_{DL}$), considers each state extracted from the GSS as a *root state* than limits the number of descendants states generated starting from a root state that have been stored in the LSS. A private descendant counter is used by each thread to keep track of the number of descendant pushed in the LSS; each descendant generated is put in the LSS while the counter is under the limit, in the GSS otherwise. Since each thread can manage a single root state, each time it pulls a state from the GSS the descendant counter is reset. It is worth to note that differently for the strategy using a depth limit for the LSS, the one adopted in $VF3P_{DL}$ does not limit directly the maximum size of the LSS, but the maximum number of states explored autonomously by a thread instead. Finally, the last policy is named *LSS with Active Work Sharing* ($VF3P_{AWS}$). It does not limit the maximum number of states in the LSS, but forces each thread to share some states periodically. Indeed, after having explored a given number of states a thread has to check if there are some inactive threads waiting for work. If so, the thread moves from its LSS to the GSS an amount of states equal to the number inactive workers. A summary of all the considered policies is shown in Table 1.

4 Experiments

In our experiments we have compared the four parallel tree-search strategies discussed in Sect. 3 and the Glasgow algorithm [1], a recent constraint programming method designed for multi-core architecture. The comparison has been performed considering two performance measures widely adopted with parallel algorithms: the *speed-up* (Sp), that is the improvement of the execution time evaluated as the ratio between the run time T_s of the most efficient sequential algorithm and T_p, and the *efficiency* (Ef) obtained as the speed-up on the number of CPUs. The computation of these measures requires a sequential algorithm to be used as baseline, we have selected VF3-Light due to its proven efficiency in solving the subgraph isomorphism and because, except for Glasgow, the compared parallel algorithms have been derived from it.

To benchmark the algorithms we have used six challenging datasets selected from the MIVIA LDG database, composed of large and dense Erdős and Rényi graphs; in particular, we have used two datasets of unlabelled random graphs having a density of 0.2, 0.3 respectively and target graph size ranging form 500 to 3000 nodes and four labelled random graphs of density 0.2 and 0.3 and size from 500 to 10000. As discussed in [7], due to its complexity, these datasets are suitable to stress the algorithms.

Table 2. Speed-up of the parallel algorithms over different target graph size and number of CPU cores employed.

Dataset		$VF3P$		$VF3P_{LS}$		$VF3P_{DL}$		$VF3P_{AWS}$		Glasgow	
All the solutions											
		8 core	16 core	8 core	16 core	8 core	16 core	8 core	16 core	8 core	16 core
No Lab	$\eta = 0.2$	5.16	8.77	**6.23**	**11.74**	5.67	10.33	1.71	3.06	2.68	2.96
	$\eta = 0.3$	5.08	10.59	7.16	12.66	6.16	11.49	4.04	8.54	**11.14**	**13.45**
Unif	$\eta = 0.2$	3.77	**5.23**	**3.86**	3.76	3.80	4.43	1.63	1.55	2.02e−03	2.07e−03
	$\eta = 0.3$	4.72	8.60	**5.81**	5.63	5.44	**9.41**	3.35	5.07	2.78e−03	3.10e−03
NonUni	$\eta = 0.2$	**3.76**	**4.61**	2.65	2.60	3.51	3.70	1.53	1.41	1.64e−03	1.73e−03
	$\eta = 0.3$	4.75	8.66	4.16	3.95	**5.62**	**9.32**	3.56	5.31	2.35e−03	2.68e−03
First solution											
Dataset		$VF3P$		$VF3P_{LS}$		$VF3P_{DL}$		$VF3P_{AWS}$		Glasgow	
		8 core	16 core	8 core	16 core	8 core	16 core	8 core	16 core	8 core	16 core
No Lab	$\eta = 0.2$	0.56	0.94	1.49	2.46	**4.37**	**8.07**	2.74	3.82	0.46	0.48
	$\eta = 0.3$	0.55	1.15	4.15	**11.42**	**5.47**	9.62	0.90	1.48	3.02	3.40
Unif	$\eta = 0.2$	0.53	0.69	**1.17**	**1.24**	0.59	0.56	0.50	0.66	9.38e−04	9.05e−04
	$\eta = 0.3$	0.52	0.94	**4.15**	**4.75**	1.49	1.98	0.68	1.06	1.41e−03	1.42e−03
NonUni	$\eta = 0.2$	0.59	0.68	**0.71**	**0.70**	0.58	0.52	0.35	0.40	8.20e−04	8.47e−04
	$\eta = 0.3$	0.52	0.93	**1.6**	1.50	1.43	**1.70**	0.57	0.77	1.26e−03	1.33e−03

Differently from [7], in this paper we are interested in analyzing both the communication overhead and the capability of each policy to effectively balance the work among the threads from the beginning of the search process. Therefore, we have performed the experiments considering a thread pool of 8 and 16 threads respectively; then we have initialized the GSS by pushing only the empty state, so letting thread that picks the first state to distribute the initial work to the others. The different strategies have been compared using the best configuration of parameters among all the experiments performed. In particular, we set the limit of the LSS depth in $VF3P_{LS}$ to a tenth of the size of the pattern and the limit to the number of descendants in $VF3P_{LS}$ to the size of the pattern. As for $VF3P_{AWS}$, we checked for inactive threads every 10 states.

The experiments have been performed on a server equipped with four Intel(R) Xeon(R) Gold 5220, 72 cores in total, and 1 Tb of Ram. The hyperthreading has been deactivated and 64 cores have been isolated so as to avoid the operating system to allocate user processes, the left cores have been used to measure the performance indices. In order to get unbiased measures of the speed-up and the efficiency, each algorithm has a subset of cores reserved for the whole execution time and each thread in the pool has been pinned to a core of such a set.

In Tables 2 and 3 we show the results in terms of speed-up and efficiency respectively; due to space we have reported the average value computed over all the graphs of each dataset. In every table we have highlighted the performance achieved by the algorithms while searching for all the solutions and the first one only. Despite the two problems are similar, the former requires to explore a considerable higher number of states.

Before discussing the results it is important to point out that, since all the four parallel algorithms proposed are derived from VF3-Light [6], given a couple of graphs they all generates the same amount of states following the same order

Table 3. Efficiency of the parallel algorithms over different target graph size and number of CPU cores employed.

Dataset		VF3P		$VF3P_{LS}$		$VF3P_{DL}$		$VF3P_{AWS}$		Glasgow	
		8 core	16 core	8 core	16 core	8 core	16 core	8 core	16 core	8 core	16 core
						All the solutions					
No Lab	$\eta=0.2$	0.64	0.54	**0.78**	**0.73**	0.70	0.65	0.21	0.19	0.33	0.18
	$\eta=0.3$	0.63	0.66	0.90	0.79	0.77	0.72	0.50	0.53	**1.39**	**0.84**
Unif	$\eta=0.2$	0.47	**0.33**	0.19	0.15	**0.48**	0.28	0.20	0.10	2.53e−04	1.29e−04
	$\eta=0.3$	0.59	0.54	**0.73**	0.35	0.68	**0.59**	0.42	0.32	3.48e−04	1.94e−04
NonUni	$\eta=0.2$	**0.47**	**0.29**	0.33	0.16	0.44	0.23	0.19	0.09	2.05e−04	1.08e−04
	$\eta=0.3$	0.59	0.54	0.52	0.25	**0.70**	**0.58**	0.45	0.33	2.94e−04	1.67e−04
						First solution					
No Lab	$\eta=0.2$	0.07	0.05	0.19	0.15	**0.54**	**0.50**	0.34	0.23	0.05	0.03
	$\eta=0.3$	0.07	0.07	0.52	**0.71**	**0.68**	0.60	0.11	0.09	0.38	0.21
Unif	$\eta=0.2$	0.07	0.04	**0.15**	**0.08**	0.07	0.04	0.06	0.04	1.17e−04	5.66e−05
	$\eta=0.3$	0.06	0.06	**0.52**	**0.30**	0.19	0.12	0.08	0.07	1.76e−04	8.86e−05
NonUni	$\eta=0.2$	0.07	**0.04**	**0.09**	**0.04**	0.07	0.03	0.04	0.03	1.02e−04	5.29e−05
	$\eta=0.3$	0.06	0.06	**0.20**	0.09	0.18	**0.11**	0.07	0.05	1.58e−04	8.31e−05

with respect to the pattern graph. This is due to the fact that the proposed algorithms do not add any new heuristics to visit the state space but an effective way to parallelize the exploration. Therefore, we can reasonably consider the performance gap among them only related to the policy they use.

As shown in Table 2 for all the analyzed strategies the speed-up grows almost linearly w.r.t the number of threads, among them the best two are $VF3P_{LS}$ and $VF3P_{DL}$. A similar result is evident while looking at the efficiency, in Table 3. In particular, when we analyze the results achieved on the unlabelled graphs, the improvement in efficiency is between the 5% and the 15% while searching for all the solutions and it is higher than the 60% when the task is to find the first solution. Dealing with labelled graphs is generally less challenging, but also in this case the efficiency improvement is 25% and 35% while searching for all the solutions on the graphs that are more dense.

Looking at the whole picture, it is not the LSS itself to provide a real advantage to the algorithms, but the way it is used together with the GSS. Both for $VF3P_{LS}$ and $VF3P_{DL}$ the use of the LSS has reduced the communication overhead, but the main difference between them and $VF3P_{AWS}$ is how the LSS affects the ability of the algorithm to balance the workload among the thread. Indeed, the latter has achieved a very poor performance because, for most of the time, only a subset of threads are fully involved in the exploration. A proof of that is the very low efficiency achieved by $VF3P_{AWS}$ also in the most challenging situation (enumerating all the solution on unlabelled graphs). Therefore, we can deduce that setting explicitly a limit to the capacity of the LSS allows the algorithm to share the work among the thread more efficiently.

As for the memory requirements, we have reported the ratio between the average memory usage peak of the parallel algorithm and that of the baseline. Analyzing the results in Table 4 it is clear that, despite the lowest memory footprint is achieved by $VF3P$ and $VF3P_{LS}$, the difference among the algorithms is

Table 4. Average peak of memory usage while searching for all the solutions. The memory usage of the baseline $VF3L$ is reported in megabytes and for each parallel algorithms we show the ratio between the memory usage of the parallel algorithm and the baseline.

		$VF3L$					
		No Lab		Unif		NonUni	
		$\eta = 0.2$	$\eta = 0.3$	$\eta = 0.2$	$\eta = 0.3$	$\eta = 0.2$	$\eta = 0.3$
		57.56	48.16	551.12	726.58	539.09	747.96

Dataset		$VF3P$		$VF3P_{LS}$		$VF3P_{DL}$		$VF3P_{AWS}$		Glasgow	
		8 core	16 core	8 core	16 core	8 core	16 core	8 core	16 core	8 core	16 core
Unl	$\eta = 0.2$	11.20	21.77	11.53	21.61	15.51	25.24	11.67	21.98	11.29	**20.84**
	$\eta = 0.3$	13.64	26.67	14.28	**26.45**	22.51	35.42	13.68	26.49	17.36	30.16
Uni	$\eta = 0.2$	2.10	3.19	2.11	**3.14**	2.45	**3.14**	2.13	3.16	1.61	3.06
	$\eta = 0.3$	1.83	**2.64**	1.86	**2.64**	3.09	3.70	1.85	2.63	1.34	2.59
NonUni	$\eta = 0.2$	2.13	3.23	2.14	3.18	2.29	**3.15**	2.15	3.21	1.73	3.44
	$\eta = 0.3$	1.81	2.60	1.82	**2.58**	2.41	3.07	1.82	**2.58**	1.47	2.39

minimal except for $VF3P_{LS}$ that in the case of unlabeled graphs. Since the algorithms uses the same data structures and generates the same amount of states, this difference can be justified by the fact that, in the average, more states are stored in the LSSs of the threads used by $VF3P_{LS}$.

Considering Glasgow, it has outperformed all the strategies only in the case of unlabelled graphs with density $\eta = 0.3$, in all the other situations it showed very poor performance. It is worth to point out that, in the case of labelled graphs the results are referred to only half of the instances considered, the ones that the algorithm has been able to complete; since it has a significant impact on the average peak of memory usage reported in Table 4 we have highlighted only the best results achieved by the other algorithms.

5   Conclusions

In this paper we have compared four different policies to manage the communication among the threads while solving the subgraph isomorphism using parallel tree-search based algorithms derived from the state-of-the-art sequential algorithm VF3-Light. In addition to the strategy proposed in [7] we have introduced two novel approaches, $VF3P_{LS}$ and $VF3P_{AWS}$, so as to further extend the analysis and evaluate the most effective way to balance the communication overhead with a good level of work sharing among the threads. These four algorithms have been compared with Glasgow, another very recent parallel algorithm based on constraint programming. Two strategies, $VF3P_{DL}$ and $VF3P_{LS}$, have been proved to the most effective on six different database of large and dense random graphs. The experimentation can be extended to other databases, for providing a better characterization of the situations in which these strategies give the best results.

References

1. Archibald, B., Dunlop, F., Hoffmann, R., McCreesh, C., Prosser, P., Trimble, J.: Sequential and parallel solution-biased search for subgraph algorithms. In: Rousseau, L.-M., Stergiou, K. (eds.) CPAIOR 2019. LNCS, vol. 11494, pp. 20–38. Springer, Cham (2019). https://doi.org/10.1007/978-3-030-19212-9_2
2. Bi, F., Chang, L., Lin, X., Qin, L., Zhang, W.: Efficient subgraph matching by postponing Cartesian products. In: Proceedings of the 2016 International Conference on Management of Data. Association for Computing Machinery (2016). https://doi.org/10.1145/2882903.2915236
3. Bonnici, V., Giugno, R., Pulvirenti, A., Shasha, D., Ferro, A.: A subgraph isomorphism algorithm and its application to biochemical data. BMC Bioinform. 14, 1–13 (2013)
4. Carletti, V., Foggia, P., Saggese, A., Vento, M.: Challenging the time complexity of exact subgraph isomorphism for huge and dense graphs with VF3. IEEE Trans. Pattern Anal. Mach. Intell. 40, 804–818 (2018)

5. Carletti, V., Foggia, P., Vento, M., Jiang, X.: Report on the first contest on graph matching algorithms for pattern search in biological databases. In: Liu, C.-L., Luo, B., Kropatsch, W.G., Cheng, J. (eds.) GbRPR 2015. LNCS, vol. 9069, pp. 178–187. Springer, Cham (2015). https://doi.org/10.1007/978-3-319-18224-7_18
6. Carletti, V., Foggia, P., Greco, A., Saggese, A., Vento, M.: The VF3-light subgraph isomorphism algorithm: when doing less is more effective. In: Bai, X., Hancock, E.R., Ho, T.K., Wilson, R.C., Biggio, B., Robles-Kelly, A. (eds.) S+SSPR 2018. LNCS, vol. 11004, pp. 315–325. Springer, Cham (2018). https://doi.org/10.1007/978-3-319-97785-0_30
7. Carletti, V., Foggia, P., Ritrovato, P., Vento, M., Vigilante, V.: A parallel algorithm for subgraph isomorphism. In: Conte, D., Ramel, J.-Y., Foggia, P. (eds.) GbRPR 2019. LNCS, vol. 11510, pp. 141–151. Springer, Cham (2019). https://doi.org/10.1007/978-3-030-20081-7_14
8. Carletti, V., Foggia, P., Vento, M.: Performance comparison of five exact graph matching algorithms on biological databases. In: Petrosino, A., Maddalena, L., Pala, P. (eds.) ICIAP 2013. LNCS, vol. 8158, pp. 409–417. Springer, Heidelberg (2013). https://doi.org/10.1007/978-3-642-41190-8_44
9. Carletti, V., Foggia, P., Vento, M.: VF2 plus: an improved version of VF2 for biological graphs. In: Liu, C.-L., Luo, B., Kropatsch, W.G., Cheng, J. (eds.) GbRPR 2015. LNCS, vol. 9069, pp. 168–177. Springer, Cham (2015). https://doi.org/10.1007/978-3-319-18224-7_17
10. Conte, D., Foggia, P., Sansone, C., Vento, M.: Thirty years of graph matching in pattern recognition. Int. J. Pattern Recogn. Artif. Intell. **18**, 265–298 (2004)
11. Cordella, L., Foggia, P., Sansone, C., Vento, M.: A (sub)graph isomorphism algorithm for matching large graphs. IEEE Trans. Pattern Anal. Mach. Intell. **26**, 1367–1372 (2004)
12. Foggia, P., Percannella, G., Vento, M.: Graph matching and learning in pattern recognition in the last ten years. Int. J. Patt. Recogn. Artif. Intell. **28**, 1450001 (2014)
13. Han, W., Lee, J.h., Lee, J.: TurboISO: towards ultrafast and robust subgraph isomorphism search in large graph databases. In: SIGMOD pp. 337–348 (2013)
14. Jenkins, J., Arkatkar, I., Owens, J.D., Choudhary, A., Samatova, N.F.: Lessons learned from exploring the backtracking paradigm on the GPU. In: Jeannot, E., Namyst, R., Roman, J. (eds.) Euro-Par 2011. LNCS, vol. 6853, pp. 425–437. Springer, Heidelberg (2011). https://doi.org/10.1007/978-3-642-23397-5_42
15. Jin, X., Lai, L.: MPMatch: a multi-core parallel subgraph matching algorithm (2019). https://doi.org/10.1109/ICDEW.2019.000-6
16. McCreesh, C., Prosser, P.: A parallel, backjumping subgraph isomorphism algorithm using supplemental graphs. In: Pesant, G. (ed.) CP 2015. LNCS, vol. 9255, pp. 295–312. Springer, Cham (2015). https://doi.org/10.1007/978-3-319-23219-5_21
17. Shang, H., Zhang, Y., Lin, X., Yu, J.X.: Taming verification hardness: An efficient algorithm for testing subgraph isomorphism. Proc. VLDB Endow. **1**(1), 364–375 (2008)
18. Solnon, C.: AllDifferent-based filtering for subgraph isomorphism. Artif. Intell. **174**, 850–864 (2010)
19. Sun, Z., Wang, H., Wang, H., Shao, B., Li, J.: Efficient subgraph matching on billion node graphs. Proc. VLDB Endow. (2012). https://doi.org/10.14778/2311906.2311907

20. Tran, H.-N., Kim, J., He, B.: Fast subgraph matching on large graphs using graphics processors. In: Renz, M., Shahabi, C., Zhou, X., Cheema, M.A. (eds.) DASFAA 2015. LNCS, vol. 9049, pp. 299–315. Springer, Cham (2015). https://doi.org/10.1007/978-3-319-18120-2_18

21. Ullmann, J.R.: Bit-vector algorithms for binary constraint satisfaction and subgraph isomorphism. ACM J. Exp. Algorithmics **15** (2011). https://doi.org/10.1145/1671970.1921702. Association for Computing Machinery

22. Vento, M.: A long trip in the charming world of graphs for pattern recognition. Pattern Recogn. **48**, 291–301 (2014)

23. Xu, Q., Jeon, H., Annavaram, M.: Graph processing on GPUS: where are the bottlenecks. In: 2014 IEEE International Symposium on Workload Characterization (2014)

24. Zeng, L., Zou, L., Özsu, M.T., Hu, L., Zhang, F.: GSI: GPU-friendly subgraph isomorphism. In: 2020 IEEE 36th International Conference on Data Engineering (ICDE), pp. 1249–1260 (2020)

Multimedia Analysis and Understanding

Multiple-Image Super-Resolution Using Deep Learning and Statistical Features

Jakub Nalepa[1,2](✉) [ID], Krzysztof Hrynczenko[1] [ID], and Michal Kawulok[1,2] [ID]

[1] Silesian University of Technology, Gliwice, Poland
{jnalepa,michal.kawulok}@ieee.org
[2] KP Labs, Gliwice, Poland

Abstract. Capturing, transferring, and storing high-resolution images has become a serious issue in a wide range of fields, in which these processes are costly, time-consuming, or even infeasible. As obtaining low-resolution images may be easier in practice, enhancing their spatial resolution is currently an active research area and encompasses both single- and multiple-image super-resolution techniques. In this paper, we propose a deep learning approach for multiple-image super-resolution that is independent from the number of available low-resolution images of the scene. It is in contrast to other deep networks which are crafted to deal with input stacks of a constant size, hence are not applicable once the number of low-resolution images varies. The experiments showed that our technique not only outperforms other single- and multiple-image super-resolution algorithms, but also it is lightweight and delivers instant operation, thus can be deployed in hardware-constrained environments.

Keywords: Super-resolution · Deep learning · Statistical features.

1 Introduction

Super-resolution reconstruction (SRR) is a process of enhancing the spatial resolution of an input image (or a stack of images presenting the same scene). In single-image SR (SISR), we rely solely on one low-resolution (LR) image to perform the reconstruction, whereas multiple-image SR (MISR) takes advantage of using a set of input images capturing the same spatial area, although with subtle differences like sub-pixel shifts. SRR plays an increasingly important role in many applications from various fields, including medical imaging, Earth observation, video processing and more, in which acquiring and storing high-resolution (HR) image data is costly, time-consuming, or even impossible in practice [8].

The most basic SISR techniques exploit various interpolation methods [17], hence are computationally lightweight. Edge-based methods are based on the

This research was supported by the National Science Centre, Poland, under Research Grant 2019/35/B/ST6/03006, and partially by European Space Agency (DeepSent). JN was supported by the Silesian University of Technology funds (02/080/BKM20/0012).

© Springer Nature Switzerland AG 2021
A. Torsello et al. (Eds.): S+SSPR 2020, LNCS 12644, pp. 261–271, 2021.
https://doi.org/10.1007/978-3-030-73973-7_25

assumption that edges are the most important factor when it comes to the perception of image quality, but they often fail to effectively reconstruct texture or fine-grained details. Finally, statistical algorithms utilize image properties and benefit from the gradient distribution to recover a HR image. There exist example-based machine learning-powered approaches which use LR-HR image pairs to learn mapping functions that are later exploited to super-resolve LR images. In this context, convolutional neural networks (CNNs) were initially introduced, commonly coupled with interpolation techniques. Dong et al. proposed Super-Resolution Convolutional Neural Network (SRCNN) [3], later redesigned to accelerate its operation [4], and followed by other deep architectures [10].

In MISR, we build upon a premise that each LR image was derived from its HR counterpart using a given imaging model [8], therefore SR may be considered as a task of reversing this degradation process. Although there are methods that exploit image processing and analysis approaches to super-resolve stacks of LR images [5], deep learning algorithms for MISR have recently emerged as a data-driven alternative which might capture fine-grained image characteristics during the reconstruction [8,13]. Unfortunately, the existent deep architectures are designed to work with fixed-size LR stacks, and cannot be seamlessly used to super-resolve different numbers of LRs without re-training. It hampers the exploitation of such methods in practice, as the number of available LRs (alongside their quality) can vary across separate acquisitions.

We tackle the problem of MISR using deep learning, and propose a deep model that is independent from the number of LRs that are fed as an input image stack for SRR (Sect. 2). We benefit from statistical features extracted from the co-registered LRs, which are later stacked together with a single LR during the reconstruction. This approach makes our network applicable over LR stacks of variable sizes. The experiments, backed up with thorough quantitative and qualitative analysis, showed that not only does our method deliver high-quality and plausible SR images and outperform other SISR and MISR techniques, but it is also lightweight and infers very fast (Sect. 3). Therefore, it could be deployed in constrained environments, e.g., on board an imaging satellite, as a pre-processing step to enhance the spatial resolution of captured scenes.

2   Method

In this section, we discuss our deep CNN for MISR which benefits from the statistical features extracted from an input stack of images (we refer to this architecture as StatNet). In StatNet, we abstract from the number of images, hence we can super-resolve image stacks of any size (greater or equal to two, as we utilize feature maps that would be empty for a single-image case) without updating the backbone deep model and re-training it. The architecture can be separated into a couple of building blocks, including the statistical block (Sect. 2.1), extraction block (Sect. 2.2), and recursive block (Sect. 2.3).

Input LR images

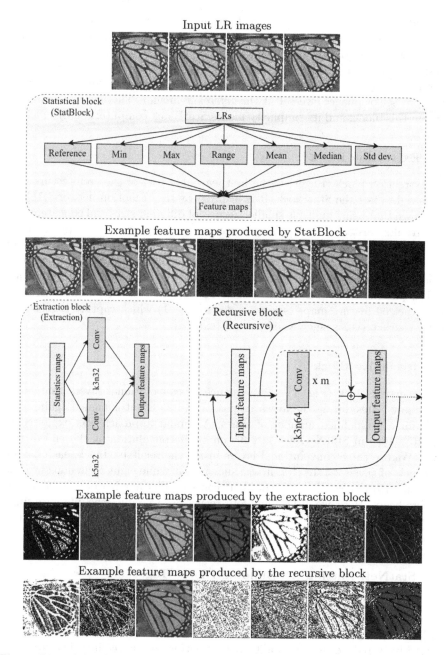

Fig. 1. Pivotal StatNet building blocks, alongside example feature maps (k—kernel size, n—the number of kernels).

2.1 Statistical Block

The statistical block (StatBlock, Fig. 1) transforms any number of input images into six feature maps (the minimum, maximum, range, mean, median, and standard deviation of the pixel values at a given position within the stack of co-registered input images; see examples in Fig. 1). These feature maps are stacked together (always in the same order), and coupled with a random image[1] from the input set of images (we call it the *reference* image). There are no learnable parts in this block, and its simplicity makes StatBlock computationally efficient.

2.2 Extraction Block

The extraction block (Fig. 1) is the first learnable part of the architecture that comes right after the StatBlock. It is inspired by the Inception network [15]—it was shown that using multiple smaller kernels of different sizes and concatenating their results, instead of using larger kernels, can capture higher-quality deep features, and improve the performance of the deep model. In StatNet, an input to the extraction block is processed by two convolutional layers (both with 32 kernels, one having kernels of size 3×3, and the other of size 5×5) that are stacked parallel to each other, and their outputs are concatenated. Therefore, we extract 64 feature maps (see examples in Fig. 1) which capture higher-level object characteristics, such as edges and textures.

2.3 Recursive Block

The recursive block (Fig. 1) is a sequence of convolutional layers that is re-used many times during the forwards pass, i.e., its output is passed again as an input for a pre-defined number of times. A similar approach was exploited in LapSRN [10]—in StatNet, we do not employ deconvolution at the end of this block. We use three convolutional layers inside the recursive block, each having 64 kernels of size 3×3. In [10], it was shown that while eight recursions provide the best performance, the difference between a smaller number of recursions was not significant. Here, we utilize only four recursions to reduce the inference time. The feature maps produced by the recursive block are harder to interpret, but we can observe that some of them resemble the object being reconstructed.

2.4 StatNet: Putting All Building Blocks Together

StatNet (Fig. 2) is built as a sequence of the aforementioned blocks (StatBlock, extraction, and recursive blocks). The feature maps produced be the recursive block are then aggregated by two convolutional layers (with 3×3 kernels) to progressively reduce dimensionality before performing upscaling. The first one combines 64 feature maps into 32, which are processed by the second convolution that outputs four feature maps. All of the convolutional layers are followed by

---

[1] In this work, we exploit the first image in the stack.

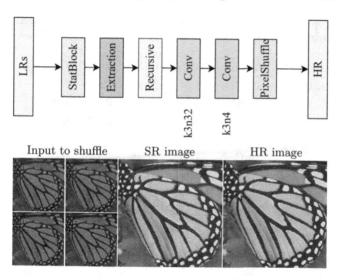

Fig. 2. The StatNet architecture (k—kernel size, n—the number of kernels), with example pixel shuffle layer's inputs, resulting SR image, and the corresponding HR image.

leaky ReLUs (LReLUs), with a negative slope of 0.2, as suggested in [11]. It was proven that LReLU is capable of increasing the network's performance, and accelerating the training process, when compared to ReLU [16]. The output of the last convolutional layer is processed by the pixel shuffle layer [14], which does the final upscaling. An example of the input pixel-shuffle feature maps, alongside the resulting SR image and the corresponding HR image are presented in Fig. 2.

3   Experiments

To confront StatNet with the state of the art and investigate its abilities, we compare it with four deep learning SR algorithms: SRCNN [3], FSRCNN [4], LapSRN [10] (all SISR), and MFCNN [6] (MISR). Additionally, the bicubic interpolation [9] is adopted to serve as a MISR baseline. Since SRCNN, FSR-CNN, and LapSRN are the SISR methods, we generate four SR images (for each LR image separately), and take the highest-quality SR—according to a given metric—in the comparisons. To quantify the super-resolution quality[2], we utilized the mean absolute error (MAE, the smaller the value, the better, thus we want to minimize it: ↓), peak signal-to-noise ratio (PSNR↑), structural similarity index measure (SSIM↑), and gradient-based structural similarity index measure (GSSIM↑). Each of these metrics is calculated and averaged over all images for each test dataset. The experiments ran on a personal computer with an AMD Ryzen 5 1600 Six-Core Processor (16 GB DDR4 2400 MHz), and NVIDIA GeForce GTX 1060 GB.

---

[2] For more details on the metrics used for evaluating SR algorithms, see [2].

Table 1. The quantitative comparison of all methods across all datasets. The green bold values show the best score, while the black bold are the second best scores.

Dataset	Algorithm	MAE↓	PSNR↑	SSIM↑	GSSIM↑
Set5	SRCNN	7.03	27.99	0.80	0.46
	FSRCNN	6.80	28.26	0.81	0.49
	LapSRN	**6.25**	**28.84**	**0.83**	**0.52**
	Bicubic	7.29	28.16	0.78	0.47
	MFCNN	17.62	21.48	0.74	0.46
	StatNet	4.81	31.20	0.88	0.61
Set14	SRCNN	8.99	26.04	0.72	0.43
	FSRCNN	9.13	25.25	0.73	0.44
	LapSRN	**8.72**	25.48	**0.75**	**0.48**
	Bicubic	8.81	**26.49**	0.73	0.47
	MFCNN	19.89	20.72	0.67	0.39
	StatNet	6.73	27.68	0.82	0.58
BSD100	SRCNN	9.55	25.62	0.69	0.40
	FSRCNN	9.41	25.36	0.70	0.42
	LapSRN	**8.92**	26.09	**0.72**	0.45
	Bicubic	9.20	**26.16**	0.71	**0.46**
	MFCNN	17.29	21.32	0.65	0.35
	StatNet	7.07	27.92	0.80	0.58
Manga109	SRCNN	7.80	25.09	0.82	0.51
	FSRCNN	8.31	22.39	0.82	0.53
	LapSRN	**6.96**	24.91	**0.86**	**0.58**
	Bicubic	8.45	26.56	0.77	0.44
	MFCNN	31.19	16.90	0.77	0.47
	StatNet	5.50	**25.68**	0.89	0.62
Urban100	SRCNN	10.79	23.91	0.73	0.48
	FSRCNN	10.75	23.18	0.75	0.50
	LapSRN	**10.36**	23.26	**0.77**	**0.55**
	Bicubic	11.13	**24.23**	0.70	0.45
	MFCNN	21.96	19.22	0.64	0.38
	StatNet	7.66	25.40	0.84	0.64

In this work, we focus on 2× SR, and utilize DIV2K [1] to randomly sample 1280 training and 320 validation patches, and Set5, Set14, BSD100, Manga109, and Urban100 [7,12] are used as test sets which were never seen during the training process. As all of these datasets are single-image, we simulate the low-resolution stacks of images for each HR (original) image by applying the following pixel shifts: $(1,0)$, $(0,1)$, and $(1,1)$ in the HR domain, which results in sub-pixel shifts in the LR domain, due to downsampling via bicubic interpolation.

Table 2. Training and inference times for each investigated algorithm.

Algorithm →	Training time					
	SRCNN	FSRCNN	LapSRN	Bicubic	MFCNN	StatNet
Total (min)	31.00	60.76	116.70	–	63.92	29.16
Avg/epoch (sec)	13.78	12.53	42.44	–	28.43	33.01
Dataset ↓	Inference time (seconds)					
Set5	0.08	0.07	0.52	0.07	0.04	0.70
Set14	0.42	0.29	2.84	0.04	0.21	0.39
BSD100	2.93	2.86	14.71	0.23	1.88	2.71
Manga109	13.15	7.26	87.58	1.65	6.27	12.41
Urban100	9.92	5.54	63.07	1.10	4.81	9.29

Additionally, we inject the Gaussian (with zero mean) and impulsive noise (using a random noise map sampled from the uniform distribution) into LR images. For simplicity, we convert all images to grayscale.

The quantitative results, obtained over all test datasets, are gathered in Table 1. We can appreciate that StatNet outperforms other techniques in the majority of cases (over all metrics). It is observable that LapSRN consistently delivers high-quality SR images too, but it trains and infers much slower (Table 2), as the number of its trainable parameters, together with the required multiply-accumulate operations (Table 3) are significantly larger when compared with StatNet. Therefore, StatNet not only elaborates higher-quality SR images, but it is also more compact and easier to deploy in constrained execution environments where real-time inference and memory/energy frugality are pivotal, e.g., in Earth observation applications. In Fig. 3, we render example reconstruction results obtained using all investigated algorithms. StatNet produces more plausible images with well-restored fine-grained image details (e.g., *Lena* and *Fish*). Although there are images with visible artifacts (*Manga*), StatNet was able to accurately restore detailed edge information in such cases.

Table 3. The number of trainable parameters, and required multiply-accumulate operations (MACs; the lower, the better) for each network.

Algorithm	# parameters	MACs (M)
SRCNN	$0.81 \cdot 10^4$	407.88
FSRCNN	$12.64 \cdot 10^4$	1317.01
LapSRN	$25.15 \cdot 10^4$	87557.99
MFCNN	$17.28 \cdot 10^4$	8594.19
StatNet	$13.18 \cdot 10^4$	23647.94

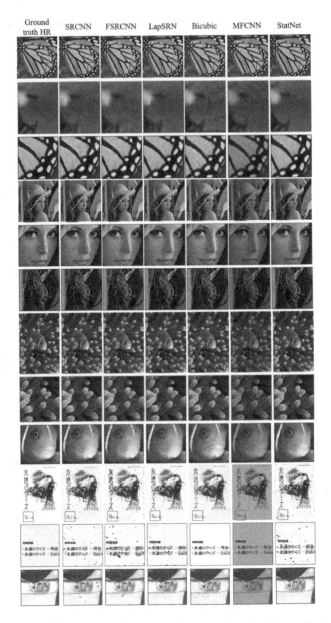

Fig. 3. Examples of SR images (three rows for each image) obtained using all methods for *Butterfly* (from the Set5 dataset), *Lena* (Set14), *Fish* (BSD100), and *Manga* (Manga109). We zoom some interesting parts of the images in blue and red rectangles. (Color figure online)

Table 4. Evaluating StatNet using different numbers of input LR images (two and three initial images from the input stack, and all images (StatNet$_2$–StatNet$_4$). The green bold values show the best score, while the black bold are the second best scores.

Dataset	Algorithm	MAE↓	PSNR↑	SSIM↑	GSSIM↑
Set5	StatNet$_2$	15.70	16.98	0.56	0.34
	StatNet$_3$	**6.12**	**27.46**	**0.84**	**0.55**
	StatNet$_4$	4.81	31.20	0.88	0.61
Set14	StatNet$_2$	15.39	19.77	0.55	0.37
	StatNet$_3$	**10.38**	**23.89**	**0.76**	**0.52**
	StatNet$_4$	6.73	27.68	0.82	0.58
BSD100	StatNet$_2$	14.47	20.59	0.53	0.36
	StatNet$_3$	**9.68**	**24.22**	**0.74**	**0.51**
	StatNet$_4$	7.07	27.92	0.80	0.58
Manga109	StatNet$_2$	18.49	16.19	0.54	0.33
	StatNet$_3$	**13.98**	**18.02**	**0.76**	**0.50**
	StatNet$_4$	5.50	25.68	0.89	0.62
Urban100	StatNet$_2$	18.19	17.55	0.54	0.38
	StatNet$_3$	**11.01**	**21.55**	**0.77**	**0.54**
	StatNet$_4$	7.66	25.40	0.84	0.64

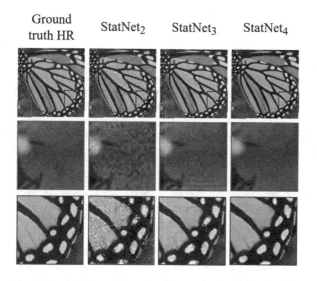

Fig. 4. The *Butterfly* image (Set5) reconstructed using StatNet with two and three initial images from the input stack, and all images (StatNet$_2$–StatNet$_4$).

To better understand the abilities of StatNet in the context of super-resolving image stacks of varying sizes, we processed test benchmarks[3] by exploiting two and three initial images from the generated LR stacks, and compared the resulting SR reconstructions with those obtained using all available LRs (Table 4). The results obtained for two-image LR stacks are significantly worse than those elaborated for three- and four-image cases. However, Fig. 4 shows that the SR images still present valid reconstruction, and increasing the number of LRs directly leads to notably better SR images. Therefore, reducing the number of LR images adversely affect the super-resolution performance of StatNet, but it can still be successfully deployed in such scenarios, and can produce valid results.

4   Conclusion

We introduced a deep learning algorithm for MISR, referred to as StatNet, which benefits from statistical features extracted from an input LR image stack, and is independent from the number of available LRs. Our experiments, performed over several benchmark datasets and coupled with quantitative and qualitative analysis, showed that StatNet outperforms other SISR and MISR approaches, also including different deep learning techniques. The sensitivity analysis revealed that StatNet can effectively super-resolve LR stacks of different sizes, it is lightweight, and train and infer very fast, hence it could be deployed e.g., on board imaging satellites where memory- and energy-frugality, together with real-time processing are critical issues.

References

1. Agustsson, E., Timofte, R.: NTIRE 2017 challenge on single image super-resolution: Dataset and study. In: Proceedings of IEEE CVPR, pp. 126–135 (2017)
2. Benecki, P., Kawulok, M., Kostrzewa, D., Skonieczny, L.: Evaluating super-resolution reconstruction of satellite images. Acta Astronautica **153**, 15–25 (2018)
3. Dong, C., Loy, C.C., He, K., Tang, X.: Learning a deep convolutional network for image super-resolution. In: Fleet, D., Pajdla, T., Schiele, B., Tuytelaars, T. (eds.) ECCV 2014. LNCS, vol. 8692, pp. 184–199. Springer, Cham (2014). https://doi.org/10.1007/978-3-319-10593-2_13
4. Dong, C., Loy, C.C., Tang, X.: Accelerating the super-resolution convolutional neural network. In: Leibe, B., Matas, J., Sebe, N., Welling, M. (eds.) ECCV 2016. LNCS, vol. 9906, pp. 391–407. Springer, Cham (2016). https://doi.org/10.1007/978-3-319-46475-6_25
5. Farsiu, S., Robinson, M.D., Elad, M., Milanfar, P.: Fast and robust multiframe super resolution. IEEE Trans. Image Process. **13**(10), 1327–1344 (2004)
6. Greaves, A., Winter, H.: Multi-frame video super-resolution using convolutional neural networks. Stanford University Course Project Reports (2017)
7. Huang, J.B., Singh, A., Ahuja, N.: Single image super-resolution from transformed self-exemplars. In: Proceedings of IEEE CVPR, pp. 5197–5206 (2015)

---

[3] Note that we did not re-train StatNet.

8. Kawulok, M., Benecki, P., Piechaczek, S., Hrynczenko, K., Kostrzewa, D., Nalepa, J.: Deep learning for multiple-image super-resolution. IEEE Geosci. Remote Sens. Lett. **17**(6), 1062–1066 (2020)

9. Keys, R.: Cubic convolution interpolation for digital image processing. IEEE Trans. Acoust. Speech Signal Process. **29**(6), 1153–1160 (1981)

10. Kim, J., Kwon Lee, J., Mu Lee, K.: Accurate image super-resolution using very deep convolutional networks. In: Proceedings of IEEE CVPR, pp. 1646–1654 (2016)

11. Kim, J., Kwon Lee, J., Mu Lee, K.: Deeply-recursive convolutional network for image super-resolution. In: Proceedings of IEEE CVPR, pp. 1637–1645 (2016)

12. Matsui, Y., Ito, K., Aramaki, Y., Fujimoto, A., Ogawa, T., Yamasaki, T., Aizawa, K.: Sketch-based manga retrieval using manga109 dataset. Multimedia Tools Appl. **76**(20), 21811–21838 (2016). https://doi.org/10.1007/s11042-016-4020-z

13. Molini, A.B., Valsesia, D., Fracastoro, G., Magli, E.: DeepSUM: Deep neural network for super-resolution of unregistered multitemporal images. IEEE Trans. Geosci. Remote Sens. **58**(5), 3644–3656 (2019)

14. Shi, W., et al.: Real-time single image and video super-resolution using an efficient sub-pixel convolutional neural network. In: Proceedings of IEEE CVPR, pp. 1874–1883 (2016)

15. Szegedy, C., Vanhoucke, V., Ioffe, S., Shlens, J., Wojna, Z.: Rethinking the inception architecture for computer vision. In: Proceedings of IEEE CVPR, pp. 2818–2826 (2016)

16. Xu, B., Wang, N., Chen, T., Li, M.: Empirical evaluation of rectified activations in convolutional network. arXiv preprint arXiv:1505.00853 (2015)

17. Yang, C.-Y., Ma, C., Yang, M.-H.: Single-image super-resolution: a benchmark. In: Fleet, D., Pajdla, T., Schiele, B., Tuytelaars, T. (eds.) ECCV 2014. LNCS, vol. 8692, pp. 372–386. Springer, Cham (2014). https://doi.org/10.1007/978-3-319-10593-2_25

Unsupervised Semantic Discovery Through Visual Patterns Detection

Francesco Pelosin[✉], Andrea Gasparetto, Andrea Albarelli,
and Andrea Torsello

Ca' Foscari University, Venice, Italy
{francesco.pelosin,andrea.gasparetto,andrea.albarelli,
andrea.torsello}@unive.it

Abstract. We propose a new fast fully unsupervised method to discover semantic patterns. Our algorithm is able to hierarchically find visual categories and produce a segmentation mask where previous methods fail. Through the modeling of what is a visual pattern in an image, we introduce the notion of "semantic levels" and devise a conceptual framework along with measures and a dedicated benchmark dataset for future comparisons. Our algorithm is composed by two phases. A filtering phase, which selects semantical hotsposts by means of an accumulator space, then a clustering phase which propagates the semantic properties of the hotspots on a superpixels basis. We provide both qualitative and quantitative experimental validation, achieving optimal results in terms of robustness to noise and semantic consistency. We also made code and dataset publicly available.

Keywords: Visual-pattern-detection · Semantic-discovery · Cosegmentation

1   Introduction

The extraction of semantic categories from images is a fundamental task in image understanding [5,18,29]. While the task is one that has been widely investigated in the community, most approaches are supervised, making use of labels to detect semantic categories [2]. Comparatively less effort has been put to investigate automatic procedures which enable an intelligent system to learn autonomously extrapolating visual semantic categories without any *a priori* knowledge of the context.

We observe the fact that in order to define what a visual pattern is, we need to define a scale of analysis (objects, parts of objects etc.). We call these scales *semantic levels* of the real world. Unfortunately most influential models arising from deep learning approaches still show a limited ability over scale invariance [13,25] which instead is common in nature. In fact, we don't really care much about scale, orientation or partial observability in the semantic world. For us, it is way more important to preserve an "internal representation" that matches reality [6,17].

© Springer Nature Switzerland AG 2021
A. Torsello et al. (Eds.): S+SSPR 2020, LNCS 12644, pp. 272–281, 2021.
https://doi.org/10.1007/978-3-030-73973-7_26

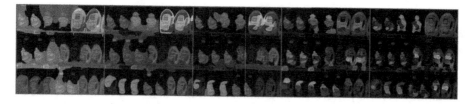

Fig. 1. A real world example of unsupervised segmentation of a grocery shelf. Our method can automatically discover both low-level coherent patterns (brands and logos) and high-level compound objects (detergents) by controlling the semantic level of the detection and segmentation process.

Our method leverages repetitions (Fig. 1) to capture the internal representation in the real world and then extrapolates categories at a specific semantic level. We do this without continuous geometrical constraints on the visual pattern disposition, which is common among other methodologies [8, 10, 21, 22].

We also do not constrain ourselves to find only one visual pattern, which is another very common assumption. Indeed, what if the image has more than one visual pattern? One can observe that this is *always* the case. Each visual repetition can be hierarchically decomposed in its smaller parts which, in turn, repeat over different semantic levels. This peculiar observation allow our work to contribute to the community as follows:

- A new pipeline able to capture semantic categories with the ability to hierarchically span over semantic levels.
- A better conceptual framework to evaluate analogous works through the introduction of the semantic levels notion along with a new metric.
- A new benchmark dataset of 208 labelled images for visual repetition detection.

Code, dataset and notebooks are public and available at: https://git.io/JT6UZ.

2   Related Works

Several works have been proposed to tackle visual pattern discovery and detection. While the paper by Leung and Malik [11] could be considered seminal, many other works build on their basic approach, working by detecting contiguous structures of similar patches by knowing the window size enclosing the distinctive pattern.

One common procedure in order to describe what a pattern is, consists to first extract descriptive features such as SIFT to perform a clustering in the feature space and then model the group disposition over the image by exploiting geometrical constraints, as in [21] and [4], or by relying only on appearance, as in [7, 14, 27].

The geometrical modeling of the repetitions usually is done by fitting a planar 2-D lattice, or a deformation of it [20], through RANSAC procedures as in [21,23] or even by exploiting the mathematical theory of crystallographic groups as in [15]. Shechtman and Irani [24], also exploited an active learning environment to detect visual patterns in a semi-supervised fashion. For example Cheng et al. [3] use input scribbles performed by a human to guide detection and extraction of such repeated elements, while Huberman and Fattal [9] ask the user to detect an object instance and then the detection is performed by exploiting correlation of patches near the input area.

Recently, as a result of the new wave of AI-driven Computer Vision, a number of Deep Leaning based approaches emerged, in particular Lettry et al. [10] argued that filter activation in a model such as AlexNet can be exploited in order to find regions of repeated elements over the image, thanks to the fact that filters over different layers show regularity in the activations when convolved with the repeated elements of the image. On top of the latter work, Rodríguez-Pardo et al. [22] proposed a modification to perform the texture synthesis step.

A brief survey of visual pattern discovery in both video and image data, up to 2013, is given by Wang et al. [28], unfortunately after that it seems that the computer vision community lost interest in this challenging problem. We point out that all the aforementioned methods look for *only one* particular visual repetition except for [14] that can be considered the most direct competitor and the main benchmark against which to compare our results.

3   Method Description

3.1   Features Localization and Extraction

We observe that any visual pattern is delimited by its contours. The first step of our algorithm, in fact, consists in the extraction of a set C of contour *keypoints* indicating a position $\mathbf{c}_j$ in the image. To extract keypoints, we opted for the Canny algorithm, for its simplicity and efficiency, although more recent and better edge extractor could be used [16] to have a better overall procedure.

A descriptor d_j is then computed for each selected $\mathbf{c}_j \in C$ thus obtaining a *descriptor set* $\mathcal{D}$. In particular, we adopted the DAISY algorithm because of its appealing dense matching properties that nicely fit our scenario. Again, here we can replace this module of the pipeline with something more advanced such as [19] at the cost of some computational time.

3.2   Semantic Hot Spots Detection

In order to detect self-similar patterns in the image we start by associating the k most similar descriptors for each descriptor $\mathbf{d}_j$. We can visualize this data structure as a star subgraph with k endpoints called *splash* "centered" on descriptor $\mathbf{d}_j$. Figure 2(a) shows one.

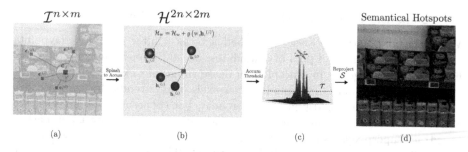

$\mathcal{I}^{n \times m}$ $\mathcal{H}^{2n \times 2m}$ Semantical Hotspots

$\mathcal{H}_w = \mathcal{H}_w + g(w, h_i^{(j)})$

(a)   (b)   (c)   (d)

Fig. 2. (a) A splash in the image space with center in the keypoint c_j. (b) $\mathcal{H}$, with the superimposed splash at the center, you can note the different levels of the vote ordered by endpoint importance i.e. descriptor similarity. (c) 3D projection showing the gaussian-like formations and the thresholding procedure of $\mathcal{H}$. (d) Backprojection through the set $\mathcal{S}$.

Splashes potentially encode repeated patterns in the image and similar patterns are then represented by similar splashes. The next step consists in separating these splashes from those that encode noise only, this is accomplished through an accumulator space.

In particular, we consider a 2-D *accumulator space* $\mathcal{H}$ of size double the image. We then superimpose each splash on the space $\mathcal{H}$ and cast k votes as shown in Fig. 2(b). In order to take into account the noise present in the splashes, we adopt a gaussian vote-casting procedure $g(\cdot)$. Similar superimposed splashes contribute to similar locations on the accumulator space, resulting in peak formations (Fig. 2(c)). We summarize the voting procedure as follows:

$$\mathcal{H}_{\mathbf{w}} = \mathcal{H}_{\mathbf{w}} + g(\mathbf{w}, \mathbf{h}_i^{(j)}) \tag{1}$$

where $\mathbf{h}_i^{(j)}$ is the i-th splash endpoint of descriptor $\mathbf{d}_j$ in accumulator coordinates and $\mathbf{w}$ is the size of the gaussian vote. We filter all the regions in $\mathcal{H}$ which are above a certain *threshold* τ, to get a set $\mathcal{S}$ of the locations corresponding to the peaks in $\mathcal{H}$. The τ parameter acts as a coarse filter and is not a critical parameter to the overall pipeline. A sufficient value is to set it to $0.05 \cdot max(\mathcal{H})$. Lastly, in order to visualize the semantic hotspots in the image plane we map splash locations between $\mathcal{H}$ and the image plane by means of a *backtracking structure* $\mathcal{V}$.

In summary, the key insight here is that similar visual regions share similar splashes, we discern noisy splashes from representative splashes through an auxiliary structure, namely an accumulator. We then identify and backtrack in the image plane the semantic hotspots that are candidate points part of a visual repetition.

3.3 Semantic Categories Definition and Extraction

While the first part previously described acts as a filter for noisy keypoints allowing to obtain a good pool of candidates, we now transform the problem of finding visual categories in a problem of dense subgraphs extraction.

We enclose semantic hotspots in superpixels, this extends the semantic significance of such identified points to a broader, but coherent, area. To do so we use the SLIC [1] algorithm which is a simple and one of the fastest approaches to extract superpixels as pointed out in this recent survey [26]. Then we choose the cardinality of the *superpixels* $\mathcal{P}$ to extract. This is the second and most fundamental parameter that will allow us to span over different semantic levels.

Once the superpixels have been extracted, let $\mathcal{G}$ be an *undirected weighted graph* where each node correspond to a superpixel $p \in \mathcal{P}$. In order to put edges between graph nodes (i.e. two superpixels), we exploit the splashes origin and endpoints. In particular the strength of the connection between two vertices in $\mathcal{G}$ is calculated with the number of splashes endpoints falling between the two in a mutual coherent way. So to put a weight of 1 between two nodes we need exactly 2 splashes endpoints falling with both origin and end point in the two candidate superpixels.

With this construction scheme, the graph has clear dense subgraphs formations. Therefore, the last part simply computes a partition of $\mathcal{G}$ where each connected component correspond to a cluster of similar superpixels. In order to achieve such objective we optimize a function that is maximized when we partition the graph to represent so. To this end we define the following *density score* that given G and a set K of connected components captures the optimality of the clustering:

$$s(G, K) = \sum_{k \in K} \mu(k) - \alpha \, |K| \tag{2}$$

where $\mu(k)$ is a function that computes the average edge weight in a undirected weighted graph.

The first term, in the score function, assign a high vote if each connected component is dense. While the second term acts as a regulator for the number of connected components. We also added a weighting factor α to better adjust the procedure. As a proxy to maximize this function we devised an *iterative algorithm* reported in Algorithm 1 based on graph corrosion and with temporal complexity of $O(|E|^2 + |E|\,|V|)$. At each step the procedure corrupts the graph edges by the minimum edge weight of G. For each corroded version of the graph that we call *partition*, we compute s to capture the density. Finally the algorithm selects the corroded graph partition which maximizes the s and subsequently extracts the node groups.

In brevity we first enclose semantic hotspots in superpixels and consider each one as a node of a weighted graph. We then put edges with weight proportional to the number of splashes falling between two superpixels. This results in a graph with clear dense subgraphs formations that correspond to superpixels clusters i.e. *semantic categories*. The semantic categories detection translates in the extraction of dense subgraphs. To this end we devised an iterative algorithm

Algorithm 1. Semantic categories extraction algorithm

Require: G weighted undirected graph
$\quad i = 0$
$\quad s^* = -\inf$
$\quad K^* = \emptyset$
$\quad$**while** G_i is not fully disconnected **do**
$\qquad i = i + 1$
$\qquad$Compute G_i by corroding each edge with the minimum edge weight
$\qquad$Extract the set K_i of all connected components in G_i
$\qquad s(G_i, K_i) = \sum_{k \in K_i} \mu(k) - \alpha |K_i|$
$\qquad$**if** $s(G_i, K_i) > s^*$ **then**
$\qquad\quad s^* = s(G_i, K_i)$
$\qquad\quad K^* = K_i$
$\quad$**return** s^*, K^*

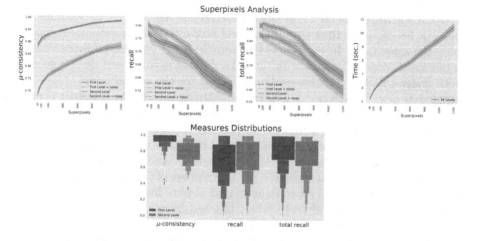

Fig. 3. (top) Analysis of measures as the number of superpixels $|\mathcal{P}|$ retrieved varies. The rightmost figure shows the running time of the algorithm. We repeated the experiments with the noisy version of the dataset but report only the mean since variation is almost equal to the original one. (bottom) Distributions of the measures for the two semantic levels, by varying the two main parameters r and $|\mathcal{P}|$.

based on graph corrosion where we let the procedure select the corroded graph partition that filters noisy edges and let dense subgraphs emerge. We do so by maximizing score that captures the density of each connected component.

4   Experiments

Dataset. As we introduced in Sect. 1 one of the aims of this work is to provide a better comparative framework for visual pattern detection. To do so we created a public dataset by taking 104 pictures of store shelves. Each picture has been

[14]   [10]   Ours

Fig. 4. Qualitative comparison between [10, 14] and our algorithm. Our method detects and segments more than one pattern and does not constrain itself to a particular geometrical disposition.

took with a 5mpx camera with approximatively the same visual conditions. We also rectified the images to eliminate visual distortions.

We manually segmented and labeled each repeating product in two different semantic levels. In the **first semantic level** *products made by the same company* share the same label. In the **second semantic level** visual repetitions consist in the *exact identical products*. In total the dataset is composed by 208 ground truth images, half in the first level and the rest for the second one.

μ-consistency. We devised a new measure that captures the semantic consistency of a detected pattern that is a proxy of the average precision of detection.

In fact, we want to be sure that all pattern instances fall on similar ground truth objects. First we introduce the concept of semantic consistency for a particular pattern $\mathbf{p}$. Let $\mathbf{P}$ be the set of patterns discovered by the algorithm. Each pattern $\mathbf{p}$ contains several instances $\mathbf{p}_i$. $\mathbf{L}$ is the set of ground truth categories, each ground truth category l contain several objects instances l_i. Let us define $\mathbf{t}_p$ as the vector of ground truth labels touched by all instances of $\mathbf{p}$. We say that $\mathbf{p}$ is consistent if all its instances $\mathbf{p}_i, i = 0 \ldots |\mathbf{p}|$ fall on ground truth regions sharing the same label. In this case $\mathbf{t}_p$ would be uniform and we consider $\mathbf{p}$ a good detection. The worst scenario is when given a pattern $\mathbf{p}$ every $\mathbf{p}_i$ falls on objects with different label l i.e. all the values in $\mathbf{t}_p$ are different.

To get an estimate of the overall consistency of the proposed detection, we average the consistency for each $\mathbf{p} \in \mathbf{P}$ giving us:

$$\mu\text{-consistency} = \frac{1}{|\mathbf{P}|} \sum_{p \in \mathbf{P}} \frac{|\text{mode}\,(\mathbf{t}_p)|}{|\mathbf{t}_p|} \tag{3}$$

Recall. The second measure is the classical recall over the objects retrieved by the algorithm. Since our object detector outputs more than one pattern we average the recall for each ground truth label by taking the best fitting pattern.

$$\frac{1}{|\mathbf{L}|} \sum_{l \in \mathbf{L}} \max_{\mathbf{p} \in \mathbf{P}} \text{ recall }(\mathbf{p}, l) \tag{4}$$

The last measure is the **total recall**, here we consider a hit if any of the pattern falls in a labeled region. In general we expect this to be higher than the recall.

We report the summary performances in Fig. 4. As can be seen the algorithm achieves a very high μ-consistency while still able to retrieve the majority of the ground truth patterns in both levels.

One can observe in Fig. 3 an inverse behaviour between recall and consistency as the number of superpixels retrieved grows. This is expected since less superpixels means bigger patterns, therefore it is more likely to retrieve more ground truth patterns.

In order to study the robustness we repeated the same experiments with an altered version of our dataset. In particular for each image we applied one of the following corruptions: Additive Gaussian Noise (*scale* $= 0.1 * 255$), Gaussian Blur ($\sigma = 3$), Spline Distortions (grid affine), Brightness ($+100$), and Linear Contrast (1.5).

Qualitative Validation. Firstly we begin the comparison by commenting on [14]. One can observe that our approach has a significant advantage in terms of how the visual pattern is modeled. While the authors model visual repetitions as geometrical artifacts associating points, we output a higher order representation of the visual pattern. Indeed the capability to provide a segmentation mask of the repeated instance region together the ability to span over different levels unlocks a wider range of use cases and applications.

As qualitative comparison we also added the latest (and only) deep learning based methodology [10] we found. This methodology is only able to find a single instance of visual pattern, namely the most frequent and most significant with respect to the filters weights. This means that the detection strongly depends from the training set of the CNN backbone, while our algorithm is fully unsupervised and data agnostic.

Quantitative Validation. We compared quantitatively our method against [14] that constitutes, to the best of our knowledge, the only work developed able to detect more than one visual pattern. We recreated the experimental settings of the authors by using the Face dataset [12] as benchmark achieving 1.00 precision vs. 0.98 of [14] and 0.77 in recall vs. and 0.63. We considered a miss on the object retrieval task, if more than 20% of a pattern total area falls outside from the ground truth. The parameter used were $|\mathcal{C}| = 9000, k = 15, r = 30, \tau = 5, |\mathcal{P}| = 150$. We also fixed the window of the gaussian vote to be 11×11 pixels throughout all the experiments.

5   Conclusions

With this paper we introduced a fast and unsupervised method addressing the problem of finding semantic categories by detecting consistent visual pattern repetitions at a given scale. The proposed pipeline hierarchically detects self-similar regions represented by a segmentation mask.

As we demonstrated in the experimental evaluation, our approach retrieves more than one pattern and achieves better performances with respect to competitors methods. We also introduce the concept of *semantic levels* endowed with a dedicated dataset and a new metric to provide to other researchers tools to evaluate the consistency of their approaches.

Acknowledgments. We would like to express our gratitude to Alessandro Torcinovich and Filippo Bergamasco for their suggestions to improve the work. We also thank Mattia Mantoan for his work to produce the dataset labeling.

References

1. Achanta, R., Shaji, A., Smith, K., Lucchi, A., Fua, P., Süsstrunk, S.: SLIC superpixels compared to state-of-the-art superpixel methods. IEEE Trans. Pattern Anal. Mach. Intell. **34**, 2274–2282 (2012)
2. Chen, L.-C., Zhu, Y., Papandreou, G., Schroff, F., Adam, H.: Encoder-decoder with Atrous separable convolution for semantic image segmentation. In: Ferrari, V., Hebert, M., Sminchisescu, C., Weiss, Y. (eds.) ECCV 2018. LNCS, vol. 11211, pp. 833–851. Springer, Cham (2018). https://doi.org/10.1007/978-3-030-01234-2_49
3. Cheng, M., Zhang, F., Mitra, N.J., Huang, X., Hu, S.: RepFinder: finding approximately repeated scene elements for image editing. ACM Trans. Graph. (2010)
4. Chum, O., Matas, J.: Unsupervised discovery of co-occurrence in sparse high dimensional data. In: The Twenty-Third IEEE Conference on Computer Vision and Pattern Recognition, CVPR 2010 (2010)
5. Cordts, M., et al.: The cityscapes dataset for semantic urban scene understanding. In: Conference on Computer Vision and Pattern Recognition CVPR (2016)
6. DiCarlo, J.J., Zoccolan, D., Rust, N.C.: How does the brain solve visual object recognition? Neuron **73**, 415–434 (2012)
7. Doubek, P., Matas, J., Perdoch, M., Chum, O.: Image matching and retrieval by repetitive patterns. In: 20th International Conference on Pattern Recognition, ICPR 2010 (2010)
8. He, K., Zhang, X., Ren, S., Sun, J.: Deep residual learning for image recognition. In: IEEE Conference on Computer Vision and Pattern Recognition, CVPR (2016)
9. Huberman, I., Fattal, R.: Detecting repeating objects using patch correlation analysis. In: 2016 IEEE Conference on Computer Vision and Pattern Recognition, CVPR (2016)
10. Lettry, L., Perdoch, M., Vanhoey, K., Gool, L.V.: Repeated pattern detection using CNN activations. In: 2017 IEEE Winter Conference on Applications of Computer Vision, WACV 2017 (2017)
11. Leung, T., Malik, J.: Detecting, localizing and grouping repeated scene elements from an image. In: Buxton, B., Cipolla, R. (eds.) ECCV 1996. LNCS, vol. 1064, pp. 546–555. Springer, Heidelberg (1996). https://doi.org/10.1007/BFb0015565
12. Li, F., Fergus, R., Perona, P.: Learning generative visual models from few training examples: an incremental Bayesian approach tested on 101 object categories. Comput. Vis. Image Underst. **106**, 59–70 (2007)
13. Li, Y., Chen, Y., Wang, N., Zhang, Z.: Scale-aware trident networks for object detection. In: IEEE International Conference on Computer Vision, ICCV (2019)

14. Liu, J., Liu, Y.: GRASP recurring patterns from a single view. In: IEEE Conference on Computer Vision and Pattern Recognition (2013)

15. Liu, Y., Collins, R.T., Tsin, Y.: A computational model for periodic pattern perception based on frieze and wallpaper groups. IEEE Trans. Pattern Anal. Mach. Intell. **26**, 354–371 (2004)

16. Liu, Y., et al.: Richer convolutional features for edge detection. IEEE Trans. Pattern Anal. Mach. Intell. **41**, 1939–1946 (2019)

17. Logothetis, N.K., Sheinberg, D.L.: Visual object recognition. Ann. Rev. Neurosci. **19**, 577–621 (1996)

18. Mottaghi, R., et al.: The role of context for object detection and semantic segmentation in the wild. In: IEEE Conference on Computer Vision and Pattern Recognition, CVPR (2014)

19. Ono, Y., Trulls, E., Fua, P., Yi, K.M.: LF-Net: learning local features from images. In: Bengio, S., Wallach, H.M., Larochelle, H., Grauman, K., Cesa-Bianchi, N., Garnett, R. (eds.) Advances in Neural Information Processing Systems 31, NeurIPS (2018)

20. Park, M., Brocklehurst, K., Collins, R.T., Liu, Y.: Deformed lattice detection in real-world images using mean-shift belief propagation. IEEE Trans. Pattern Anal. Mach. Intell. **31**, 1804–1816 (2009)

21. Pritts, J., Chum, O., Matas, J.: Rectification, and segmentation of coplanar repeated patterns. In: 2014 IEEE Conference on Computer Vision and Pattern Recognition, CVPR (2014)

22. Rodríguez-Pardo, C., Suja, S., Pascual, D., Lopez-Moreno, J., Garces, E.: Automatic extraction and synthesis of regular repeatable patterns. Comput. Graph. **83**, 33–41 (2019)

23. Schaffalitzky, F., Zisserman, A.: Geometric grouping of repeated elements within images. Shape, Contour and Grouping in Computer Vision. LNCS, vol. 1681, pp. 165–181. Springer, Heidelberg (1999). https://doi.org/10.1007/3-540-46805-6_10

24. Shechtman, E., Irani, M.: Matching local self-similarities across images and videos. In: 2007 IEEE Computer Society Conference on Computer Vision and Pattern Recognition (CVPR 2007). IEEE Computer Society (2007)

25. Singh, B., Davis, L.S.: An analysis of scale invariance in object detection snip. In: IEEE Conference on Computer Vision and Pattern Recognition CVPR, pp. 3578–3587 (2018)

26. Stutz, D., Hermans, A., Leibe, B.: Superpixels: an evaluation of the state-of-the-art. Comput. Vis. Image Underst. **166**, 1–27 (2018)

27. Torii, A., Sivic, J., Okutomi, M., Pajdla, T.: Visual place recognition with repetitive structures. IEEE Trans. Pattern Anal. Mach. Intell. (2015)

28. Wang, H., Zhao, G., Yuan, J.: Visual pattern discovery in image and video data: a brief survey. Wiley Interdiscip. Rev. Data Min. Knowl. Discov. **4**, 24–37 (2014)

29. Zhou, B., Zhao, H., Puig, X., Fidler, S., Barriuso, A., Torralba, A.: Scene parsing through ADE20K dataset. In: IEEE Conference on Computer Vision and Pattern Recognition CVPR (2017)

Deep Residual Neural Network for Child's Spontaneous Facial Expressions Recognition

Abdul Qayyum[1] and Imran Razzak[2(✉)]

[1] University of Burgundy, Dijon, France
abdul.qayyum@u-bourgogne.fr
[2] Deakin University, Geelong, Australia
imran.razzak@deakin.edu.au

Abstract. Early identification of deficits in emotion recognition and expression skills may prevent low social functioning in adulthood. Deficits in young children's ability to recognize facial expressions can lead to impairments in social functioning. Kids may need extra help learning to read facial expressions. Most of the earlier efforts consider the problem of emotion recognition in adults; however, ignore the child's emotions, especially in an unconstrained environment. In this paper, we present progressive light residual learning to classify spontaneous emotion recognition in children. Unlike earlier residual neural network, we reduce the skip connection at the earlier part of the network and increase gradually as the network go deeper. The progressive light residual network can explore more feature space due to limiting the skip connection locally, which makes the network more vulnerable to perturbations which help to deal with overfitting problem for smaller data. Experimental results on benchmark children emotions dataset show that the proposed approach showed a considerable gain in performance compared to the state of the art methods.

Keywords: Emotion recognition · Child expression · Facial expression · Spontaneous expression · Residual learning

1 Introduction

From childhood to our lifespan, emotions are crucial for social interactions and essential for communication. Emotions play a critical role in how we live our lives, and we adapt our behaviour according to our feelings and reactions. How we engage with others in our day to day lives to affecting the decisions we make and our ability to understand What we feel and show our emotions are factors of integration in the society at all era of our age. It is important to remember, however, that no emotion is an island. Instead, the many emotions we experience are nuanced and complex, working together to create the rich and varied fabric of our emotional life [13–15]. At times, it may seem like these emotions rule us.

A. Torsello et al. (Eds.): S+SSPR 2020, LNCS 12644, pp. 282–291, 2021.
https://doi.org/10.1007/978-3-030-73973-7_27

The choices we make, the actions we take, and our perceptions are all influenced by the emotions we are experiencing at any given moment.

Facial expressions represent the most effective way to convey information on one's emotional state to others. Facial expression recognition (FER) is not only important for social interactions but also effective communication. Deficits in young children's ability to recognize facial expressions can lead to impairments in social functioning. Peer-rated popularity and academic achievement directly correlated with the ability to recognize emotional expressions of others. For decades, researchers have been fascinated by how humans respond to, detect, and interpret emotional facial expressions. Much of the research in this area has relied on controlled stimulus sets of adults posing various facial expressions.

Face expression recognition based on six common expressions of happiness, anger, disgust, fear, surprise, and sadness has already existed since long. Many works focus on developing novel methods, and there are several publicly available datasets for common facial expression. However, research for spontaneous/natural basic emotions is quite limited, especially in children's expression, while, it has been proved that spontaneous facial expressions differ substantially from posed expression. This may be because the spontaneous facial expression is further challenging compared to relatively still and exhibit posed expressions due to large variations in face appearance and pose. Generally, real-world applications are spontaneous interactions such as meeting, person to person interactions, personality, debates summarizing etc. Thus, there is a need of joint head-pose analysis for facial expression identification. Lai et al. presented GAN-based face localization by localizing the front face image and preserving the expression and identifying characteristics [11]. The discriminator is used to differentiate the facial images from real images. Similarly, Zhang et al. presented GAN based approach to generated images with different facial expression under arbitrary pose [21]. In order to learn the identity-invariant representation, Chen et al. presented privacy-preserving representation-learning variational GAN by combining VAE, and GAN [3] which is explicitly disentangled from identity information and generative for facial synthesis expression preserving the image. Yang et al. presented identity-adaptive generation that generates the face images of the same subject with variable facial expressions using cGANs followed by facial expression recognition for single identify [20]. The recurrent neural network can drive information in sequential data as nodes connections form directed graph along temporal sequence; thus, it exploits the fact that features vectors are connected semantically. Long short term memory (LSTM) can handle variable length with less computational complexity. Chanti and Caplier applied 2D grid convolution that encode the spatial correlation followed by LSTM that model the temporal relationship of facial expression sequence [2]. Instead of 2D kernels along the time axis, Tran et al. applied 3D convolutional kernels with shared weights [18], which has been widely applied on dynamic facial expression recognition. Vielzeuf et al. extend it by applying weighted 3D convolutional based on the structural score by extracting windows from consecutive frames in each facial expression sequence [19]. Cascaded deep network based on CNN and LSTM has also been applied to involve the time-varying spontaneous facial expression sequential data [5,10].

Recently, Khan et al. presented LIRIS-CSE first ever spontaneous children's facial expression dataset. The recurrent neural network can drive information in sequential data as nodes connections form directed graph along temporal sequence; thus, it exploits the fact that features vectors are connected semantically. Long short term memory (LSTM) can handle variable length with less computational complexity. Chanti and Caplier applied 2D grid convolution that encode the spatial correlation followed by LSTM that model the temporal relationship of facial expression sequence [2]. Instead of 2D kernels along time axis, Tran et al. applied 3D convolutional kernels with shared weights [18] which has been widely applied on dynamic facial expression recognition [1,4,5,22]. Vielzeuf et al. extend it by applying weighted 3D convolutional based on the structural score by extracting windows from consecutive frames in each facial expression sequence [19]. Cascaded deep network based on CNN and LSTM has also been applied to involve the time-varying spontaneous facial expression sequential data [5,10].

Early identification of deficits in emotion recognition and expression skills may prevent poor social functioning in adulthood, thus, intending to identify the child's facial expression, we presented a novel end to end framework for spontaneous children's facial expression. We reduce the skip connection at the earlier part of the network and increase gradually as the network go deeper. The progressive light residual network can explore more feature space due to limiting the skip connection locally, which makes the network more vulnerable to perturbations which help to deal with overfitting problem for smaller data. To validate the performance, we experimented on benchmark publicly available dataset that showed considerable improvement in classification performance.

2   Spontaneous Child's Expression Recognition

Though automatic FER has made substantial progress in the past few decades, occlusion-robust and pose-invariant FER issues have received relatively less attention, especially in real-world scenarios. Early identification of deficits in emotion recognition and expression skills may prevent poor social functioning in adulthood. In this work, we consider the problem of spontaneous facial expression recognition in children. We present progressive light residual network that can explore more feature space due to limiting the skip connection locally, which makes the network more vulnerable to perturbations which help to deal with the over-fitting problem for smaller data. Unlike earlier residual neural network, we reduce the skip connection at the earlier part of the network and increase gradually as the network go deeper. In the following discussion, we first provide the proposed progressive light residual learning, followed by its application for spontaneous facial expression recognition in children. Figure 1 describes the proposed framework for emotion classification in children.

2.1   Progressive Residual Learning

Many modern neural network architectures with over parameterized regime have been used for emotion classification. Recent work showed that network, where the hidden units are polynomially smaller in size, showed better performance than over parameterized models. In DenseNet, each layer obtains additional inputs from all preceding layers and passes on its own feature-maps to all subsequent layers. In results, each layer is receiving a "collective knowledge" from all preceding layers. However, this results in a complex network, whereas ResNet is over parameterize, and it is natural to expect the training loss to have numerous global optima that perfectly interpolate the training data (Fig. 2).

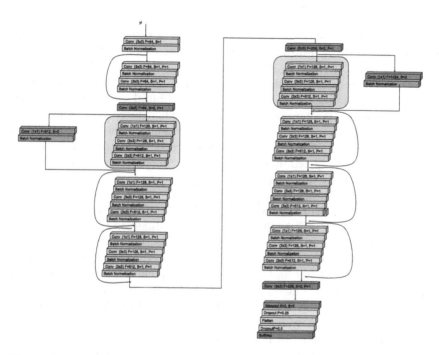

Fig. 1. Architectures of proposed three stage progressive light residual network

In this study, we develop the progressive residual network with residual learning (PLResNe) by leveraging multistage deep residual learning with additional transition layers to overcome the challenge as mentioned above. Unlike earlier residual networks and its counter network [16], we have reduced the number of residual connection at the earlier part of the network, resulting in better representation comparatively. The proposed progressive residual-based architecture consisting of 3 blocks, is shown in Fig. 1. PLResNe also considers the features of the use of all complexity levels. Notice that the depth of network (number of blocks) varies and depends on the dataset's complexity and size. Large datasets require more number of blocks where a complex task requires a small size of

Fig. 2. Spontaneous facial expressions

initial blocks. The first stage consists of only one residual unit with identify function followed by a transition block, which acts as a projection shortcut and 2nd stage consist of three residual units. This enables the network to extract text features with semantic information by limiting the jump over at an earlier stage. Thus, in the case of a complex task, the initial stages consist of fewer residual skip units. We have increased each residual block's size and numbers with the network's growth, which forces the network to learn basic features with semantic information at the initial stage and high-level features at later stages. In addition to this, it speeds up the learning by reducing the impact of vanishing gradients, as there are only fewer layers to propagate through in each stage as well as allows the network to capture the shallow information but not deep at earlier stages thus it help to prevent over-fitting for smaller data.

The proposed PLResNe is divided into several stages and each stage consists of the variable number of residual unit of different structure. Figure 1 shows the proposed progressive light residual learning. Notice that the first stage consists of the smaller residual unit and less number of skip connection. The progressive light residual network built on constructs known from pyramidal cells in the cerebral cortex and constructed small units in each block. It uses residual skip connection to jump over layers with each small unit. Thus, the resultant network is less complicated and make it possible to learn hundreds or even thousands of layers without training loss and still achieves better performance. Figure 1 shows the proposed progressive light network architecture with residual learning (PLResNet). Thus the error signal can be easily propagated to earlier blocks more directly. Unlike ResNet, we have divided each individual block into several subunits and adopted the residual learning thus our proposed PLResNet results in learning collective knowledge from all preceding layers as well as substantial deeper than ResNet and its counterpart. Notice that the error signal can be easily propagated at stage level directly. This is a kind of implicit in-depth supervision as earlier units can get direct supervision from the next stage. PLResNe has a much smaller size than its counter ResNet.

2.2 Child Facial Expression Recognition

In childhood and in adolescence, we learn to discriminate and produce facial expression to gain the level of full potential. Children and adolescents with emotion recognition skills showed better performance in their studies and excel in social interaction comparatively [7,8,12]. Studies showed that understanding children's emotions is linked to various outcomes and is very important for their cognitive development [17]. Early identification of deficits in emotion recognition and expression skills may prevent poor social functioning in adulthood. In this work, we consider the problem of spontaneous facial expression recognition in children. We present an end-to-end classification of emotion recognition that does not rely on the pre-trained model and showed significant performance compared to pre-trained models. In this work, we applied progressive light residual learning for the task of children facial expression recognition. Figure 1 shows the proposed framework of children facial expression recognition. The input images are forwarded to the proposed network. Figure 1 shows the initial stage structure. It consists of two residual units, followed by a transition layer. The progressive light residual network offers several advantages over its counter network. For example, it has low number of parameters, thus faster comparatively. Limiting the residual skip connection at the earlier part of the network helps to learn better features.

Table 1. Network parameter

Parameters	Values
Number of epochs	35
Validation frequency	300
Initial learn rate	1×10^{-4}
Momentum	0.9

Table 2. Evaluation (Precision, Recall, F1 Score) of proposed progressive light residual network (%)

	Precision	Recall	F1-score	Support
Disgust	1.00	1.00	1.00	97
Fear	0.99	0.98	0.99	529
Happy	1.00	0.99	0.99	778
Sad	0.99	1.00	0.99	512
Surprise	0.98	0.99	0.98	546
Macro avg	0.99	0.99	0.99	2462
Weighted avg	**0.9941**	**0.9925**	**0.9915**	2462

3   Experiment

This section presents the experimental results and evaluation of the proposed progressive residual network on benchmark dataset LIRIS-CSE [9], which consists of 26,000 frames of kids emotions. As kids were recorded in constraints free environment i.e. there was no restriction on kids applied during dataset collection while they were watching special build images, thus, their expression are natural and spontaneous. To generalize the performance of the proposed approach, we have performed 10-fold cross-validation.

Progressive light residual architecture consists of multiple residual units with at each stage. The initial blocks are much smaller than the later blocks, which force the network to learn better feature representation. For example, stage-1 contains of 2x convolution layers of with the kernel size of 64. Stage-2 consists of 3x residual units and each unit consists of 3x convolution layers with the kernel size of 128, 128 and 512 respectively. Similarly, stage 3 consist of 4x residual units and each unit consists of 4x convolution layers with the kernel size of 128, 128 and 512 respectively. The number of stages varies and depends upon dataset size and complexity of the task. To learn best possible features, we have used fewer residual units in earlier stages such as one unit and 2 units in stage 1 and stage 2 respectively. However, we have increased the number of residual units in later stages such as four residual units in stage 3 and 6 in stage 4. We initialize the weights as in [6] and train the network from scratch. To find the best parameters, we have trained the network on a different set of parameters for each block. We have set the momentum of (0.9) and learning rate to (1×10^{-4}). We have stopped the training if there is no improvement in the validation set's error rate for 35 epochs. Table 1 describes the training parameters (Table 2).

Table 3. Comparative analysis of proposed progressive light residual learning with its counter network (DenseNet, ResNet, SqueezNet, MobileNet and Inception)

	Average Accu	Average precision	Average recall	Average F1 score
Progressive light residual	**99.06**	**99.16**	**99.22**	**99.19**
SqueezNet	86.92	89.25	82.20	84.46
DensNet121	87.65	86.95	86.94	86.78
ResNet101	90.37	89.98	84.28	86.41
Inception_V3	89.27	90.26	88.83	89.45
MobileNet-v2	87.32	85.40	88.68	86.71
Deep-CNN [9]	77.23	69.43	77.88	81.44

Table 4. Comparative evaluation (Precision, Recall, F1-Score) of proposed progressive light residual network

Model	Precision	Recall	F1-score
Proposed network	**0.9941**	**0.9925**	**0.9915**
DenseNet-Freeze	0.94	0.95	0.94
ResNet-Freeze	0.97	0.97	0.97
Inception-Freeze	0.97	0.97	0.97
MobileNet-Freeze	0.95	0.95	0.95
DenseNet-Fine-tuned	0.88	0.88	0.88
ResNet-Fine-tuned	0.90	0.89	0.89
Inception-Fine-tuned	0.90	0.90	0.90
MobileNet-Fine-tuned	0.88	0.87	0.87
Khan et al. [9]	0.81	0.82	0.83

We have performed experiments with different network parameters and network variations. To generalize the results, we performed 10-fold cross-validation. We have used different evaluation metrics such as the area under the curve (AUC), sensitivity, specificity, precision and F-score. Table 3 describes the emotion recognition results for each class. We have accuracy classify the disgust emotions followed by 'happy' that showed 1 (precision), 0.99 (recall) and 0.99 (F1 score). A similar trend can be noticed for other emotions. Results show that our proposed approach showed significant emotion classification performance. We can notice performance is slightly poor for 'fear' expression (Table 4).

Table 5. Comparative analysis of proposed progressive light residual learning with its counter network (DenseNet, ResNet, SqueezNet, MobileNet using Freeze based fine tuning)

	Average Accu	Average precision	Average recall	Average F1 score
Progressive light residual	**99.06**	**99.16**	**99.22**	**99.19**
SqueezNet	86.92	89.25	82.20	84.46
DensNet121	87.65	86.95	86.94	86.78
ResNet101	90.37	89.98	84.28	86.41
Inception_V3	89.27	90.26	88.83	89.45
MobileNet-v2	87.32	85.40	88.68	86.71
Deep-CNN [9]	77.23	69.43	77.88	81.44

Tables 3 and 5 showed that our proposed progressive light residual network achieve best performance 0.99 (precision), 0.99 (recall), and 0.99 (F1 score) on 2462 images. Confusion matrix showed that disgust and happy showed the highest performance with no false rate. We have also applied transfer learning using several states of the art methods, especially counter networks such as ResNet, MobileNet, Inception and DenseNet. We have considered both freeze and fine turning in transfer learning. Proposed progressive light residual learning showed significant improvement results compared to the baseline state of the art methods.

4   Conclusion

In this paper, we present progressive light residual learning to classify spontaneous emotion recognition in children. Unlike earlier residual neural network, we reduce the skip connection at the earlier part of the network and increase gradually as the network go deeper. The progressive light residual network can explore more feature space due to limiting the skip connection locally, which makes the network more vulnerable to perturbations which help to deal with overfitting problem for smaller data. Experimental results on benchmark children emotions dataset show that our proposed network showed the significantly better performance to 99.19% in comparison to 89.45%, 86.41 and 86.71% for Inception, ResNet and MobileNet respectively. The gain in performance compared to Resnet and its variants showed the robustness of the proposed network for complex and relativity small dataset problems.

References

1. Abbasnejad, I., Sridharan, S., Nguyen, D., Denman, S., Fookes, C., Lucey, S.: Using synthetic data to improve facial expression analysis with 3D convolutional networks. In: Proceedings of the IEEE International Conference on Computer Vision Workshops, pp. 1609–1618 (2017)
2. Al Chanti, D.A., Caplier, A.: Deep learning for spatio-temporal modeling of dynamic spontaneous emotions. IEEE Trans. Affect. Comput. (2018). https://doi.org/10.1109/TAFFC.2018.2873600
3. Chen, J., Konrad, J., Ishwar, P.: VGAN-based image representation learning for privacy-preserving facial expression recognition. In: Proceedings of the IEEE Conference on Computer Vision and Pattern Recognition Workshops, pp. 1570–1579 (2018)
4. Cano Montes, A., Hernández Gómez, L.A.: Audio-visual emotion recognition system for variable length spatio-temporal samples using deep transfer-learning. In: Abramowicz, W., Klein, G. (eds.) BIS 2020. LNBIP, vol. 389, pp. 434–446. Springer, Cham (2020). https://doi.org/10.1007/978-3-030-53337-3_32
5. Fan, Y., Lu, X., Li, D., Liu, Y.: Video-based emotion recognition using CNN-RNN and C3D hybrid networks. In: Proceedings of the 18th ACM International Conference on Multimodal Interaction, pp. 445–450 (2016)
6. He, K., Zhang, X., Ren, S., Sun, J.: Delving deep into rectifiers: surpassing human-level performance on ImageNet classification. In: Proceedings of the IEEE International Conference on Computer Vision, pp. 1026–1034 (2015)

7. Jones, S.M., Brown, J.L., Hoglund, W.L.G., Aber, J.L.: A school-randomized clinical trial of an integrated social-emotional learning and literacy intervention: impacts after 1 school year. J. Consult. Clin. Psychol. **78**(6), 829 (2010)
8. Jones, S.M., Brown, J.L., Aber, J.L.: Two-year impacts of a universal school-based social-emotional and literacy intervention: an experiment in translational developmental research. Child Dev. **82**(2), 533–554 (2011)
9. Khan, R.A., Crenn, A., Meyer, A., Bouakaz, S.: A novel database of children's spontaneous facial expressions (LIRIS-CSE). Image Vis. Comput. **83**, 61–69 (2019)
10. Kim, D.H., Baddar, W.J., Jang, J., Ro, Y.M.: Multi-objective based spatio-temporal feature representation learning robust to expression intensity variations for facial expression recognition. IEEE Trans. Affect. Comput. **10**(2), 223–236 (2017)
11. Lai, Y.-H., Lai, S.-H.: Emotion-preserving representation learning via generative adversarial network for multi-view facial expression recognition. In: 2018 13th IEEE International Conference on Automatic Face and Gesture Recognition (FG 2018), pp. 263–270. IEEE (2018)
12. McDowell, D.J., O'Neil, R., Parke, R.D.: Display rule application in a disappointing situation and children's emotional reactivity: relations with social competence. Merrill-Palmer Q. (1982-) **46**, 306–324 (2000)
13. Qayyum, A., Razzak, I., Mumtaz, W.: Hybrid deep shallow network for assessment of depression using electroencephalogram signals. In: Yang, H., Pasupa, K., Leung, A.C.-S., Kwok, J.T., Chan, J.H., King, I. (eds.) ICONIP 2020. LNCS, vol. 12534, pp. 245–257. Springer, Cham (2020). https://doi.org/10.1007/978-3-030-63836-8_21
14. Razzak, I., Blumenstein, M., Guandong, X.: Multiclass support matrix machines by maximizing the inter-class margin for single trial EEG classification. IEEE Trans. Neural Syst. Rehabil. Eng. **27**(6), 1117–1127 (2019)
15. Razzak, I., Hameed, I.A., Xu, G.: Robust sparse representation and multiclass support matrix machines for the classification of motor imagery EEG signals. IEEE J. Trans. Eng. Health Med. **7**, 1–8 (2019)
16. Razzak, I., Naz, S.: Unit-vise: deep shallow unit-vise residual neural networks with transition layer for expert level skin cancer classification. IEEE/ACM Trans. Comput. Biol. Bioinform. (2020). https://doi.org/10.1109/TCBB.2020.3039358
17. Sprung, M., Münch, H.M., Harris, P.L., Ebesutani, C., Hofmann, S.G.: Children's emotion understanding: a meta-analysis of training studies. Dev. Rev. **37**, 41–65 (2015)
18. Tran, D., Bourdev, L., Fergus, R., Torresani, L., Paluri, M.: Learning spatiotemporal features with 3D convolutional networks. In: Proceedings of the IEEE International Conference on Computer Vision, pp. 4489–4497 (2015)
19. Vielzeuf, V., Pateux, S., Jurie, F.: Temporal multimodal fusion for video emotion classification in the wild. In: Proceedings of the 19th ACM International Conference on Multimodal Interaction, pp. 569–576 (2017)
20. Yang, H., Zhang, Z., Yin, L.: Identity-adaptive facial expression recognition through expression regeneration using conditional generative adversarial networks. In: 2018 13th IEEE International Conference on Automatic Face and Gesture Recognition (FG 2018), pp. 294–301. IEEE (2018)
21. Zhang, F., Zhang, T., Mao, Q., Xu, C.: Joint pose and expression modeling for facial expression recognition. In: Proceedings of the IEEE Conference on Computer Vision and Pattern Recognition, pp. 3359–3368 (2018)
22. Zhao, J., Mao, X., Zhang, J.: Learning deep facial expression features from image and optical flow sequences using 3D CNN. Vis. Comput. **34**(10), 1461–1475 (2018)

Multi-layer PCA Network for Image Classification

Mubarakah Alotaibi$^{(\boxtimes)}$ ⓘ and Richard C. Wilson$^{(\boxtimes)}$

University of York, York, UK
{mmma512,richard.wilson}@york.ac.uk

Abstract. PCANet is a simple deep learning baseline for image classification, which learns the filters banks by PCA instead of stochastic gradient descent (SGD) in each layer. It shows a good performance for image classification tasks with only a few parameters and no backpropagation procedure. However, PCANet suffers from two main problems. The first problem is the features explosion which limits its depth to two layers. The second issue is the binarization process which leads to discriminative information loss. To handle these problems, we adopted CNN-like convolution layers to learn the PCA filter-bank and reduce the number of dimensions. We also used second-order pooling with z-score normalization to replace the histogram descriptor. The late fusion method is used to combine the class posteriors generated each layer. The proposed network has been tested on image classification tasks including MNIST, Cifar10, Cifar100 and Tiny ImageNet databases. The experimental results show that our model achieves better performance than standard PCANet and is competitive with some CNN methods.

Keywords: PCANet · PCANet+ · PCANet II · CNN · Spatial pyramid pooling · Second-order pooling · Fusion Neural Network

1 Introduction

Convolutional neural networks (CNNs) have witnessed immense success in image classification since Alexnet won the ImageNet Large Scale Visual Recognition Challenge (ILSVRC) in 2012 [19]. Some of the famous CNNs include VGGNet [23], ResNet [24] and in general, architectures have become complicated with more layers. For example, one common realization of ResNet has 152 layers. The filters learning process in CNNs relies on optimization via gradient descent using backpropagation, making the whole process unexplainable and computationally expensive, particularly with more layers.

As a response to this, PCANet [2] was proposed in 2015 as a simple baseline for image classification problems. PCANet is trained using a closed-form non-iterative and unsupervised procedure and is an order of magnitude faster to train than a traditional CNN. Although the performance is not state-of-the-art, it is remarkably effective for such a simple architecture. PCANet has led to a family

ⓒ Springer Nature Switzerland AG 2021
A. Torsello et al. (Eds.): S+SSPR 2020, LNCS 12644, pp. 292–301, 2021.
https://doi.org/10.1007/978-3-030-73973-7_28

of related techniques, and the network proposed in this paper is broadly part of this family. While PCANet achieved good performance in various benchmarks, it suffers from some problems:

- The network is shallow, potentially causing a loss of performance on more complex datasets.
- Due to the way PCANet convolves the images and uses histogram-pooling, there is an explosion in the number of features with more layers or filters.
- Feature binarization restricts the filter responses to only 1's, and 0's, leading to discriminative-information loss.
- Since filter learning is unsupervised, later layers may not preserve important classification information from earlier layers.

In this paper, we propose a new network with some innovations. Firstly, we generate feature maps from all network layers and use *late fusion* to combine class posteriors. This introduces an element of supervised learning while preserving the single-pass strategy of PCANet. We are then able to preserve essential features from early network layers. Secondly, to improve the amount of discriminative information, we use *second-order pooling* which has recently become popular in CNNs and is used in PCANet-II [11]. We refined the approach by using z-score normalization to provide more stable and informative features. Finally, we adopt multi-channel convolution layers typically used in CNNs (but not in PCANet) and PCANet+ [1]. This helps to reduce the size of the feature maps. The performance of this new network is good and comparable to old NNs-based architectures but not the recent architectures.

2 Related Work

PCANet [2] is a simple feedforward network which does not require backprop- agation to learn the filter-bank. Instead, they adopted the PCA eigenvectors learnt from stacking patches of images to be their candidate filters. After several convolutional layers, the features are encoded employing a binary hashing fol- lowed by block-wise histograms. In fact, this network generates (unsupervised) features, and the classification is left to a final stage. The filters were applied on a one-channel basis, i.e. each filter operates on one channel and produces one new grayscale image. As a result, the number of channels at the next level is channels × filters and rises quickly. The histogram pooling also generates many features, leading to a feature explosion if more layers are added.

The design of PCANet has led to other related variants. For example, DCT- Net [20], LBP-Net [21] and ICANet [22] changed the type of filters used by PCANet while pursuing with the same structure. PCANet-II [11] also used the same PCANet architecture but replaced the histogram-pooling with the second- order pooling to reduce the number of features. PCANet+ [1] tried to change the PCANet filters' topology by proposing PCA filter ensemble learning. Other research such as [23] and [31] attempted to learn a non-linear representation of the PCA filters using kernel methods.

3   Multi-layer PCA Network Structure

Assume we have a set of N training samples $X^{(0)}$ that feed into the network, where $X^{(0)} = \{\{X_i\}_1^N : X_i \in \mathbb{R}^{m \times n \times d}\}$ and $d \in \{1, 3\}$ represents the grayscale and the coloured images respectively. Figure 1 describes the architecture of the network. The first layers are convolution layers, where the input to each convolution layer $(X^{(L-1)})$ is the previous layer's output. The following equation defines the output of each convolution layer (defined in the next section):

$$X^L = W^L * X^{(L-1)} \tag{1}$$

Where X^L represents the output of the layer (L), W^L is the PCA-based filters learned for the current layer (L), and $(*)$ is the convolution operator.

For every convolution layer's output, we extract the features using the second-order pooling and then, optionally, apply multi-level spatial pyramid pooling (SPP) as in [2]. Finally, we run a classifier for each layer which outputs the class posteriors and send the fused posteriors to a classifier for final prediction. Details about the network structure are discussed in the following sections.

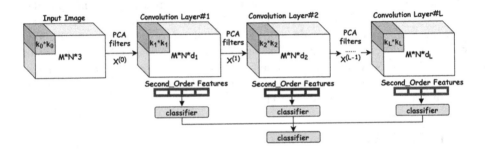

Fig. 1. Multi-layer PCA network architecture

3.1   PCA Convolution Layers

The PCA filters are calculated as in [1]. For N samples $X^{L-1} = \{X_i\}_1^N$, where $X_i \in \mathbb{R}^{m \times n \times d_{L-1}}$, and d_{L-1} is the number of filters in the layer (L-1), or $d_{L-1} \in \{1, 3\}$ for the input layer, we find the PCA filters as the following:

1. Extract all overlapping patches of size $k_L \times k_L \times d_{L-1}$ in each image X_i and subtract the mean-patch, where k_L represents the filter's size in layer (L).
2. Calculate $\bar{X}_i \in \mathbb{R}^{k_L^2 d_{L-1} \times N \tilde{m} \tilde{n}}$ by concatenating all the vectorized zero-mean patches from all sample images, where $\tilde{m} = (m - k_L) + 1$ and $\tilde{n} = (n - k_L) + 1$, m and n are the width and the height of the image.
3. Solve the following equation to find d_L principle components of $(\bar{X}^{L-1} \bar{X}^{L-1^T})$.

$$\min_{V \in R^{(k_L \times k_L) \times d_L}} ||\bar{X}^{L-1} - VV^T \bar{X}^{L-1}||_F^2, \ V^T V = I_{d_L-1} \tag{2}$$

Where $I_{d_{L-1}}$ is the identity matrix of size $d_{L-1} \times d_{L-1}$.

4. The PCA filters can be expressed as the following:

$$W_s^L = \max_{k_L \times k_L \times d_{L-1}} q_s, \ s = 1, 2, \ldots, d_L. \tag{3}$$

Where $\max_{k_L \times k_L \times d_{L-1}} (v)$ is the function that maps the vector $v \in \mathbb{R}^{k_L^2 d_{L-1}}$ to tensor $W \in \mathbb{R}^{k_L \times k_L \times d_{L-1}}$, q_s is the sth principal eigenvector of $\bar{X}^{L-1}\bar{X}^{L-1^T}$ and d_L is the number of filters chosen for the layer (L).

5. The output of each convolution layer is obtained by convolving each filter with the sample images:

$$X_i^L = \bar{X}_i^{L-1} * W_s^L \in \mathbb{R}^{m \times n \times d_L} \tag{4}$$

where $s = 1, 2, \ldots, d_L$ and $\bar{X}_i^{L-1}$ is zero-padded to get the same image size.

3.2   Second-Order Features

Figure 2 illustrates the mechanism of calculating second-order features from each convolution layer's output. So, for each output $X_i^L \in \mathbb{R}^{m \times n \times d_L}$, where $i = [1, 2, \ldots, N]$, N is the number of samples, and d_L is the number of responses in layer L. We first divide the tensors into patches of the same size and normalize each patch using z-score normalization. Then, we find the channel-wise covariance matrix for each patch, representing the second-order features related to that position in the image (determined by the patch size). Details about covariance calculation have been discussed in the following subsection.

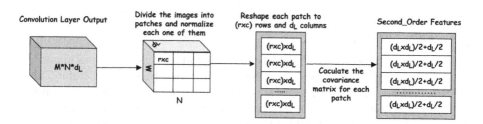

Fig. 2. Second order features for every convolution output

Covariance Computation. Assume $X \in \mathbb{R}^{M \times d}$ is sampled from the normal distribution, where M is the number of instances, and d is the number of dimensions. The following equation defines the covariance matrix of X.

$$\Sigma = \frac{1}{M} \sum_{k=1}^{M} (x_k - \mu)^T (x_k - \mu) \tag{5}$$

Where μ is the sample mean.

The covariance matrix Σ is a symmetric positive definite matrix that forms a Riemannian manifold [5]. Most of the classifiers, such as SVM, deals with Euclidean data so direct use of covariance matrix entries for features is not ideal. We should first map Σ to the Euclidean tangent space using log matrix $\log \Sigma$ [5], or square-root matrix $\Sigma^{1/2}$ [3,4,6]. Empirically, the square-root matrix seems to produce better performance than logarithmic matrix [3].

The covariance matrix is difficult to estimate robustly when the number of samples is small compared to the number of dimensions [4,7], and so the square-root covariance cannot be accurately calculated directly. Therefore, we used the following equation as in [7] to estimate the covariance matrix.

$$\widetilde{\Sigma} = U \mathrm{diag}(\delta_i : i = 1, 2, \ldots, d) U^T \ , \ \delta_i = \sqrt{\left(\frac{1-\alpha}{2\alpha}\right)^2 + \frac{\lambda_i}{2\alpha}} - \frac{1-\alpha}{2\alpha} \qquad (6)$$

Where U is the orthogonal matrix consisting of the eigenvectors, $(\lambda_i, i = 1 \ldots d)$ are the eigenvalues in decreasing order, α is regularizing parameter and set to be $\frac{1}{2}$ in all experiments as in [4]. This gives $\widetilde{\Sigma}$ as a regularized estimate of $\Sigma^{1/2}$. After computing the estimated covariance matrix, we combine the mean and the estimated covariance using the following positive-definitive matrix as in [7].

$$\mathcal{N}(\mu, \widetilde{\Sigma}) \sim \begin{pmatrix} \widetilde{\Sigma} + \mu\mu^T & \mu \\ \mu^T & 1 \end{pmatrix} \qquad (7)$$

Where $\widetilde{\Sigma}$ and μ are the estimated covariance and the sample mean, respectively. Because this matrix is symmetric, the number of features for each convolution layer is $(d_L + 1)(d_L + 2)/2 - 1\times$ the number of patches.

3.3 Late Fusion

The intermediate activations' outputs could provide informative clues about the images, including local parts, boundaries, and low-level textures. Therefore, integrating information from all layers is essential for better performance and reliable prediction [8]. Generally, there are two standard fusion methods, namely early fusion and late fusion [9,10]. The difference between the two methods is explained in Fig. 3. The early fusion works in the features level, where we fuse the features first using one of the methods discussed in Table 1 and send the fused features to a classifier to predict the classes' labels. On the other hand, the late fusion works by running a classifier each level, combine the posteriors using one of the methods described in Table 1 and send them to the primary classifier for the final prediction. The researchers [9,10] showed that the late fusion method could provide comparable or better performance than the early fusion. We used the late fusion method in our experiments, where the class posteriors were averaged to predict the final results. This produces a large reduction in the number of features used in the final classifier.

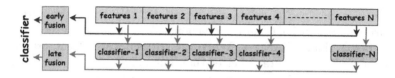

Fig. 3. Late fusion versus early fusion

Table 1. Fusion methodology

Fusion method	Definition
Concatenation	$f = [f_1, f_2, \ldots, f_N]$, where $f_i \in \mathbb{R}^{d_i}$ $f \in \mathbb{R}^{\sum_{k=1}^{N} d_k}$
Max	$f_k = \max_{i=1}^{d}(f_k^i)$, where $k = 1, \ldots, N$ and $f \in \mathbb{R}^N$
Sum	$f_k = \sum_{i=1}^{d}(f_k^i)$, where $k = 1, \ldots, N$ and $f \in \mathbb{R}^N$

4 Experiments and Results

We investigated several architectures to test our model in 4 benchmarks: CIFAR-10 [29], CIFAR-100 [29], MNIST [26] and Tiny ImageNet [30]. Table 2 presents the best configurations we found for the four datasets. Each entry gives the receptive field size and the number of output filters for that layer. The filter size for all of our experiments has fixed to 3 × 3. The classifier we ran for every convolution layer is the linear discriminant analysis (LDA), and the posteriors generated of the LDA classifiers were averaged and sent to SVM [28] to produce the final prediction. More details about these configurations and their accuracies have been discussed in the following subsections.

Table 2. Configurations for CIFAR-10/100, MNIST and Tiny ImageNet

CIFAR-10	CIFAR-100	MNIST	Tiny ImageNet
Input Image 32 × 32 × 3		Input Image 28 × 28	Input Image 64 × 64 × 3
3 × 3 conv-27		3 × 3 conv-9	3 × 3 conv-27
[3 × 3 conv-50] ×7	[3 × 3 conv-50] ×4	[3 × 3 conv-40] ×8	2 × 2 max-pooling, stride = 2
			[3 × 3 conv-70] ×3

4.1 Experiment on CIFAR-10 Database

CIFAR-10 database [29] consists of 50,000 coloured images for training and 10,000 images for testing. Each class contains objects that come with different angles and poses. The model used for testing CIFAR-10 as described in Table 2 consists of 8 layers with 27 filters for the first layer and 50 for the rest. We divided the feature maps of each layer into patches of size 8 × 8 with a stride = 1. Each convolution layer's second-order features are reduced using 3-levels SPP of 4 × 4, 2 × 2 and 1 × 1 subregions. The number of the first convolution layer features is 8508, and 27825 from the remaining layers.

Table 3 compares our model's accuracy with PCANet-2 and the current state of the art results (without data augmentation). The accuracy of our model is 4.58% better than PCANet-2. While the current state of the art methods achieved around 12% accuracy better than our model, our model is competitive with some simpler deep learning methods and learns features in a one-pass closed-form algorithm. So, the result is still promising.

Table 3. Comparison of the accuracy (%) of some methods on the CIFAR-10/CIFAR-100 database with no data augmentation

Method	CIFAR-10	CIFAR-100
Stochastic pooling [12]	84.87	57.49
Maxout network [13]	88.32	61.43
Network in network [14]	89.59	64.32
ALL-CNN [15]	90.2	–
Fractal network [16]	89.82	64.66
110 ResNet reported by [17,18]	86.82	55.26
ResNet stochastic depth [17]	–	62.20
164-ResNet(pre-activation) reported by [18]	–	64.42
Dense network(k = 24) [18]	94.08	76.58
Dense network-BC (k = 24)[18]	94.81	80.36
PCANet-2 [2]	77.14	51.62
PCANet-2 (combined) [2]	78.67	–
Multi-layer PCANet (ours)	81.72	57.86

4.2   Experiment on CIFAR-100 Database

CIFAR-100 database [29] is similar to CIFAR-10 but with 100 classes. The model we used in this experiment is identical to the one we used for CIFAR-10 but with 5 convolutional layers. We choose the number of layers to be five because there is no improvement in the accuracy when using eight layers as in CIFAR-10.

Table 3 compares the results achieved by our model with PCANet and other neural networks-based models (without data augmentation) including Residual network, Fractal network [16] network in network [14] and dense network [18]. The PCANet result has been achieved by running the same model used for CIFAR-10 [2], but with the CIFAR-100 database. The results show that our 5-layers PCANet accuracy is 6.24% better than the 2-layers PCANet and 2.60% better than ResNet with 110 layers, and about the same error rate as the stochastic pooling method [12]. The dense-network achieved the best performance with an error rate of 22% less than our model. Again our method improves on PCANet and is competitive with some older CNN-based methods, although it is not as good as more recent ones.

4.3   Experiment on the MNIST Database

The MNIST dataset [26] consists of 60,000 training examples and 10,000 test samples organized into 10 classes. Testing on this dataset was also performed without data augmentation. As described in Table 2, we used 9 convolutional layers, with 9 filters for the first layer and 40 for the rest. The performance reported here is the last convolution layer's accuracy, where we divided the 40 feature-maps into patches of size = 7×7 and stride = 1. The number of dimensions of the last layer has reduced using 3-level SPP of 16, 4 and 1 bins. The classifier used here is also the LDA, and the number of dimensions is 18060.

Table 4 compares the results obtained in this study with those obtained by PCANet [2]. The performance of our model is equivalent to those of PCANet-2, LDANet-2 and PCANet-1 (k = 13). Our interpretation of this situation is that the accuracy was good, and could not improve much by adding more layers.

Table 4. Comparison of cthe MNIST database with no data augmentation

Method	Accuracy(%)
PCANet-1 [2]	99.06
PCANet-2 [2]	99.34
LDANet-1 [2]	99.02
LDANet-2 [2]	99.38
PCANet-1 (k = 13) [2]	99.38
Multi-layer PCANet (ours)	99.40

4.4   Experiment on Tiny ImageNet

Tiny ImageNet database [30] consists of 100,000 training images divided into 200 categories. The validation and the test sets contain 10,000 images each, with 50 images per class. The test set is not labelled, and the results here are reported on the validation set. The model used for this experiment, as described in Table 2 consists of 4 convolution layers and one max-pooling layer with 27 filters generated from the first layer and 70 filters produced by the next layers. We introduced the max-pooling layer here to reduce the size of the output images. We divided the output images into patches of size 16×16 with stride = 1. The first layer's covariance features were pooled using 3-level SPP with 16, 4 and 1 bins. The next convolution layers have been connected to 2-level SPP with 4 and 1 bins. The number of features generated = 8508 from the first layer and 9450 from the remaining convolution layers.

Table 5 displays our model's accuracy compared to PCANet-2, ResNet34 and ResNet-50 reported by [27] without data augmentation. We tried to choose the best parameters to run PCANet-2 on 1TB memory. Therefore, PCANet-2 was trained with filter size k1 = k2 = 5, the number of filters L1 = 30, L2 = 8, and the block size = 16×16 with overlapping ratio = 0.5. To the best of our knowledge, we obtained the best error rate with no data augmentation.

Table 5. Comparison of the accuracy of some methods on the Tiny ImageNet database with no data augmentation

Method	ResNet-34 [27]	ResNet-50 [27]	PCANet-2	Multi-layer PCANet (ours)
Accuracy(%)	33.50	26.20	30.00	40.87

5 Conclusions

In this paper, we have presented some new refinements to the PCANet family of image classification methods to improve the performance and reduce the number of features. We introduced late fusion to preserve the information from all layers, adopted multi-channel convolutional layers and used second-order pooling with z-score normalisation. These substantially reduce the number of generated features and allow us to use deeper networks. We have shown that this offers improved performance over PCANet and results which are competitive with some simpler CNN architectures. We believe this is promising for a method where the features are unsupervised, but this is also a weakness of the architecture because we cannot learn which features are important for classification. In future work, we intend to study supervised convolutional layers where the filters are learnt with simple closed-form solutions in the same spirit as PCANet.

References

1. Low, C.Y., Teoh, A.B.-J., Toh, K.-A.: Stacking PCANet+: an overly simplified convnets baseline for face recognition. IEEE Signal Process. Lett. **24**(11), 1581–1585 (2017)
2. Chan, T.-H., et al.: PCANet: a simple deep learning baseline for image classification? IEEE Trans. Image Process. **24**(12), 5017–5032 (2015)
3. Yu, K., Salzmann, M.: Statistically-motivated second-order pooling. In: Proceedings of the European Conference on Computer Vision (ECCV) (2018)
4. Wang, Q., et al.: Deep CNNs meet global covariance pooling: better representation and generalization. arXiv preprint arXiv:1904.06836 (2019)
5. Carreira, J., Caseiro, R., Batista, J., Sminchisescu, C.: Semantic segmentation with second-order pooling. In: Fitzgibbon, A., Lazebnik, S., Perona, P., Sato, Y., Schmid, C. (eds.) ECCV 2012. LNCS, vol. 7578, pp. 430–443. Springer, Heidelberg (2012). https://doi.org/10.1007/978-3-642-33786-4_32
6. Mao, Y., Wang, R., Shan, S., Chen, X.: COSONet: compact second-order network for video face recognition. In: Jawahar, C.V., Li, H., Mori, G., Schindler, K. (eds.) ACCV 2018. LNCS, vol. 11363, pp. 51–67. Springer, Cham (2019). https://doi.org/10.1007/978-3-030-20893-6_4
7. Wang, Q., et al.: RAID-G: robust estimation of approximate infinite dimensional Gaussian with application to material recognition. In: Proceedings of the IEEE Conference on Computer Vision and Pattern Recognition (2016)
8. Liu, Yu., Guo, Y., Georgiou, T., Lew, M.S.: Fusion that matters: convolutional fusion networks for visual recognition. Multimedia Tools Appl. **77**(22), 29407–29434 (2018). https://doi.org/10.1007/s11042-018-5691-4

9. Ergun, H., et al.: Early and late level fusion of deep convolutional neural networks for visual concept recognition. Int. J. Semant. Comput. **10**(03), 379–397 (2016)
10. Ebersbach, M., Herms, R., Eibl, M.: Fusion methods for ICD10 code classification of death certificates in multilingual corpora. In: CLEF (Working Notes), September 2017
11. Fan, C., et al.: PCANet-II: when PCANet meets the second order pooling. IEICE Trans. Inf. Syst. **101**(8), 2159–2162 (2018)
12. Zeiler, M.D., Fergus, R.: Stochastic pooling for regularization of deep convolutional neural networks. arXiv preprint arXiv:1301.3557 (2013)
13. Goodfellow, I., et al.: Maxout networks. In: International Conference on Machine Learning. PMLR (2013)
14. Lin, M., et al.: Network in network. arXiv preprint arXiv:1312.4400 (2013)
15. Springenberg, J.T., Dosovitskiy, A., Brox, T., Ried-miller, M.: Striving for simplicity: the all convolutional net. arXiv preprint arXiv:1412.6806 (2014)
16. Larsson, G., Maire, M., Shakhnarovich, G.: FractalNet: ultra-deep neural networks without residuals. arXiv preprint arXiv:1605.07648 (2016)
17. Huang, G., Sun, Yu., Liu, Z., Sedra, D., Weinberger, K.Q.: Deep networks with stochastic depth. In: Leibe, B., Matas, J., Sebe, N., Welling, M. (eds.) ECCV 2016. LNCS, vol. 9908, pp. 646–661. Springer, Cham (2016). https://doi.org/10.1007/978-3-319-46493-0_39
18. Huang, G., et al.: Densely connected convolutional networks. In: Proceedings of the IEEE Conference on Computer Vision and Pattern Recognition (2017)
19. Krizhevsky, A., et al.: ImageNet classification with deep convolutional neural networks. In: Advances in Neural Information Processing Systems (2012)
20. Ng, C.J., Teoh, A.B.J.: DCTNet: a simple learning-free approach for face recognition. In: 2015 Asia-Pacific Signal and Information Processing Association Annual Summit and Conference (APSIPA). IEEE (2015)
21. Xi, M., et al.: Local binary pattern network: a deep learning approach for face recognition. In: 2016 IEEE international conference on Image processing. IEEE (2016)
22. Zhang, Y., Geng, T., Wu, X., Zhou, J., Gao, D.: ICANet: a simple cascade linear convolution network for face recognition. EURASIP J. Image Video Process. **2018**(1), 1–7 (2018). https://doi.org/10.1186/s13640-018-0288-4
23. Wu, D., et al.: Kernel principal component analysis network for image classification. arXiv preprint arXiv:1512.06337 (2015)
24. Simonyan, K., Zisserman, A.: Very deep convolutional networks for large-scale image recognition. arXiv preprint arXiv: 1409.1556 (2014)
25. Zeiler, M.D., Fergus, R.: Visualizing and understanding convolutional networks. In: Fleet, D., Pajdla, T., Schiele, B., Tuytelaars, T. (eds.) ECCV 2014. LNCS, vol. 8689, pp. 818–833. Springer, Cham (2014). https://doi.org/10.1007/978-3-319-10590-1_53
26. LeCun, Y., et al.: Gradient-based learning applied to document recognition. Proc. IEEE **86**(11), 2278–2324 (1998)
27. Abai, Z., Rajmalwar, N.: DenseNet models for tiny imagenet classification. arXiv preprint arXiv:1904.10429 (2019)
28. Fan, R.-E., et al.: LIBLINEAR: a library for large linear classification. J. Mach. Learn. Res. **9**, 1871–1874 (2008)
29. Krizhevsky, A., et al.: Learning multiple layers of features from tiny images (2009)
30. https://tiny-imagenet.herokuapp.com/
31. Qaraei, M., et al.: Randomized non-linear PCA networks. Inf. Sci. **545**, 241–253 (2021)

A Multimodal Fusion Model Based on Hybrid Attention Mechanism for Gesture Recognition

Yajie Li[1,2] , Yiqiang Chen[1,2], Yang Gu[2], and Jianquan Ouyang[1(✉)]

1 Xiangtan University, Xiangtan 411105, China
oyjq@xtu.edu.cn
2 Institute of Computing Technology, Chinese Academy of Sciences,
Beijing 100190, China

Abstract. Gesture recognition based on multimodal information plays a significant role in the field of human-computer interaction. In recent years, although many researchers devoted themselves to the related work in this field, the correlation and complementarity of multimodal information have not been explored and utilized fully. Consequently, this paper proposes a multimodal fusion network based on the hybrid attention mechanism for gesture recognition, where: 1. the cross-attention mechanism is introduced to fuse and enhance multi-dimensional features mutually, such as video and audio features; 2. the single-attention mechanism is employed to balance the correlation and redundancy between one-dimensional representation and multi-dimensional representation, such as skeleton and video features. The proposed network aims to excavate the relationship between modalities from different perspectives, fuse various information in different fusion stages, and achieve high accuracy of recognition. The method is evaluated on the publicly available datasets, ChaLearn Montalbano dataset, and obtains 95.97% accuracy when fusing video, skeleton, and audio modalities, which outperforms state-of-the-art approaches.

Keywords: Gesture recognition · Multimodal fusion · Hybrid attention mechanism

1 Introduction

Gesture recognition designed to explain human action is essential in the field of human-computer interaction (HCI) [21], which has a wide range of applications, such as virtual reality, intelligent navigation, sign language recognition, and other aid systems [3]. In consequence, researchers have done many works in the field of gesture recognition, which can be broadly divided into two categories, according to the number of modalities. One category is based on one modality, and the other is based on multimodal information.

© Springer Nature Switzerland AG 2021
A. Torsello et al. (Eds.): S+SSPR 2020, LNCS 12644, pp. 302–312, 2021.
https://doi.org/10.1007/978-3-030-73973-7_29

For the gesture recognition methods based on the single modality, most approaches employ the extracted video features. Pigou et al. [13] proposed a model with temporal convolutions for the gesture and sign recognition task. Simonyan et al. [15] proposed a two-stream convolutional architecture including spatial and temporal networks for action recognition. However, gesture recognition with high accuracy is a challenging task because of individual differences in tempos and styles of gestures, the complex observation conditions, tiny gestures that are hard to spot, and variation real-time performance [11].

Researchers have introduced multiple modality data to alleviate the bias easily generated in the single modal and improve the robustness of the model. Besides, the machine learning-based algorithms have made a tremendous impact in lots of fields, such as computer vision [7], natural language processing [18] and bioinformatics [14]. As a result, the Multimodal Machine Learning (MML) [2] is a hot topic of current research, in which multimodal fusion is an essential procedure [6]. Most studies have focused on hybrid multimodal fusion that allows fusing multimodal information on different layers of the model [19]. Some works are fusing multimodel information by concatenating low-level or high-level features extracted by pre-trained models [1,12,17]. Some works are fusing multimodel information by using attention mechanism [6,9,22].

For the gesture recognition methods based on multimodal information, Zhang et al. [20] learned spatiotemporal features of RGB, depth, and optical flow data by using 3DCNN and LSTM for gesture recognition. Besides, Molchanov et al. [10] employed a recurrent three-dimensional convolutional neural network that can detect and classify the dynamic hand gestures from multimodal data simultaneously. Neverova et al. [11] proposed a multimodal fusion method named ModDrop for gesture detection based on multimodal deep learning. Li et al. [8] introduced a multimodal fusion selection method based on stochastic regularization and can get high accuracy in gesture recognition. However, the simple concatenation may limit the ability of the model to dynamically determine the relevance of each modality feature about the different task, because different modalities may carry diverse task-relevant information. And most approaches have not explored the correlations between multiple modalities fully.

This paper focuses on the research of multimodal fusion in the field of gesture recognition. We propose a multimodal fusion modal based on the hybrid attention mechanism. And two attention mechanisms are presented: the cross-attention mechanism for fusing multi-dimensional representation; the single-attention mechanism for fusing one-dimensional information and multi-dimension-al information. We employ our methods in different fusion stages. In video sub-modalities fusion stages, video and audio modalities, we apply the cross-attention method. In video and skeleton fusion stages, audio and skeleton fusion stages, we employ the single-attention method. By employing the attention mechanism, the most meaningful feature information for the task from modalities can dynamically receive a stronger weight. Besides, the model can detect interference and dynamically reduce the impact of unimportant information. By the experiments on the ChaLearn Montalbano dataset, we get 95.97% accuracy when fusing video, skeleton, and audio modalities, which is higher than state-of-the-art approaches.

2   Methodology

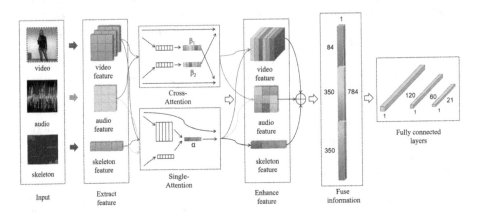

Fig. 1. The architecture of our network

In this section, we will illustrate our multimodal fusion method that is based on the hybrid attention mechanism for gesture recognition. We employ the same extracted feature method as [11], and finally get the video feauture ($V \in \mathbb{R}^{C \times H_v \times W_v}$), audio feature ($A \in \mathbb{R}^{1 \times H_a \times W_a}$), and skeleton feature ($S \in \mathbb{R}^{1 \times W_s}$). Then we use our method to fuse and enhance the extracted features of three modalities and concatenate enhanced features to form a fusion layer with 784 dimensions. We send the fusion layer to the fully connected module and get the output that is one of 21 gesture categories. The fully connected module contains three layers with 120, 60, and 21 dimensions. And the network architecture is shown in Fig. 1. The hybrid attention mechanism consists of two mechanisms. For the mutual enhancement between multi-dimensional representations, we apply the cross-attention mechanism, which will be introduced in Sect. 3.1. For the enhancement between the multi-dimensional representation and the one-dimensional representation, we employ the single-attention mechanism and will be presented in Sect. 3.2.

2.1   The Cross-Attention Mechanism

The cross-attention mechanism is proposed to perform feature recalibration, through using global information to selectively emphasize informative features and depress less useful ones. The cross-attention can make multi-dimensional data from different modalities refer to each other and enhance the meaningful channel characteristics between modalities, as shown in Fig. 2. We use the multi-dimensional feature maps extracted from the modality and denote it as $X \in \mathbb{R}^{C \times H \times W} = [x_1, x_2, ..., x_c], x_c \in \mathbb{R}^{H \times W}$. To exploit the channel dependencies, we squeeze the feature maps across their spatial dimensions ($H \times W$) by using the global average pooling operation and get the channel descriptors $Z \in \mathbb{R}^C = [z_1, z_2, ..., z_c], z_c \in \mathbb{R}$, as the Eq. 1. We capture channel-wise dependencies and obtain the attention weights, denoted $\beta \in \mathbb{R}^C$, according to the Eq. 2.

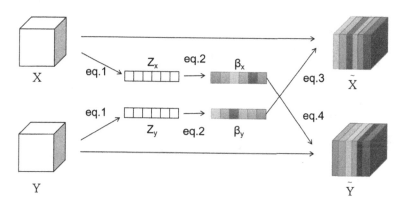

Fig. 2. The cross-attention mechanism

$$z_c = \frac{1}{H \times W} \sum_{h=1}^{H} \sum_{w=1}^{W} x_c(h, w), \tag{1}$$

$$\beta = softmax\left(W_2 Relu\left(W_1 Z\right)\right), \tag{2}$$

where W_1 and W_2 are the trainable parameters. As shown in Fig. 2, we use the same way to get β_x and β_y for different modalities.

Then we rescale the feature map of other modalities to make full use of the channel descriptors aggregated in the prior step and enhance the information from the channel dimension. According to the Eq. 3 and Eq. 4, we apply the β_x to excite the feature map $Y \in \mathbb{R}^{C \times H \times W}$, and use the β_y to activate the feature map $X \in \mathbb{R}^{C \times H \times W}$, which can achieve mutual reinforcement between different modalities.

$$\tilde{Y} = F\left(\beta_x Y\right) = \beta_x Y, \tag{3}$$

$$\tilde{X} = F\left(\beta_y X\right) = \beta_y X, \tag{4}$$

where the $F(*)$ refers to channel-wise multiplication function.

In this way, we can employ the correlation of highly correlated modalities with multi-dimensional representation to enhance each other from channel dimension and improve the ability of the model to represent features.

2.2 The Single-Attention Mechanism

For the multi-dimensional representation $X \in \mathbb{R}^{C \times H \times W_1}$ and one-dimensional information $Y \in \mathbb{R}^{1 \times W_2}$, we employ the single-attention mechanism to improve the representation of multi-dimensional data X, which is shown in Fig. 3. We first transform the representation $X \in \mathbb{R}^{C \times H \times W_1}$ into $X_r \in \mathbb{R}^{C \times (H \times W_1)}$. And then we calculate the attention weights by the following formula:

$$X_x \in \mathbb{R}^{C \times K} = Relu\left(W_x X_r + b_x\right), \tag{5}$$

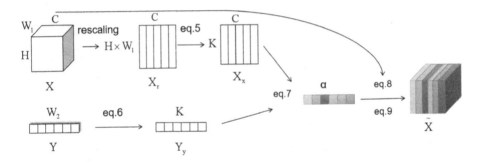

Fig. 3. The single-attention mechanism

$$Y_y \in \mathbb{R}^{1 \times K} = Relu\left(W_y Y + b_y\right), \tag{6}$$

$$\alpha = \tanh\left(X_x Y_y^T\right), \tag{7}$$

where W_x and W_y are the trainable parameters, b_x and b_y are the bias and the T is transpose operation. The weight value $\alpha \in \mathbb{R}^C$, which can be interpreted as the attention scores for every channel. The final output is obtained by rescaling the multi-dimensional representation with the value α, according to the Eq. 8 and Eq. 9:

$$\tilde{x}_c = F(x_c, \alpha_c) = x_c \alpha_c, \tag{8}$$

$$\tilde{X} = [\tilde{x}_1, \tilde{x}_2, ..., \tilde{x}_c], \tag{9}$$

where $\tilde{X}$ means the latest representation after activation by using attention score, and the $F(x_c, \alpha_c)$ refers to channel-wise multiplication between the scalar $\alpha \in \mathbb{R}^C$ and the feature maps $X \in \mathbb{R}^{C \times H \times W}$.

By this method, we can apply the one-dimensional feature to enhance the multi-dimensional information from channels, which makes use of the relationship between various modalities with different dimensional features sufficiently.

Through the above attention mechanisms, the cross-reference between multi-dimensional data can be achieved, the connection between various modalities information can be enhanced, which improves the effectiveness of multimodal data fusion and alleviates the interference between them.

3   Experiments

3.1   Datasets and Method of Data Processing

ChaLearn Montalbano Dataset. We evaluate our method on the ChaLearn Montalbano dataset [5], which is the preprocessed version of the multimodal gesture recognition dataset, Chalearn 2014 Looking at People Challenge track 3. This dataset is composed of four modalities: RGB video data, depth video data, skeleton data, and audio data, and contains 20 Italian gesture categories executed by 20 performers and one non-gesture category. Besides, we divide the gesture instances into the training set and the test set according to the ratio of 9:1. Aiming to focus on the research of multimodal fusion, we use the same method of data processing and feature extracting with ModDrop [11]. Training network for every modality, removing the output layer and employing the extracted features by hidden layers for fusion.

Table 1. Parameters configuration of data preprocessing on Montalbano dataset

Video	Skeleton	Audio
Input-(36 * 36 * 5)	Input-915	Input-(40 * 9)
CONV3D-(25 * 5 * 5 * 3)	FC1-700	CONV2D-(25 * 5 * 5)
Maxpooling-(2 * 2 * 3)	FC2-400	Maxpooling-(1 * 1)
CONV2D-(25 * 5 * 5)	FC3-350	FC1-700
Maxpooling-(1 * 1)	Output-21	FC2-350
FC1-900	–	Output-21
FC2-450	–	–
Output-21	–	–

According to [11], the specific feature extraction methods for video, skeleton, and audio modalities are shown in Table 1. In particular, for video modality, every hand is localized relied on skeleton data and is extracted feature respectively, rather than for the whole gesture. And the parameters of the right and left-hand network of video modality are shared. The extracted feature operation of the RGB modality is the same as the depth modality. We first fuse the RGB and depth modalities, then fuse the right and left hands modalities in our experiments, as shown in the Fig. 4.

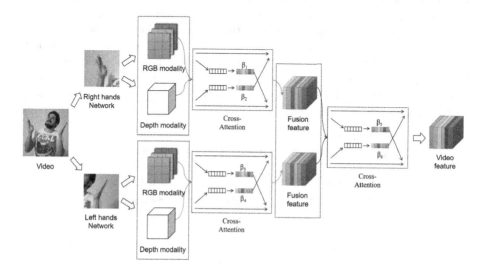

Fig. 4. The fusion of different video modalities

3.2   Training Details

The training details of our method are described as follows. The learning is completed with 100 epochs and the batch size is 512 for the Montalbano dataset. And the cross-entropy function is selected as the loss function. The stochastic gradient descent method is used as the optimization algorithm to train the networks with learning rate decay, weight decay, and the momentum method. The initial learning rate is 0.01 and the decay parameter of the learning rate is 0.9995. The weight decay rate is 0.0002, and the parameter of momentum is 0.8. The first two layers of the fully connected module use the tanh function as activation function, each followed by a dropout [16] layer of rate 0.5. We employ accuracy as the evaluation metric.

3.3   Experiment Results

We explore the hybrid attention method on ChaLearn Montalbano dataset and choose ModDrop [11], Modout [8] and Block [4] as baselines. ModDrop is a multi-scale multimodal fusion method based on the deep learning algorithm for gesture recognition. Modout is a multimodal fusion model based on stochastic regularization for gesture recognition. And Block is a new multimodal fusion based on the block-superdiagonal tensor decomposition. We do three groups of the experiment that fusing different numbers of modalities. Since the video modality is the main data of this dataset, every group of experiments contains the video modality. The experiments of various video modalities fusion are presented firstly. Then the results of video and skeleton modalities fusion, video and audio modalities fusion are discussed. Finally, the experiments of video, audio, and skeleton modalities fusion are introduced.

The Results of Different Video Modalities Fusion. We explore the cross-attention method on the situation fusing various video modalities. We use the cross-attention mechanism between the RGB modality and depth modality (dubb-ed CD), and employ the same method between the right hand and left hand (dubbed RL), since the representation of these four modalities is multi-dimensional. The results are shown in Table 2. From the results, adding our method on two fusion models of video modalities can get higher accuracy than other baselines, 89.56%, and 89.45% respectively.

Table 2. The accuracy of fusing different video modalities

Methods	ModDrop	Modout	Block	Ours
CD	89.33%	87.82%	88.99%	**89.56%**
RL	89.33%	89.45%	89.31%	**89.45%**

Table 3. The accuracy of fusing two modalities

Methods	ModDrop	Modout	Block	Ours
VS	94.17%	93.79%	94.20%	**94.45%**
VA	91.65%	91.62%	**93.48%**	92.05%

The Results of Two Modalities Fusion. We explore the hybrid attention method on the situation fusing two modalities. Because the video features and audio features are multi-dimensional, the skeleton features are one-dimensional, we use the single-attention mechanism on fusing video and skeleton modalities (dubbed VS) and employ the cross-attention mechanism on fusing video and audio modalities (dubbed VA). The results are shown in Table 3.

When fusing video and skeleton data, the accuracy of our approach is the highest in the table, 94.45%, which outperforms other methods. However, when fusing video and audio data, our method has a 92.05% accuracy that exceeds the baselines except for Block. The proposed method has a slight fluctuation in this situation, because of the presence of noise in the audio modality.

The Results of Three Modalities Fusion. We also test the performance of the hybrid attention mechanism when having video, skeleton, and audio modalities. We use the cross-attention for fusing RGB and depth modalities (dubbed CD), right and left modalities (dubbed RL), video and audio modalities (dubbed VA), respectively, since their features are multi-dimensional. Besides, we employ the single-attention on the skeleton and video modalities fusion (dubbed VS), skeleton and audio modalities fusion (dubbed SA), respectively. The related results are shown in Table 4. From the results, at each fusion stage in the table, our method outperforms the others.

Table 4. The accuracy of fusing three modalities

Methods	ModDrop	Modout	Block	Ours
CD	95.15%	94.68%	94.80%	**95.68%**
RL	95.05%	95.11%	94.74%	**95.77%**
VS	95.00%	93.97%	93.91%	**95.77%**
SA	94.91%	94.22%	94.37%	95.35%
VA	95.10%	92.25%	95.60%	95.62%

Finally, We also explored the effects of using different fusion methods at different fusion stages. For example, the cross-attention is used on the RGB and depth, right and left hands fusion stages at the same time (dubbed CDRL). The single-attention is employed on the skeleton and video fusion stages, the cross-attention is used on the RGB and depth, right and left hands fusion stages at the

310   Y. Li et al.

same time (dubbed CDRLVS). Similarly, we can get the CDRLSA, CDRLVA, CDRLVSA. And CDRLVSA means that using fusion methods on every fusion level when fusing all modalities. These codes represent the fusion methods used on the corresponding fusion stages and the results are shown in Table 5.

Table 5. The accuracy of multiple combinations on three modalities fusion

Methods	ModDrop	Modout	Block	Ours
CDRL	95.13%	94.54%	95.02%	96.05%
CDRLVS	95.10%	92.51%	94.34%	95.97%
CDRLSA	94.91%	94.40%	94.80%	95.02%
CDRLVA	95.12%	93.19%	94.48%	96.00%
CDRLVSA	95.17%	94.91%	94.45%	**95.97%**

As can be seen from Table 5, the better results can be obtained by using our method on several different fusion stages at the same time, with the accuracy ranging from 95.97% to 96.05% except for CDRLSA. Among them, when using the hybrid attention mechanism on every fusion stage (CDRLVSA), we have 95.97% accuracy, which outperforms ModDrop 0.80%, exceed the Modout by 1.06%, and is more than Block with 1.52%. The results illustrate that our approach can make use of their relationship, excavate the useful information between different modalities sufficiently, and achieve mutual reinforcement of multimodal.

4   Conclusion

In this work, a multimodal fusion network based on the hybrid attention mechanism for gesture recognition is proposed. We explore different patterns for the fusion of different modalities. Using the cross-attention approach for the multi-dimensional features can achieve mutual enhancement between the multiple modalities with a high correlation. Employing the single-attention method between one-dimensional representation and multi-dimensional information can reinforce the useful parts of the multi-dimensional feature according to the one-dimensional feature. Our approach balance the relationship between complementation and redundancy of multimodal, optimize the tradeoff between the representation ability and complexity of the fusion model, and achieve fine interactions between modalities. In addition, our approach can be applied to different fusion stages and its effectiveness is also verified from the relevant experimental results.

Acknowledgement. This work is supported by the National Key Research and Development Plan of China (No. 2017YFB1002802) and the Natural Science Foundation of China (No. 61902377).

References

1. Anastasopoulos, A., Kumar, S., Liao, H.: Neural language modeling with visual features. Comput. Lang. (2019)
2. Baltrusaitis, T., Ahuja, C., Morency, L.: Multimodal machine learning: a survey and taxonomy. IEEE Trans. Pattern Anal. Mach. Intell. **41**(2), 423–443 (2019)
3. Bastos, I.L.O., Melo, V.H.C., Robson Schwartz, W.: Multi-loss recurrent residual networks for gesture detection and recognition. In: 2019 32nd SIBGRAPI Conference on Graphics, Patterns and Images (SIBGRAPI), pp. 170–177 (2019)
4. Benyounes, H., Cadene, R., Thome, N., Cord, M.: Block: bilinear superdiagonal fusion for visual question answering and visual relationship detection. Comput. Vis. Pattern Recogn. (2019)
5. Escalera, S., et al.: ChaLearn looking at people challenge 2014: dataset and results. In: Agapito, L., Bronstein, M.M., Rother, C. (eds.) ECCV 2014. LNCS, vol. 8925, pp. 459–473. Springer, Cham (2015). https://doi.org/10.1007/978-3-319-16178-5_32
6. Hori, C., Hori, T., Lee, T.Y., Sumi, K., Hershey, J.R., Marks, T.K.: Attention-based multimodal fusion for video description (2017)
7. LeCun, Y., Bengio, Y., Hinton, G.: Deep learning. Nature **521**(7553), 436–444 (2015)
8. Li, F., Neverova, N., Wolf, C., Taylor, G.W.: ModOut: learning multi-modal architectures by stochastic regularization, pp. 422–429 (2017)
9. Lu, P., Li, H., Zhang, W., Wang, J., Wang, X.: Co-attending free-form regions and detections with multi-modal multiplicative feature embedding for visual question answering. arXiv: Computer Vision and Pattern Recognition (2017)
10. Molchanov, P., Yang, X., Gupta, S., Kim, K., Tyree, S., Kautz, J.: Online detection and classification of dynamic hand gestures with recurrent 3D convolutional neural networks, pp. 4207–4215 (2016)
11. Neverova, N., Wolf, C., Taylor, G., Nebout, F.: ModDrop: adaptive multi-modal gesture recognition. IEEE Trans. Pattern Anal. Mach. Intell. **38**(8), 1692–1706 (2016)
12. Perezrua, J., Vielzeuf, V., Pateux, S., Baccouche, M., Jurie, F.: MFAS: multimodal fusion architecture search, pp. 6966–6975 (2019)
13. Pigou, L., Van Herreweghe, M., Dambre, J.: Gesture and sign language recognition with temporal residual networks, pp. 3086–3093 (2017)
14. Rifaioglu, A.S., Atas, H., Martin, M.J., Cetin-Atalay, R., Atalay, V., Doğan, T.: Recent applications of deep learning and machine intelligence on in silico drug discovery: methods, tools and databases. Briefings Bioinf. **20**(5), 1878–1912 (2019)
15. Simonyan, K., Zisserman, A.: Two-stream convolutional networks for action recognition in videos, pp. 568–576 (2014)
16. Srivastava, N., Hinton, G.E., Krizhevsky, A., Sutskever, I., Salakhutdinov, R.: DropOut: a simple way to prevent neural networks from overfitting. J. Mach. Learn. Res. **15**(1), 1929–1958 (2014)
17. Wang, H., Meghawat, A., Morency, L., Xing, E.P.: Select-additive learning: improving generalization in multimodal sentiment analysis, pp. 949–954 (2017)
18. Young, T., Hazarika, D., Poria, S., Cambria, E.: Recent trends in deep learning based natural language processing [review article]. IEEE Comput. Intell. Mag. **13**(3), 55–75 (2018)
19. Zhang, C., Yang, Z., He, X., Deng, L.: Multimodal intelligence: representation learning, information fusion, and applications. IEEE J. Sel. Topics Signal Process. **14**(3), 478–493 (2020)

20. Zhang, L., Zhu, G., Shen, P., Song, J.: Learning spatiotemporal features using 3D CNN and convolutional LSTM for gesture recognition, pp. 3120–3128 (2017)
21. Zhu, G., Zhang, L., Shen, P., Song, J., Shah, S.A.A., Bennamoun, M.: Continuous gesture segmentation and recognition using 3D CNN and convolutional LSTM. IEEE Trans. Multimedia **21**(4), 1011–1021 (2018)
22. Zhuang, B., Wang, W., Shinozaki, T.: Investigation of attention-based multimodal fusion and maximum mutual information objective for DSTC7 track3. In: In DSTC7 at AAAI 2019 workshop (2019)

StegColNet: Steganalysis Based on an Ensemble Colorspace Approach

Shreyank N. Gowda[1]([✉]) and Chun Yuan[2]

[1] University of Edinburgh, Edinburgh, UK
[2] Tsinghua University, Beijing, China

Abstract. Image steganography refers to the process of hiding information inside images. Steganalysis is the process of detecting a steganographic image. We introduce a steganalysis approach that uses an ensemble color space model to obtain a weighted concatenated feature activation map. The concatenated map helps to obtain certain features explicit to each color space. We use a levy-flight grey wolf optimization strategy to reduce the number of features selected in the map. We then use these features to classify the image into one of two classes: whether the given image has secret information stored or not. Extensive experiments have been done on a large scale dataset extracted from the Bossbase dataset. Also, we show that the model can be transferred to different datasets and perform extensive experiments on a mixture of datasets. Our results show that the proposed approach outperforms the recent state of the art deep learning steganalytical approaches by 2.32% on average for 0.2 bits per channel (bpc) and 1.87% on average for 0.4 bpc.

Keywords: Steganalysis · Color spaces · Greywolf optimization · Concatenated feature maps

1 Introduction

Steganography is a means of covert communication in which secret information is embedded into some form of digital media, such as an image, video or text file [3]. Usually, this form of embedding is done such that there is no apparent perceptible change in the embedding file. In multimedia security, steganography forms a critical research topic [4]. The difference between steganography and cryptography is that in cryptography data is encrypted and although difficult to break, raises a doubt in the mind of an attacker about the presence of secret information. Steganography, on the other hand, aims to reduce the risk of being detected. In general, images are considered as the embedding medium due to minute changes in an image being imperceptible to the human eye [4]. There are three main properties that a steganographic algorithm should possess: security, robustness, and capacity. In case of an image steganographic algorithm, security would mean how securely the algorithm can hide information, i.e., how little visual change is caused on an image using an image steganography algorithm.

© Springer Nature Switzerland AG 2021
A. Torsello et al. (Eds.): S+SSPR 2020, LNCS 12644, pp. 313–323, 2021.
https://doi.org/10.1007/978-3-030-73973-7_30

Robustness refers to the invariability of the steganographic algorithm when an image is subject of different transforms such as scaling, resizing, rotation, etc. The capacity for a steganographic algorithm represents the amount of data that can be embedded in an image before there is a noticeable visual change in the image [5]. Steganalysis is the process of detecting if a given image has information hidden in it or not [27]. In this regard, we can convert this problem into that of a simple classification problem. To detect if an image is embedded with information we propose the use of an ensemble color space model. Recently, it was seen an ensemble colorspace model [1] obtained excellent results on large scale image classification datasets such as imagenet [2]. Based on [1] we propose a novel steganalysis approach.

Steganalysis is the process of detecting if a given image has information hidden in it or not. In this regard, we can convert this problem into that of a simple classification problem. To detect if an image is embedded with information, we propose the use of an ensemble color space model.

We do the following:

- We use a colorspace approach to determine if an image is hiding information or not. We use ColorNet [1] and take the final activation map from each colorspace.
- We use weighted averaging to obtain a single feature map from all the individual feature maps that are generated by each colorspace. It was seen [1] that each color space had features explicit to themselves and this would help us detect minute changes in the image.
- We then use a levy-flight grey wolf optimization method (meta-heuristic approach) to select a smaller subset of features. Using these features, we classify the given image into one of two classes: containing concealed information or not.

1.1 Steganography

Most steganography algorithms can be expressed in Fig. 1. An image is broken down to it's RGB (Red Green Blue) channels and pixels in the individual channels are modulated with some cost function 'C' which embeds information into that channel. The most straightforward steganography algorithm is the LSB (Least Significant Bit) algorithm. Here, as the name suggests the least significant bit is taken, and one bit of information is stored (either as a 1 or a 0).

Steganography algorithms can be classified broadly into four categories: 1) cover image size 2) embedding domain-based algorithms 3) nature of retrieval based algorithms 4) adaptive steganographic algorithms. In the case of 2-D images, the information is embedded onto the 2-D plane of the cover image. This embedding can be done over transform domain coefficients (such as discrete cosine transforms, Fourier transforms, etc.) or on the spatial domain (an example is LSB). The 3-D approaches essentially follow the same general procedure. However, the procedure is repeated on multiple planes (for instance RGB

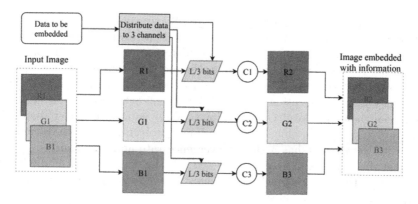

Fig. 1. Pipeline of a standard steganography algorithm (Color figure online)

in a color image has 3 planes that can embed information). Image steganography on 3-D images can be made in either geometrical domain [5], representation domain [6] or topological domain [7].

Some of the transform-based steganographic algorithms include discrete Fourier transform (DFT) [9], discrete cosine transform (DCT), discrete wavelet transform [10], complex wavelet transform [11] among others. Here, frequency coefficients obtained after applying transforms are used to hide secret bits. Along with the security being improved, these algorithms are robust to image compression, cropping, scaling, etc. Off late, machine learning approaches have been proposed such as SVM (Support Vector Machine)[12], genetic algorithm approaches [13], neural network-based steganography [14]. Though these approaches are black-box approaches, they have shown good results.

1.2 Steganalysis

Steganalysis is the method of trying to either determine a stego image (image where information is hidden) or extract the secret information. Our method deals with the former. We treat the problem at hand to be a classification problem, wherein, each image either contains some hidden information or not.

There are two basic approaches to steganalysis: signature steganalysis and statistical steganalysis. Signature steganalysis is the method wherein patterns, or signatures relevant to various steganographic algorithms are searched for. The presence of a pattern indicating that secret information is being hidden in the image. The quintessential process here is the repetition of patterns due to embedded secret information. The statistical approach searches for mathematical results to determine if the information is being hidden. Signature steganalysis is further classified into specific embedding [16] and universal blind steganalysis [15]. Specific embedding approaches are impractical because we need to know what steganography approach has been used to embed information. Hence, universal blind steganalysis [8,17] is preferred. These approaches help in the extraction of high dimensional features. However, the curse of dimensionality occurs.

Hence, a need to reduce feature size occurs. Some commonly used algorithms to do the same include wrappers, filters, etc. Filters are less complex; however, they perform poorly. Wrapper methods evaluate feature subset using predictive models [18]. However, wrappers are complex and time-consuming.

To overcome this, meta-heuristic approaches have been deployed. These approaches solve optimization problems by utilizing natural phenomena [19,20]. It was seen that Grey Wolf Optimization (GWO) performed better than other metaheuristic approaches for solving non-linear problems in a multi-dimensional space [19]. However, it has a slow convergence rate and gets trapped in local optima at times. It has been seen that GWO can be optimized by modifying it's parameter **A** to obtain a quick convergence rate, better convergence precision and higher agility for global searching.

2   Proposed Approach

2.1   Overall Architecture and Effect of Using Color Spaces

We consider steganalysis as a 2 class classification problem. The overall architecture is described in Fig. 2. The experimental analysis along with details regarding training set etc. are explained in the next section. Recently, the effect of color spaces on image classification has been explored [1]. It was seen that individual color spaces inherited classification features explicitly to themselves. This helped us ponder about the ability to extract information in an image where there is secret information being embedded. Colornet [1] being an ensemble model, that could extract features specific to each colorspace, was an excellent choice to utilize to help us in determining if an image could have information hidden in it. The output of Colornet is a high-dimensional vector, which causes a computationally intensive execution. To reduce the number of features selected we have to use an optimization approach for feature selection. Figure 1 shows the architecture of the model.

2.2   Optimization Process for Feature Selection

Feature Selection Using LF-Grey Wolf Optimization. In GWO, the head of the pack is the α. The next level of the hierarchy is β, δ and finally followed by ω. GWO models the social hierarchy and mathematically illustrates the hunting procedure as an optimization problem. If $\mathbf{X}_p(t)$ and $\mathbf{X}(t)$ represent the position of prey and wolf at iteration 't', we can mathematically model the encircling process [19] with two coefficients **A** and **C** as shown in (1). **A** and **C** are calculated by (2).

$$\mathbf{D} = |\mathbf{C}.\mathbf{X}_p(t) - \mathbf{X}(t)|; \mathbf{X}(t+1) = \mathbf{X}_p(t) - \mathbf{A}.\mathbf{D} \qquad (1)$$

$$\mathbf{A} = 2\mathbf{a}.\mathbf{r}_1 - \mathbf{a}; \mathbf{C} = 2.\mathbf{r}_2 \qquad (2)$$

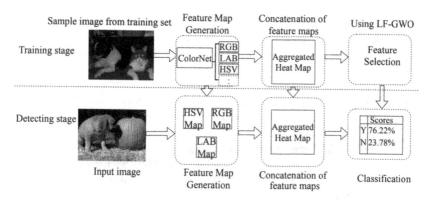

Fig. 2. Two phases involved in the overall architecture of the model: training the model using colornet and detecting stego-image using feature map aggregation

Here, $\mathbf{r_1}$ and $\mathbf{r_2}$ are random vectors in [0,1], $\mathbf{a}$ is a parameter that decreases linearly from 2 to 0 over iterations and also helps to control step size $\mathbf{D}$ of a grey wolf. Implementation of the end of the hunting process is done by decreasing the value of $\mathbf{A}$ which in turn depends on $\mathbf{a}$. Once $\mathbf{a}$ turns zero, it means that the wolves have stopped moving. The linear decrease in $\mathbf{A}$ helps to exploit search space with minimal exploration. Hence, this traps a local optimum.

The size of the aggregated feature map creates an issue in terms of the complexity of the algorithm and the overall time needed for execution. To deal with this, we propose the use of levy flight-based grey wolf optimization (LF-GWO) for feature selection based on Levy probability function in (3). Here, μ represents position parameter, γ represents scale parameter and η represents the collection of samples in the distribution. The above equation holds good for all positive values of μ and 0 otherwise. The parameter $\mathbf{A}$ is modified by the Levy flight function as $\mathbf{A} = \mathbf{L(S)}*\mathbf{r1}$. This makes $\mathbf{A}$ take up values in a non-linear decrease. $\mathbf{S}$ is the position of the wolf and $\mathbf{r1}$ is a random vector.

$$\mathbf{L}(\eta, \gamma, \mu) = \frac{\sqrt{\gamma}}{2\pi} \exp[-\frac{\gamma}{2(\eta - \mu)}] \frac{1}{(\eta - \mu)^{\frac{3}{2}}} \tag{3}$$

The reason for selection of LF-GWO is based in the statistical results obtained in [21]. It was seen that for 15 defined benchmark functions, the wilcoxon rank sum test of LF-GWO outperforms existing optimization approaches in terms of mean fitness values. For further technical analysis please refer [21].

3 Experimental Analysis

3.1 Datasets and Training

Most commonly used steganalysis datasets are the Bossbase [22] and BOWS2 [23]. Each contains 10000 grayscale images. However, the approach proposed is dependent on color, and as such, we use a dataset with color images. Hence, starting with the 10000 images of Bossbase [22] dataset, we generate a dataset by following the process done in [24]. We downsampled the full-resolution images to a size of 512×512. We then followed the process in [25], so that the training and testing scenarios were conducted in a similar environment. In [25], two datasets were created by using two demosaicing algorithms: Patterned pixel grouping (PPG) and Adaptive Homogeneity-Directed (AHD) and named BOSS-PPG-LAN and BOSS-AHD-LAN correspondingly. Further, by removing the down-sampling method, we can obtain two more datasets: BOSS-PPG-CRP and BOSS-AHD-CRP. By pairing a demosaicing algorithm with bilinear or bicubic kernels, we obtain four more datasets: BOSS-PPG-BIL, BOSS-AHD-BIL, BOSS-AHD-BIL, and BOSS-AHD-BIC.

We train our model by utilizing mini-batch stochastic gradient descent with the following parameters: learning rate: 0.0001, weight decay: 0.0005, step size: 5000, momentum: 0.75, gamma: 0.75, batch size: 32, maximum iterations: 40×10^4. Testing of the trained model was done for every 5000 iterations and accuracy in 40×104 iterations. HILL, SUNIWARD, CMD-C-SUNIWARD and CMD-C-HILL: 4 state of the art color steganography algorithms, were used as attacking targets for experimental analysis. The embedding payload was set to 0.2 bpc (bits per channel/band pixel) and 0.4 bpc. In order to select the most challenging scenarios and also follow similar conditions for result comparison, we followed the process executed in WISERNet [25].

3.2   Results Comparison

To compare our results, we considered three deep learning approaches for color steganalyzers, that are widely considered state of the art approaches: WISERNet [25], Deep Hierarchical Representations (DHR) [26] and Deep-CNN [27]. Experiments were conducted on the same datasets and using similar resources for a fair comparison. Popular steganography methods such as SUNIWARD [28], MiPOD [29], HILL [30] adopt an additive embedding distortion approach for minimizing framework [31]. Recently, CMD-C was proposed [32] by improvising the CMD approach for color images. We denote the CMD-C method using SUNIWARD and HILL as CMD-C-SUNIWARD and CMD-C-HILL respectively. Although DHR [26] and D-CNN [27] can be executed in channel-wise convolution, normal convolution and input concatenation as seen in [25], we show results only for the normal convolution as WiserNet [25] outperforms DHR and D-CNN in all cases. We also compare results with channel gradient correlation (CGC) [34].

The parameters used in terms of batch size and iterations were the same for all the comparisons. The other parameters were used as described in the original paper. Each experiment constituted 75% training images, i.e., 7500 images and 2500 images were used for testing. All experiments were performed 10 times and the average accuracy of testing was used. Table 1 compares the results of our approach with WISERNet (W-Net) [25], DHR [26], D-CNN [27], on BOSS-PPG-LAN (B-P-L), BOSS-PPG-BIC (B-P-Bc), BOSS-PPG-BIL (B-P-Bl), BOSS-AHD-BIC (B-A-Bc) and BOSS-AHD-BIL (B-A-Bl) with 0.2 bpc and Table 2 with 0.4 bpc. As can be seen, the proposed method outperforms other state of the art methods for all but one case and also the percentage increase in detection is significant when patterned pixel grouping is performed on the datasets.

Table 1. Comparison of results for CMD-C-HILL stego images with 0.2 bpc. D-CNN is executed with 30 fixed SRM kernels. The best results are represented in bold font.

Dataset	DHR	D-CNN	W-Net	CGC	Proposed
B-P-L	0.6474	0.6562	0.7139	0.7231	**0.7741**
B-P-Bc	0.6589	0.7124	0.7318	0.7278	**0.7912**
B-P-Bl	0.7611	0.7487	0.8033	0.8120	**0.8316**
B-A-Bc	0.6614	0.6627	**0.7369**	0.7168	0.7368
B-A-Bl	0.7622	0.7647	0.8022	0.7981	**0.8044**

Table 2. Comparison of results for CMD-C-HILL stego images with 0.4 bpc. D-CNN is executed with 30 fixed SRM kernels. The best results are represented in bold font.

Dataset	DHR	D-CNN	W-Net	CGC	Proposed
B-P-L	0.7568	0.7941	0.8361	0.8268	**0.8724**
B-P-Bc	0.7732	0.8068	0.8435	0.8314	**0.8814**
B-P-Bl	0.87211	0.9045	0.9169	0.9165	**0.9381**
B-A-Bc	0.7728	0.8141	0.8448	0.8412	**0.8468**
B-A-Bl	0.8738	0.9067	**0.9144**	0.9044	0.9088

Further experimental analysis is done by mixing datasets as shown in [27]. Table 3 shows how the datasets were mixed. We further label the datasets in roman numerals for simplicity to display in the comparison of steganalyzers in Table 4 and 5. BPL, BPBc, BPBl, BABc, BABl, BAL are further abbreviations of BOSS-PPG-LAN, BOSS-PPG-BIC, BOSS-PPG-BIL, BOSS-AHD-BIC, BOSS-AHD-BIL and BOSS-AHD-LAN. Similarly to Tables 1 and 2, Table 4 compares results on the above-mentioned mixture of datasets with 0.2 bpc. Table 5 compares the results with 0.4 bpc. As can be seen, the proposed method outperforms recent state of the art approaches, by a significant margin.

Table 3. Representation of mixture of datasets. ✓ implies dataset has been selected and - implies otherwise.

Name	BPL	BPBc	BPBl	BABc	BABl	BAL
Set-I	✓	✓	✓	–	–	–
Set-II	–	–	–	✓	✓	✓
Set-III	✓	–	–	–	–	✓
Set-IV	✓	✓	✓	✓	✓	✓

Table 4. Comparison of results for CMD-C-HILL stego images with 0.2 bpc on mixture of datasets. D-CNN is executed with 30 fixed SRM kernels. The best results are represented in bold font.

Dataset	DHR	D-CNN	W-Net	CGC	PVC	Proposed
Set-I	0.7237	0.7259	0.7675	0.7712	0.7734	**0.8029**
Set-II	0.7214	0.7217	0.7714	0.7710	0.7684	**0.8026**
Set-III	0.6722	0.6865	0.7284	0.7412	0.7388	**0.7648**
Set-IV	0.7164	0.7182	0.7671	0.7782	0.7684	**0.8048**

Table 5. Comparison of results for CMD-C-HILL stego images with 0.4 bpc on mixture of datasets. D-CNN is executed with 30 fixed SRM kernels. The best results are represented in bold font.

Dataset	DHR	D-CNN	W-Net	CGC	PVC	Proposed
Set-I	0.8241	0.8289	0.8594	0.8788	0.8641	**0.9041**
Set-II	0.8231	0.8417	0.8806	0.8762	0.8661	**0.9021**
Set-III	0.7812	0.7892	0.8316	0.8411	0.8421	**0.8598**
Set-IV	0.8161	0.8214	0.8893	0.8796	0.8812	**0.9013**

4   Conclusion

With recent developments of color based steganography algorithms, the need for a powerful steganalyzer is needed. We saw recently, that an ensemble model of colorspaces has a significant impact on classification results. We propose StegCol-Net as a powerful color image steganalyzer. We employ an ensemble colorspace strategy to determine if an image is protecting information or not. We use ColorNet and take the final activation map from each colorspace. We use weighted averaging to obtain a single feature map from all the feature maps that are generated by each colorspace. We then use a levy-flight grey wolf optimization method to select a smaller subset of features. Using these features, we classify the given image into one of two classes: containing concealed information or not.

References

1. Gowda, S.N., Yuan, C.: ColorNet: investigating the importance of color spaces for image classification. In: Jawahar, C.V., Li, H., Mori, G., Schindler, K. (eds.) ACCV 2018. LNCS, vol. 11364, pp. 581–596. Springer, Cham (2019). https://doi.org/10.1007/978-3-030-20870-7_36

2. Deng, J., et al.: ImageNet: a large-scale hierarchical image database. In: IEEE Conference on Computer Vision and Pattern Recognition, pp. 248–255. IEEE (2009)

3. Kahn, D.: The history of steganography. In: Anderson, R. (ed.) IH 1996. LNCS, vol. 1174, pp. 1–5. Springer, Heidelberg (1996). https://doi.org/10.1007/3-540-61996-8_27

4. Cheddad, A., Condell, J., Curran, K., Mc Kevitt, P.: Digital image steganography: survey and analysis of current methods. Signal Process. 90(3), 727–752 (2010)

5. Li, N., Hu, J., Sun, R., Wang, S., Luo, Z.: A high-capacity 3D steganography algorithm with adjustable distortion. IEEE Access 5, 24457–24466 (2017)

6. Tsai, Y.-Y.: An adaptive steganographic algorithm for 3D polygonal models using vertex decimation. Multimedia Tools Appl. 69(3), 859–876 (2012). https://doi.org/10.1007/s11042-012-1135-8

7. Cheng, Y.M., Wang, C.M.: A high-capacity steganographic approach for 3D polygonal meshes. Visual Comput. 22(9–11), 845–855 (2006)

8. Chakraborty, S., Jalal, A.S., Bhatnagar, C.: LSB based non blind predictive edge adaptive image steganography. Multimedia Tools Appl. 76(6), 7973–7987 (2016). https://doi.org/10.1007/s11042-016-3449-4

9. Jayaram, P., Ranganatha, H.R., Anupama, H.S.: Information hiding using audio steganography-a survey. Int. J. Multimedia Appl. (IJMA) 3, 86–96 (2011)

10. Kumar, V., Kumar, D.: A modified DWT-based image steganography technique. Multimedia Tools Appl. 77(11), 13279–13308 (2017). https://doi.org/10.1007/s11042-017-4947-8

11. Narasimmalou, T., Joseph, R.A.: Discrete wavelet transform based steganography for transmitting images. In: IEEE-International Conference On Advances In Engineering, Science And Management (ICAESM-2012), pp. 370–375. IEEE (2012)

12. Gowda, S.N.: Innovative enhancement of the Caesar cipher algorithm for cryptography. In: 2016 2nd International Conference on Advances in Computing, Communication and Automation (ICACCA) (Fall), pp. 1–4. IEEE, September 2016

13. Chang, C.C., Yu, Y.H., Hu, Y.C.: Hiding secret data into an AMBTC-compressed image using genetic algorithm. In Second International Conference on Future Generation Communication and Networking Symposia, vol. 3, pp. 154–157. IEEE, December 2008

14. Gowda, S.N.: Using Blowfish encryption to enhance security feature of an image. In: 6th International Conference on Information Communication and Management (ICICM), pp. 126–129. IEEE, October 2016

15. Luo, X.Y., Wang, D.S., Wang, P., Liu, F.L.: A review on blind detection for image steganography. Signal Process. 88(9), 2138–2157 (2008)

16. Fridrich, J., Goljan, M.: On estimation of secret message length in LSB steganography in spatial domain. In: Security, Steganography, and Watermarking of Multimedia Contents VI, vol. 5306, pp. 23–35. International Society for Optics and Photonics, June 2004

17. Kodovsky, J., Fridrich, J., Holub, V.: Ensemble classifiers for steganalysis of digital media. IEEE Trans. Inf. Forensics Secur. 7(2), 432–444 (2012)

18. Deng, H. and Runger, G.: Feature selection via regularized trees. In: The 2012 International Joint Conference on Neural Networks (IJCNN), pp. 1–8. IEEE, June 2012

19. Chhikara, R.R., Sharma, P., Singh, L.: An improved dynamic discrete firefly algorithm for blind image steganalysis. Int. J. Mach. Learn. Cybern. 9(5), 821–835 (2016). https://doi.org/10.1007/s13042-016-0610-3

20. Yao, X., Liu, Y., Lin, G.: Evolutionary programming made faster. IEEE Trans. Evol. Comput. 3(2), 82–102 (1999)

21. Pathak, Y., Arya, K.V., Tiwari, S.: Feature selection for image steganalysis using levy flight-based grey wolf optimization. Multimedia Tools Appl. 78(2), 1473–1494 (2018). https://doi.org/10.1007/s11042-018-6155-6

22. Gowda, S.N.: Human activity recognition using combinatorial deep belief networks. In: Proceedings of the IEEE Conference on Computer Vision and Pattern Recognition Workshops, pp. 1–6. (2017)

23. Bas, P., Filler, T., Pevný, T.: "Break our steganographic system": the ins and outs of organizing BOSS. In: Filler, T., Pevný, T., Craver, S., Ker, A. (eds.) IH 2011. LNCS, vol. 6958, pp. 59–70. Springer, Heidelberg (2011). https://doi.org/10.1007/978-3-642-24178-9_5

24. Piva, A. and Barni, M.: The first BOWS contest: break our watermarking system. In: Security, Steganography, and Watermarking of Multimedia Contents IX, vol. 6505, p. 650516. International Society for Optics and Photonics, February 2007

25. Goljan, M., Fridrich, J., Cogranne, R.: Rich model for steganalysis of color images. In: IEEE International Workshop on Information Forensics and Security (WIFS), pp. 185–190. IEEE, December 2014

26. Zeng, J., Tan, S., Liu, G., Li, B., Huang, J.: WISERNet: wider separate-then-reunion network for steganalysis of color images. IEEE Trans. Inf. Forensics Secur. 14(10), 2735–2748 (2019)

27. Ye, J., Ni, J., Yi, Y.: Deep learning hierarchical representations for image steganalysis. IEEE Trans. Inf. Forensics Secur. 12(11), 2545–2557 (2017)

28. Xu, G.: Deep convolutional neural network to detect J-UNIWARD. In: Proceedings of the 5th ACM Workshop on Information Hiding and Multimedia Security, pp. 67–73. ACM, June 2017

29. Holub, V., Fridrich, J., Denemark, T.: Universal distortion function for steganography in an arbitrary domain. EURASIP J. Inf. Secur. 2014(1), 1–13 (2014). https://doi.org/10.1186/1687-417X-2014-1

30. Sedighi, V., Cogranne, R., Fridrich, J.: Content-adaptive steganography by minimizing statistical detectability. IEEE Trans. Inf. Forensics Secur. 11(2), 221–234 (2016)

31. Li, B., Wang, M., Huang, J., Li, X.: A new cost function for spatial image steganography. In: IEEE International Conference on Image Processing (ICIP), pp. 4206–4210. IEEE, October 2014

32. Fridrich, J., Filler, T.: Practical methods for minimizing embedding impact in steganography. In: Security, Steganography, and Watermarking of Multimedia Contents IX, vol. 6505, p. 650502. International Society for Optics and Photonics, February 2007

33. Tang, W., Li, B., Luo, W., Huang, J.: Clustering steganographic modification directions for color components. IEEE Signal Process. Lett. 23(2), 197–201 (2016)

34. Qin, X., Li, B., Tan, S., Zeng, J.: A novel steganography for spatial color images based on pixel vector cost. IEEE Access 7, 8834–8846 (2019)

35. Kang, Y., Liu, F., Yang, C., Xiang, L., Luo, X., Wang, P.: Color image steganalysis based on channel gradient correlation. Int. J. Distrib. Sensor Netw. **15**(5), 1550147719852031 (2019)
36. Gowda, S.N., Yuan, C.: Using an ensemble color space model to tackle adversarial examples (2020). arXiv preprint arXiv:2003.05005
37. Gowda, S.N.: An intelligent fibonacci approach to image steganography. In: 2017 IEEE Region 10 Symposium (TENSYMP), pp. 1–4. IEEE, July 2017

Residual Multiscale Full Convolutional Network (RM-FCN) for High Resolution Semantic Segmentation of Retinal Vasculature

Tariq M. Khan[1(✉)], Antonio Robles-Kelly[1], Syed S. Naqvi[2],
and Muhammad Arsalan[3]

[1] The School of Information Technology, Faculty of Science Engineering and Built
Environment, Deakin University, Waurn Ponds, VIC 3216, Australia
tariq.khan@deakin.edu.au
[2] Department of ECE, COMSATS University Islamabad, Islamabad Campus,
Islamabad, Pakistan
[3] Division of Electronics and Electrical Engineering, Dongguk University,
30 Pildong-ro, 1-gil, Jung-gu, Seoul 100-715, Korea

Abstract. In a fundus image, Vessel local characteristics like direction, illumination and noise vary considerably, making vessel segmentation a challenging task. Methods based upon deep convolutional networks have consistently yield state of the art performance. Despite effective, of the drawbacks of these methods is their computational complexity, whereby testing and training of these networks require substantial computational resources and can be time consuming. Here we present a multi-scale kernel based on fully convolutional layers that is quite lightweight and can effectively segment large, medium, and thin vessels over a wide variations of contrast, position and size of the optic disk. Moreover, the architecture presented here makes use of these multi-scale kernels, reduced application of pooling operations and skip connections to achieve faster training. We illustrate the utility of our method for retinal vessel segmentation on the DRIVE, CHASE_DB and STARE data sets. We also compare the results delivered by our method with a number of alternatives elsewhere in the literature. In our experiments, our method always provides a margin of improvement on specificity, accuracy, AUC and sensitivity with respect to the alternative.

Keywords: Retinal vessel segmentation · Convolutional neural networks · Diabetic retinopathy

1 Introduction

Retinal fundus images contain important features often used to diagnose eye-related illnesses such as diabetic retinopathy (DR), glaucoma, age-related macular degeneration (AMD) and systemic illnesses such as arteriosclerosis and hypertension. Among these diseases, DR and AMD are the major causes of blindness

© Springer Nature Switzerland AG 2021
A. Torsello et al. (Eds.): S+SSPR 2020, LNCS 12644, pp. 324–333, 2021.
https://doi.org/10.1007/978-3-030-73973-7_31

[1,2]. Fundus images, acquired during an ophthalmic exam, are used to inspect and monitor DR and AMD disease progression. As a result, a computer-aided diagnosis system that can significantly reduce the burden on the ophthalmologists and alleviate the inter and intra observer variability is highly desired.

Here, we focus on the segmentation of retinal blood vessels. These originate from the centre of optic disc and spread over the other regions of the retina. The blood vessels are responsible for supplying blood to the entire region of the retina, whereby microaneurysm, hemorrhages and exudate lesions are formed in the retinal image due to leakages taking place and appear as bright spots in the fundus image. Recently, convolutional neural networks (CNNs) have gained significant importance in semantic segmentation [3]. Methods such as those presented in [4–6] have yielded state of the art performance. Moreover, approaches such as that in [7] are able to address the pixel-wise classification problem by mapping low resolution features produced by the encoder back to the input resolution through a decoder. The advantage of such mapping resides in the fact that they can preserve fine-grained information, which is of capital importance for effective boundary detection.

As related to retinal vessel segmentation, the authors in [8] explore a deep learning approach that focuses on the thickness of the retinal vasculature. In [9], the authors present a skip connection encoder-decoder architecture that is quite effective detecting vessel boundaries. Gu et al. [10] present a context encoder for vessel segmentation network. Yan et al. [8] introduced a joint-loss including both a pixel-wise and a segmentation-level cost. Despite the higher accuracy of these deep learning methods, there are still many problems that demand significant attention from researchers. One of the drawbacks of these methods is their computational complexity, whereby both the pre-processing and post-processing tasks needed for deep learning approaches require substantial computational resources, training and testing times.

This paper presents a residual multiscale full convolutional network (RM-FCN) for retinal vessel segmentation. The proposed method is quite lightweight compared to other methods elsewhere in the literature, with only 6 convolutional layers with 3 multi-scale fully convolutional kernels per layer. The proposed model not only is able to accurately detect thick vessels but, when applied to the thin ones, these are also segmented due to the use of our multi-scale architecture. In our networks only two max-pooling operations are required and these are paired with external skip-connections. This yields an architecture that makes use of reduced convolutional layers, multi-scale kernels and reduced application of pooling operations so as to achieve a faster training. The rest of the paper is organized as follows. Our architecture is in Sect. 2. We then present results for retinal image segmentation and compare to alternatives in Sect. 3. Finally, in Sect. 4, we conclude on the developments presented here.

2   Residual Multiscale Network

Recall that, in retinal vessel segmentation applications, the vessel size may vary considerably across patients with a variety of medical conditions. Diabetic

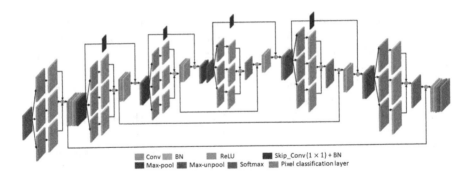

Fig. 1. Block diagram of the proposed method

retinopathy can cause the swelling of the retinal vessels and can also encourage the development of smaller, newer ones. Hypertensive retinopathy, in the other hand, can cause the shrinkage of retinal vessels. As mentioned above, here we employ multi-scale kernels to develop a neural network architecture that can cope with large size variations.

Neural networks elsewhere in the literature often employ a single sized convolutional kernel which often focuses in larger vessels and, therefore, is not quite effective for the segmentation of smaller vascular structures. This accounts for the notion that very thin vessels may not affect overly affect the overall performance in terms. This is debatable since several diagnosis in medical applications heavily rely upon small-sized vessels. Our multiscale kernels are based on 3×3, 5×5, and 7×7 convolutions for large, medium, and very small vessels, respectively. The architecture of our RM-FCN is illustrated in Fig. 1.

To construct our network, we have used multiscale convolutional blocks with important design concerns. The first of these is to keep to a minimum the use of pooling layers which are used to reduce the dimension of the feature maps. This is since these pooling operations also cause the loss of spatial information. Secondly, we employ multi-scale kernels so as to account for the large variation in retinal vessel sizes. Thirdly, we reduce the overall number of convolutions in the network. These can also be responsible for spatial information loss. Finally, we employ fine-grained information and residual skip paths to improve the segmentation results and make training more computationally efficient. Figure 2 shows the overall architecture of our proposed multi-scale convolutional blocks within the network. The network has six multi-scale convolutional blocks, where the first block is an input one, followed by two down multi-scale blocks. There is an intermediate block which connects down and up blocks. This is followed by the two up-multiscale convolutional blocks with a final output one which is equipped with a softmax loss layer.

In Fig. 2 presents the example up multi-scale convolutional block, which receives the feature map F from the pooling layer and distributes them to the convolutions C_3^A, C_5^A, C_7^A and C_1^A. Note that C_1^A is, in fact, part of the skip connection. These kernels have sizes 3×3, 5×5, 7×7, 1×1, respectively. Each of the multi-scale convolutional kernels C_3^A, C_5^A, C_7^A outputs the features F_a, F_b, F_c, respectively. These are given by

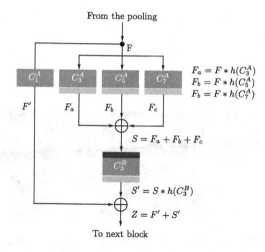

Fig. 2. Block diagram of the proposed multiscale convolutional block.

$$F_a = F * C_3^A$$
$$F_b = F * C_5^A \qquad\qquad (1)$$
$$F_c = F * C_7^A$$

which are then used to obtain S, which is given by

$$S = F_a + F_b + F_c \qquad\qquad (2)$$

Thus, S can be viewed as a combined feature map which can later be fed into a ReLU and batch normalized. This is done after an additional convolution C_3^B is applied so as to obtain the feature map S' given by

$$S' = S * C_3^B \qquad\qquad (3)$$

where S' is the multi-scale feature map. To further improve the feature map quality S' is combined with F', which arises from the skip path comprising C_1^A (a 1×1 convolutional kernel). This yield the feature map Z given by $Z = F' + S'$.

As shown it the figure, the encoder blocks generate the respective feature maps using convolutions between the input image and a multi-scale filter bank. Here, we have followed [11] and applied batch normalisation on the features followed by a ReLU. For the down sampling blocks, the resulting feature maps are fed to the a 2×2, non-overlapping max-pooling with a stride of size 2. In this manner, the down-sampled feature maps created from the final down-sampling block can be used for the up-sampling procedure. This is carried out by using the indices of the max-pooling information. In our architecture, the feature maps yielded by the down-sampling blocks are unpooled. These maps, which are sparse in nature, are augmented in the up-sampling blocks by the multi-scale filter banks. These dense feature maps are then normalized by using batch normalization. The size of the feature maps yielded by the up-sampling blocks are identical to those obtained by the respective down-sampling blocks. The only difference is in the final layer of the decoder, where a multi-channel function map is obtained as an output compared to the three-channel RGB data of the first encoder. At output, our network yields a final map where pixels are labelled as vessels or not on the basis of a soft-max classifier.

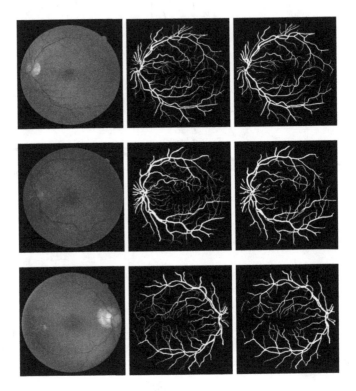

Fig. 3. Segmentation results of the RM-FCN model on three noisy test images i.e. image number 1, 3 and 8 from the DRIVE dataset. From left-to-right, we show the input images, the ground truth segmentation map and that yielded by our method.

3   Experiments

3.1   Datasets

We now turn our attention to the evaluation of our method on three publicly available retinal image databases. These are the CHASE [12][1], DRIVE [13][2] and STARE [14][3] data sets. The DRIVE dataset covers a wide age range of diabetic

---

[1] The dataset can found at https://blogs.kingston.ac.uk/retinal/chasedb1/.

[2] The dataset is widely available at https://drive.grand-challenge.org/.

[3] More information regarding the STARE project can be found at https://cecas. clemson.edu/~ahoover/stare/.

patients and consists of 20 color images for training and 20 color images for testing. The STARE dataset is a collection of 20 color retinal fundus images captured at 35° FOV with an image size of 700 × 605 pixels. Out of these 20 images, 10 images contain pathologies. Two different manual segmentation as ground truth are available. Here we employ the first experts segmentation as ground truth where available. There is no dedicated test dataset available for STARE. The CHASE dataset consists of 28 color images of 14 school children in England. Two different manual segmentation maps are available as ground truth. Again, here we employ the first experts segmentation for our experiments. The CHASE dataset doesn't contain any dedicated training or testing sets. Here we have used the first 20 images for training and the last 8 images for testing.

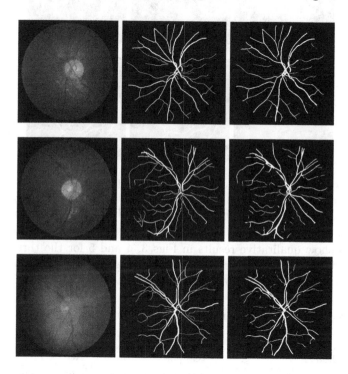

Fig. 4. Segmentation results of the RM-FCN model on images number 5, 6 and 8 from the CHASE_DB dataset. From left-to-right, we show the input images, the ground truth segmentation map and that yielded by our method.

3.2   Results and Comparison

Here we compare the results obtained by our approach on the three data sets above with those yielded by a number of alternatives. For all the methods under consideration we have used four common performance parameters. These are Sensitivity (Se), Specificity (Sp), Accuracy (Acc) and AUC. These results are shown in Tables 1, 2 and 3.

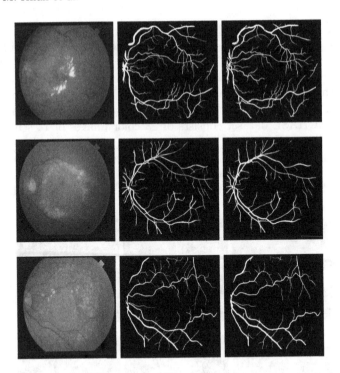

Fig. 5. The segmentation results of the RM-FCN model on the two noisy test images i.e. image number 1, 2 and 3 from the STARE dataset. From left-to-right, we show the input images, the ground truth segmentation map and that yielded by our method.

We also show qualitative results in Figs. 3, 4 and 5 for the three data sets under consideration. In all figures we show, from left-to-right the input imagery, the segmentation ground truth provided by the hand-labeled vessel maps and

Table 1. Performance comparison of our RM-FCN on DRIVE data set with respect to other methods elsewhere in the literature

Method	Year	Se	Sp	Acc	AUC
Guo *et al.* [15]	2019	0.7891	0.9804	0.9561	0.9806
Ma *et al.* [16]	2019	0.7916	0.9811	0.9570	0.9810
Wang *et al.* [17] DU-Net	2019	0.7940	0.9816	0.9567	0.9772
Wu *et al.* [18]	2019	0.8038	0.9802	0.9578	0.9821
Gu *et al.* [10] CE-Net	2019	0.8309	–	0.9545	0.9779
Arsalan *et al.* [19] VessNet	2019	0.8022	0.9810	0.9655	0.9820
Wang *et al.* [20]	2020	0.7991	0.9813	0.9581	0.9823
Yin *et al.* [21]	2020	0.8038	0.9837	0.9578	0.9846
Segnet-Basic	2020	0.7949	0.9738	0.9579	0.9720
Our method	**2020**	**0.8342**	**0.9825**	**0.9695**	**0.9830**

the results yielded by our method. From the figures, we can see that our method can cope well with thinner vessels, preserving well the fine-grained detail while being quite robust to different conditions, variations in contrast and optic disk position and size.

From Table 1, it is clear that our method's accuracy is the highest amongst the alternatives for the DRIVE data set. The second best accuracy on the Drive data set is that delivered by the method of Arsalan et al. [19]. In terms of sensitivity, on the DRIVE dataset, our method also achieve the highest value. The second best sensitivity on DRIVE dataset is that of the method in [10] (CE-Net). Similarly, the results presented in Table 2 indicate that method proposed here has the best overall performance on the CHASE data set across all the measures used. The sensitivity achieved by the Arsalan et al. [19] is the second highest in Table 2. The accuracy of Yin et al. [21] is the best among all the approaches under consideration. Finally, Table 3 shows that is also the best performing method on the STARE data set. The sensitivity achieved by the Arsalan et al. [19] is again the second highest.

Table 2. Performance comparison of our RM-FCN on CHASE_DB1 data set with respect to other methods elsewhere in the literature

Method	Year	Se	Sp	Acc	AUC
Zhang et al. [22]	2016	0.7626	0.9661	0.9452	0.9606
Khawaja et al. [23]	2019	0.7974	0.9697	0.9528	NA
Jin et al. [24]	2019	0.7595	0.9878	0.9641	0.9832
Arsalan et al. [19] VessNet	2019	0.8206	0.9800	0.9726	0.9800
Wang et al. [20]	2020	0.8186	0.9844	0.9673	0.9881
Yin et al. [21]	2020	0.7993	0.9868	0.9783	0.9869
Segnet-basic	2020	0.8190	0.9735	0.9638	0.9780
Our method	2020	**0.8463**	**0.9828**	**0.9735**	**0.9810**

Table 3. Performance comparison of our RM-FCN on STARE database with respect to other methods elsewhere in the literature

Method	Year	Se	Sp	Acc	AUC
Chen et al. [25] Deeplab v3++	2018	0.8320	0.9760	0.9650	0.9735
Jin et al. [24]	2019	0.8155	0.9752	0.9610	0.9804
Guo et al. [15]	2019	0.7888	0.9801	0.9627	0.9840
Wang et al. [17]	2019	0.8074	0.9821	0.9661	0.9812
Wu et al. [18]	2019	0.8132	0.9814	0.9661	0.9860
Arsalan et al. [19] VessNet	2019	0.8526	0.9791	0.9697	0.9883
Wang et al. [20]	2020	0.8239	0.9813	0.9670	0.9871
SegNet-Basic	2020	0.8118	0.9738	0.9543	0.9728
Our method	2020	**0.8565**	**0.9834**	**0.9739**	**0.9890**

332   T. M. Khan et al.

4 Conclusions

In this paper we have presented a residual multi-scale network for retinal vessel segmentation that employs skip connections, multiscale filters and a reduced number of pooling operations so as to segment large, medium and thin vasculature under large variations of contrast, optic disk position and size. We have illustrated the utility of the method for the task in hand by performing experiments on three publicly accessible databases, namely CHASE DB1, STARE and DRIVE. In our experiments, our network outperformed a number of state-of-the-art alternatives. For our comparison, we have used well-known measurement parameters, namely sensitivity, balanced accuracy and accuracy.

References

1. Khan, T.M., Alhussein, M., Aurangzeb, K., Arsalan, M., Naqvi, S.S., Nawaz, S.J.: Residual connection-based encoder decoder network (RCED-net) for retinal vessel segmentation. IEEE Access **8**, 131257–131272 (2020)
2. Khan, T.M., Naqvi, S.S., Arsalan, M., Khan, M.A., Khan, H.A., Haider, A.: Exploiting residual edge information in deep fully convolutional neural networks for retinal vessel segmentation. In: 2020 International Joint Conference on Neural Networks (IJCNN), pp. 1–8. IEEE (2020)
3. Khan, T.M., Abdullah, F., Naqvi, S.S., Arsalan, M., Khan, M.A., Shallow vessel segmentation network for automatic retinal vessel segmentation. In: 2020 International Joint Conference on Neural Networks (IJCNN), pp. 1–7. IEEE (2020)
4. Khan, T.M., Robles-Kelly, A., Naqvi, S.S.: A semantically flexible feature fusion network for retinal vessel segmentation. In: Yang, H., Pasupa, K., Leung, A.C.-S., Kwok, J.T., Chan, J.H., King, I. (eds.) ICONIP 2020. CCIS, vol. 1332, pp. 159–167. Springer, Cham (2020). https://doi.org/10.1007/978-3-030-63820-7_18
5. Khawaja, A., Khan, T.M., Naveed, K., Naqvi, S.S., Rehman, N.U., Nawaz, S.J.: An improved retinal vessel segmentation framework using Frangi filter coupled with the probabilistic patch based denoiser. IEEE Access **7**, 164344–164361 (2019)
6. Khan, M.A.U., Khan, T.M., Bailey, D.G., Soomro, T.A.: A generalized multi-scale line-detection method to boost retinal vessel segmentation sensitivity. Pattern Anal. Appl. **22**(3), 1177–1196 (2018). https://doi.org/10.1007/s10044-018-0696-1
7. Badrinarayanan, V., Kendall, A., Cipolla, R.: SegNet: a deep convolutional encoder-decoder architecture for image segmentation. IEEE Trans. Pattern Anal. Mach. Intell. **39**(12), 2481–2495 (2017)
8. Yan, Z., Yang, X., Cheng, K.T.: Joint segment-level and pixel-wise losses for deep learning based retinal vessel segmentation. IEEE Trans. Biomed. Eng. **65**, 1912–1923 (2018)
9. Ronneberger, O., Fischer, P., Brox, T.: U-Net: convolutional networks for biomedical image segmentation. In: Medical Image Computing and Computer-Assisted Intervention (2015)
10. Gu, Z., et al.: CE-net: context encoder network for 2D medical image segmentation. IEEE Trans. Med. Imaging **38**(10), 2281–2292 (2019)
11. Ioffe, S., Szegedy, C.: Batch normalization: accelerating deep network training by reducing internal covariate shift. In: International Conference on Machine Learning, pp. 448–456 (2015)

12. Fraz, M.M., et al.: An approach to localize the retinal blood vessels using bit planes and centerline detection. Comput. Methods Programs Biomed. **108**(2), 600–616 (2012c)
13. Staal, J., Abramoff, M.D., Niemeijer, M., Viergever, M.A., van Ginneken, B.: Ridge-based vessel segmentation in color images of the retina. IEEE Trans. Med. Imaging **23**(4), 501–509 (2004)
14. Hoover, A.D., Kouznetsova, V., Goldbaum, M.: Locating blood vessels in retinal images by piecewise threshold probing of a matched filter response. IEEE Trans. Med. Imaging **19**(3), 203–210 (2000)
15. Guo, S., Wang, K., Kang, H., Zhang, Y., Gao, Y., Li, T.: BTS-DSN: deeply supervised neural network with short connections for retinal vessel segmentation. Int. J. Med. Inf. **126**, 105–113 (2019)
16. Ma, W., Yu, S., Ma, K., Wang, J., Ding, X., Zheng, Y.: Multi-task neural networks with spatial activation for retinal vessel segmentation and artery/vein classification. In: Medical Image Computing and Computer Assisted Intervention (2019)
17. Wang, B., Qiu, S., He, H.: Dual encoding U-net for retinal vessel segmentation. In: Shen, D., et al. (eds.) MICCAI 2019. LNCS, vol. 11764, pp. 84–92. Springer, Cham (2019). https://doi.org/10.1007/978-3-030-32239-7_10
18. Wu, Y., et al.: Vessel-Net: retinal vessel segmentation under multi-path supervision. In: Medical Image Computing and Computer Assisted Intervention (2019)
19. Arsalan, M., Oqais, M., Mahmood, T., Cho, S.W., Park, K.R.: Aiding the diagnosis of diabetic and hypertensive retinopathy using artificial intelligence-based semantic segmentation. J. Clin. Med. **8**(9), 1446 (2019)
20. Wang, D., Haytham, A., Pottenburgh, J., Saeedi, O., Tao, Y.: Hard attention net for automatic retinal vessel segmentation. IEEE J. Biomed. Health Inf. **24**, 3384–3396 (2020)
21. Yin, P., Yuan, R., Cheng, Y., Wu, Q.: Deep guidance network for biomedical image segmentation. IEEE Access **8**, 116106–116116 (2020)
22. Zhang, J., Dashtbozorg, B., Bekkers, E., Pluim, J.P.W., Duits, R., Romeny, B.M.: Robust retinal vessel segmentation via locally adaptive derivative frames in orientation scores. IEEE Trans. Med. Imaging **35**(12), 2631–2644 (2016)
23. Khawaja, A., Khan, T.M., Khan, M.A.U., Nawaz, S.J.: A multi-scale directional line detector for retinal vessel segmentation. Sensors **19**(22), 4949 (2019)
24. Jin, Q., Meng, Z., Pham, T.D., Chen, Q., Wei, L., Su, R.: DUNet: a deformable network for retinal vessel segmentation. Knowl. Based Syst. **178**, 149–162 (2019)
25. Chen, L.-C., Zhu, Y., Papandreou, G., Schroff, F., Adam, H.: Encoder-decoder with Atrous separable convolution for semantic image segmentation. In: Ferrari, V., Hebert, M., Sminchisescu, C., Weiss, Y. (eds.) ECCV 2018. LNCS, vol. 11211, pp. 833–851. Springer, Cham (2018). https://doi.org/10.1007/978-3-030-01234-2_49

A Practical Hybrid Active Learning Approach for Human Pose Estimation

Sinan Kaplan[1]([✉]), Joni Juvonen[2], and Lasse Lensu[1]

[1] Computer Vision and Pattern Recognition Laboratory,
Department of Computational and Process Engineering,
School of Engineering Science, LUT University, Lappeenranta, Finland
`sinan.kaplan@student.lut.fi`
[2] Department of Future Technologies, University of Turku, Turku, Finland

Abstract. Active learning (AL) has not received much attention in deep learning (DL) for human pose estimation. In this paper, a practical hybrid active learning strategy is proposed for training a human pose estimation model, and it is tested in an industrial online environment. The conducted experiments show that the active learning strategy to select diverse samples to be annotated outperforms the baseline method with random sampling. As a result, the strategy enables a significant improvement in the performance of pose estimation.

Keywords: Active learning · Human pose estimation · Human in the loop · Artificial intelligence

1 Introduction

DL has contributed to the significant interest to machine learning and its success has been regularly seen in supervised learning tasks where a large amount of labeled data is available [9,18,35]. For applications where the amount of data is limited, methodological remedies exist in the form of data augmentation, transfer learning and few-shot learning. Even with these approaches, however, a relatively large amount of data is needed for rapid adaptation of deep models in complex domains where the initial data is limited. In applications where the acquisition of data and labelling them can be an expensive and laborious task, one may consider an iterative approach to sample and label the data. AL is a family of such techniques to appropriately select data samples to be annotated next [27]. These methods enable collaboration of a human and artificial intelligence (AI) to annotate a subset of data without resorting to fully annotating the data or purely random selection of samples.

The ultimate goal in AL is to reduce the annotation and training effort/cost while making models as accurate as possible with less amount of data. To achieve this, the data to be annotated is sampled by an acquisition function and is brought to an oracle (human annotator) to review and perform the annotation.

S. Kaplan—Supported by PintaWorks Oy.

There are two settings where AL strategies can be applied and tested: 1) online (live) and 2) offline (simulated) environments [16]. In offline environments, there is already a pool of labeled data available and the goal is to evaluate the performance of a AL strategy on this labeled data pool. Offline environments are more common than the online ones in the AL research area.

An online environment is a setting where machine learning (ML) research makes an impact on real-world problems. It connects research and applications in real life to assess whether an improvement of a particular ML model makes a difference outside the common benchmark data sets [33].

In this paper, an AL strategy is presented for an online environment involving human pose estimation [8]. Compared to object detection and image classification tasks, the annotation cost is much higher as labeling of a number of keypoints on a human body image is required. For instance, COCO-style keypoint annotation requires 18 keypoints to be labeled [7].

To use active learning in this context, a hybrid approach combining model-based uncertainty sampling and diversity sampling [2] is proposed in this study. The method takes advantage of transfer learning and approximate nearest neighbors as parts of the solution. The aims are as follows: 1) To improve the accuracy of the pose estimation model with diversely picked data samples, 2) to avoid adding another level of complexity, such as training another model for sampling, to the AL pipeline, 3) to reduce the sampling time by using approximate nearest neighbors instead of exhaustive search methods, and 4) to provide an analysis of practical challenges in the online environment.

The paper is organized as follows: Sect. 2 reviews the studies of active learning and human pose detection. Section 3 covers the proposed method in detail. Section 4 focuses on experiments with the results. The findings are further evaluated in Sect. 5 and Sect. 6 presents the conclusion[1].

2   Related Work

Human pose estimation focuses on providing reliable estimates of human poses in various applications, including person tracking and analysis of sports activities [5,8]. In the literature, it is studied under two categories: 1) top-down, and 2) bottom-up approaches. The top-down approaches first detect person candidates and then perform pose extraction. For instance, the regional multi-person pose estimation framework [11] and the cascaded pyramid network [6] fall into this category. On the other hand, the bottom-up approaches first extract features from a given image and construct bipartite graphs to produce a human pose estimate. A widely used framework in this category is OpenPose [4] that is based on part affinity fields described in the work by Cao et al. [5]. Other important studies apply long short term memory (LSTM) models [1] and different variations of convolutional neural networks (CNNs) [14,34] to improve the reliability of human pose estimation.

---

[1] Initial results of this study have been presented in the 2nd ICML 2020 Workshop on Human in the Loop Learning as a work-in-progress paper.

Active learning is widely studied in the literature and the methods vary depending on the application context [2,27]. For instance, [23] focuses on applications of active learning techniques in DL models in particular. The techniques applied for active learning can be divided into two categories: 1) pool-based strategies and 2) query synthesizing strategies [2,27].

The pool-based strategies utilize information, ensemble and uncertainty based methods to select samples from an unlabeled data pool [2,27,28]. Bayesian models are also studied among the pool-based methods [12]. As they embrace Bayesian inference principles, they provide a natural basis for the uncertainty estimation.

The strategies relying on query synthesis take advantage of generative adversarial networks (GANs) [36] and variational autoencoders (VAEs) [21,29]. The methods in this category are based on learning a latent representation for both labeled and unlabeled data for a discriminator module to classify whether a given data sample is from the labeled or unlabeled data pool. They are difficult to generalize for an online setting [16], since the data distribution is prone to change over time in online environments. Besides, this would add another level of complexity and cost by requiring an additional model to be trained on both labeled and unlabeled data pool, which this study aims to avoid.

Human Pose Estimation and Active Learning. In the field of computer vision, active learning methods are mainly studied and tested on image classification and object detection tasks [24,26]. There is a limited number of studies that cover human pose estimation tasks. Liu et al. [19] studied active learning for human pose detection. They applied uncertainty measurement and the principle of influence [10] for sampling, and convolutional pose machine (CPM) [34] was used as the pose model. One limitation of the CPM-based methods is that they provide heatmaps with the same resolution as the image size. This causes heatmaps to be diffuse making the application of non-maximum suppression (NMC) more difficult. To solve this issue, in this paper, the model provides low resolution heatmaps for each keypoint to extract peak points with NMC. In addition to that, in contrast to greedy representative sampling strategy used by Liu et al., approximate nearest neighbors approach [22] is applied to reduce sampling time (more details are given in Sects. 3 and 4).

3   Methods

The proposed AL approach for human pose estimation is shown in Fig. 1. For the development, OpenPose-plus [32] is used as the pose model, which is based on the popular OpenPose framework [4]. It is an well-suited framework for our study, since it provides real time pose estimation with simplicity and flexibility of switching between different backbones (MobileNet, Resnet18, VGG and VGGTiny) stated in the repository. By considering the given inference time, model size and accuracy in OpenPose-plus repo, the model with VGGTiny backbone is trained from scratch with initially annotated 2000 data samples.

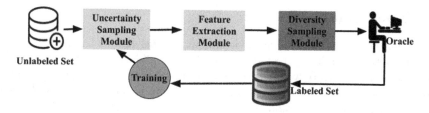

Fig. 1. An overview of the active learning procedure.

3.1 Active Learning Framework

The OpenPose-plus model provides confidence maps (heatmaps) and part affinity fields (PAFs). Heatmaps represent activations of the last layer of the model as confidence scores, whereas PAFs represent connection vectors between the keypoints. In total, there are 19 heatmaps (18 specific body points and one background). Based on the heatmap activations, AL is designed module-by-module as follows[2]: (1) **Uncertainty Sampling Module** is responsible for filtering data samples based on heatmap activations. Since the heatmaps correspond to confidence scores, a lower activation value implies higher uncertainty. (2) **Feature Extraction Module** takes filtered data from the uncertainty module and computes embedding features using the pretrained Resnet50 model on ImageNet data. Given the speed, accuracy level and amount of operations required for a single forward pass in the Resnet50 model, it is chosen to extract embedding features [3]. (3) **Diversity Sampling Module** takes the filtered data with features from the preceding module and constructs an approximate nearest neighbor tree [22] to apply diversity sampling. To shorten the sampling time on high-dimensional embedding features, an approximate nearest neighbor search is used. (4) **Oracle** is a human annotator who reviews and labels filtered data from the diversity sampling module and updates the training set. (5) **Training Module** resumes the training task with the updated training set.

As the model is deployed in an online environment, the whole procedure is an iterative process and it repeats itself depending on the model performance, the available unlabeled data size and available resources, such as the oracle and/or hardware resources, to complete the task.

3.2 Baseline Method and Evaluation Metric

To compare the effectiveness of our model for human pose estimation, *random sampling* is used as the baseline. For the evaluation of experiments, *person count accuracy vs. size of training set* is chosen to report the results for comparing the AL approach and the baseline method. Person count accuracy is a percentage of correctly detected people out of the total number of people.

---

[2] For a detailed flow of each module refer to Algorithm on   https://github.com/kaplansinan/S-SPRR2020ALpose.

4   Experiments

The experiments run in an online environment are presented in this section. First, the AL method is initially tested on the COCO validation set to qualitatively assess whether the strategy selects diverse samples after the uncertainty sampling procedure. Afterwards, experiments are run in the online environment and report results for each training session. The experiments are conducted in collaboration with PintaWorks Oy. The company provides solutions for person tracking and activity recognition for the healthcare industry.

Data and Model. The data are provided by the company and there are environment-dependent variations in the data, such as location, lighting conditions and camera angle. The data consists of grayscale images with size of 368×368 (W $\times$ H). A limited amount of data (2000 samples) was available to train the model, thus, data augmentation was used to increase the original data set. The augmentation techniques applied are as follows: rotation with limit of $[-30, 30]$ degrees, translation with limit of $[-0.62, 0.62]$ in width or height, scaling factor with range of $[0.6, 1.4]$, random brightness factor with limit of $[0.7, 1.3]$, and random contrast factor with limit of $[0.7, 1.3]$.

The OpenPose-plus model [32] is used with the VGGTiny backbone. Two different libraries were experimented with for the approximate nearest neighbor search: Annoy [31] and FLANN [20]. Based on initial tests with both of them, as a major difference, Annoy was found faster than FLANN and it was selected for the experiments.

Validation of AL Strategy. Before testing the proposed method, a qualitative assessment was performed on a benchmark set to see whether it is capable of selecting diverse samples. To do so, the trained model on the first batch of the data provided by the company is used to apply uncertainty and diversity sampling on the COCO validation set. In Fig. 2a, randomly picked samples from the COCO validation set with low and high activations from the model outputs are shown. The heatmap threshold th_{HM} was set to 0.3 and the threshold for the number of keypoints th_{KP} was set to 6 after initial tests in the development environment. After the uncertainty module, the embedding features of the size of 1×2048 are extracted for each selected sample by the feature extraction module. Next, 20% of the samples (2K images in total) are identified by the diversity sampling module to be labeled. The chosen diverse samples from the COCO validation set are shown in Fig. 2b. The figure reveals that the chosen samples are scattered across the image set. Hence, it is viable that the proposed strategy can identify diverse samples successfully.

Training Details. Tensorflow stack is used for training the models and monitoring each training session. Experiments are conducted on NVIDIA GeForce GTX 1060 with 6 GB and CUDA Development Toolkit (CUDA 10.2). The data is divided into training and validation sets at each session and the aforementioned augmentation techniques are applied during training. Early stopping criteria is used on validation set to halt the training when the model performance stops improving.

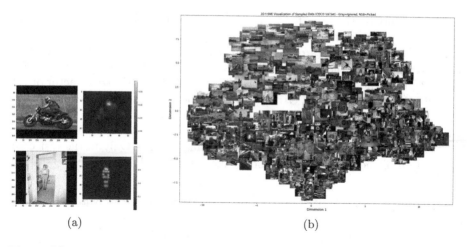

(a)   (b)

Fig. 2. (a) Examples of low and high heatmap activations. The first row shows an image with corresponding low activation values and the second row demonstrates an image with high activations. The first image will be filtered by uncertainty module for annotation.; (b) The results from diversity sampling module on the COCO validation set. The images shown in color are the ones chosen for the annotation based on approximate nearest neighbors search.

After initial tests with the training pipeline, the hyperparameters for each training session are set as follows: batch size of 4, step decay learning rate scheduler with $learningrate = 0.0001$, early stopping patience 35, and Adam optimizer [17] with $learningrate = 0.0001$, $beta_1 = 0.9$, $beta_2 = 0.999$.

Four training sessions were run for each method. It is important to note that the baseline method with random sampling was evaluated twice to provide information on the performance variation. This was necessary to see variation within the baseline method. In total, 12 training sessions were executed. At each training session, 1000 data samples are selected from the available unlabeled data pool and added to the training set after annotated by the oracle. The next training session continues from the previous one. Each training iteration takes 6–8 hours on our training setup above.

Testing Details and Results. The proposed AL method and the baseline method were evaluated on the test set, separate from the training set. The evaluation was done based on the metric given in Sect. 3.2. The comparison of the methods is shown in Fig. 3(A). The proposed method outperforms the baseline method at each iteration with a clear margin based on the person count accuracy. This is because of that the AL method can pick data samples to be annotated diversely, which leads the pose model to generalize well on the data set. In other words, the samples that the model underperforms at previous iterations are successfully selected to be annotated for the next training iteration. On the other hand, the baseline method without any guidance towards diverse samples suffers from large performance variation especially in the case of small training sets.

To test further the robustness of the proposed method under environmental variations, also test time augmentations (TTAs) was utilized on the test set. The variations in the data occur due to person size, lightning conditions, and camera noise. Thus, the following TTAs were applied: blur with kernel size of 5 (BLUR), brightness factor of 1.3 (BRIGHT) and 0.7 (DARK), scale factor of 1.4 (ZOOM IN) and 0.6 (ZOOM OUT). The evaluations on each of these augmented sets are presented in Fig. 3(B–F). One can observe the performance variation of the baseline method in most of the cases, whereas the AL approach provides consistent behaviour with the increasing training set size. Only in the test case (ZOOM OUT) both approaches produce consistent results.

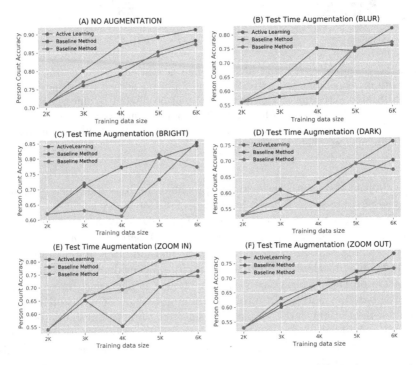

Fig. 3. Comparison of the AL method and the baseline method: (A) No Augmentations; (B–F) Test time augmentations.

5   Discussion

Data labeling is a costly and time consuming process in the development and training of human pose estimation models. For instance, COCO-style keypoint annotations for a single person takes 15–20 seconds per one image. To shorten this process in practical applications and get an accurate model with less data, AL methods can be taken into use.

In this study, an AL strategy was proposed to sample data for human pose estimation. The proposition is a hybrid procedure that makes use of both uncertainty and diversity sampling. To evaluate the method, it was compared with a baseline method based on random sampling. The experiments showed that the proposed method is able to select diverse data samples successfully and boost the performance of the human pose estimation better than the baseline method. The robustness of the method is further tested and proved in a production environment of the company.

The main issue of the proposed AL method is adversarial data samples. For instance, adversarial samples, in which an object resembles a human shape, are discarded. Since they cause high activation values, the uncertainty sampling module filters out these samples. To the best knowledge of the authors, this issue has not been not studied as part of AL in deep models and it may bring an advantage.

As this is an empirical study, it is equally important to state hidden gems and issues encountered during training process. In terms of the training procedure, the following methodologies were found useful: 1) learning rate finder [30] is an important technique to use for getting starting value for the learning rate of the optimizer, 2) weight initialization is an important factor to consider when training DL methods from scratch. In this study, better success was achieved with Glorot uniform [13] than Kaiming uniform [15], 3) Tensorflow Profiler is quite helpful to identify data loading/pipeline bottlenecks during the training, 4) One should not try an augmentation that cannot be observed in an environment, where the model is in use. It reduces the generalization capability of the model, and 5) as noted by Ruggero et al. [25], small person size and occlusions are two main issues of human pose models that were encountered in the testing and production pipeline.

Further development of the AL method will be focused on substituting the diversity sampling module with a hierarchical clustering method. Also, it can be worth using both local and global image features to perform diversity sampling. One can combine global features, such as image hashes, with the local features extracted by the feature extraction module. To improve uncertainty sampling in catching hard negatives, it can help to use average of multiple TTAs as a validation step (this is because of that true person detections likely have higher activations than adversarial ones).

6   Conclusion

In this study, an active learning strategy for human pose estimation that is implemented in an online development environment of the company was proposed. The method combines both uncertainty and diversity sampling. The uncertainty sampling is applied on the heatmap activations of the pose model, while the diversity sampling is further carried with the embedding features extracted from the pretrained feature extraction model on ImageNet. To reduce sampling time on the high dimensional embedding features, an approximate nearest neighbor method is used. The experiments revealed that the proposed AL strategy improves the performance of the pose model by introducing a smart data sampling framework.

References

1. Artacho, B., Savakis, A.: UniPose: unified human pose estimation in single images and videos. arXiv preprint arXiv:2001.08095 (2020)
2. Budd, S., Robinson, E.C., Kainz, B.: A survey on active learning and human-in-the-loop deep learning for medical image analysis. arXiv preprint arXiv:1910.02923 (2019)
3. Canziani, A., Paszke, A., Culurciello, E.: An analysis of deep neural network models for practical applications. arXiv preprint arXiv:1605.07678 (2016)
4. Cao, Z., Hidalgo, G., Simon, T., Wei, S.E., Sheikh, Y.: OpenPose: real-time multi-person 2D pose estimation using part affinity fields. arXiv preprint arXiv:1812.08008 (2018)
5. Cao, Z., Simon, T., Wei, S.E., Sheikh, Y.: Realtime multi-person 2D pose estimation using part affinity fields. In: Proceedings of the IEEE Conference on Computer Vision and Pattern Recognition, pp. 7291–7299 (2017)
6. Chen, Y., Wang, Z., Peng, Y., Zhang, Z., Yu, G., Sun, J.: Cascaded pyramid network for multi-person pose estimation. In: Proceedings of the IEEE Conference on Computer Vision and Pattern Recognition, pp. 7103–7112 (2018)
7. COCO: COCO keypoints, June 2020. http://cocodataset.org/#keypoints-2019
8. Dang, Q., Yin, J., Wang, B., Zheng, W.: Deep learning based 2D human pose estimation: a survey. Tsinghua Sci. Technol. **24**(6), 663–676 (2019)
9. Dargan, S., Kumar, M., Ayyagari, M.R., Kumar, G.: A survey of deep learning and its applications: a new paradigm to machine learning. Arch. Comput. Methods Eng. **27**(4), 1071–1092 (2019). https://doi.org/10.1007/s11831-019-09344-w
10. Dutt Jain, S., Grauman, K.: Active image segmentation propagation. In: Proceedings of the IEEE Conference on Computer Vision and Pattern Recognition, pp. 2864–2873 (2016)
11. Fang, H.S., Xie, S., Tai, Y.W., Lu, C.: RMPE: regional multi-person pose estimation. In: Proceedings of the IEEE International Conference on Computer Vision, pp. 2334–2343 (2017)
12. Gal, Y., Islam, R., Ghahramani, Z.: Deep Bayesian active learning with image data. In: Proceedings of the 34th International Conference on Machine Learning, vol. 70, pp. 1183–1192. JMLR.org (2017)
13. Glorot, X., Bengio, Y.: Understanding the difficulty of training deep feedforward neural networks. In: Proceedings of the Thirteenth International Conference on Artificial Intelligence and Statistics, pp. 249–256 (2010)
14. Golda, T., Kalb, T., Schumann, A., Beyerer, J.: Human pose estimation for real-world crowded scenarios. In: 2019 16th IEEE International Conference on Advanced Video and Signal Based Surveillance (AVSS), pp. 1–8. IEEE (2019)
15. He, K., Zhang, X., Ren, S., Sun, J.: Delving deep into rectifiers: surpassing human-level performance on ImageNet classification. In: Proceedings of the IEEE International Conference on Computer Vision, pp. 1026–1034 (2015)
16. Kagy, J.F., Kayadelen, T., Ma, J., Rostamizadeh, A., Strnadova, J.: The practical challenges of active learning: lessons learned from live experimentation. arXiv preprint arXiv:1907.00038 (2019)

17. Kingma, D., Ba, J.: Adam: a method for stochastic optimization. In: International Conference on Learning Representations, December 2014
18. LeCun, Y., Bengio, Y., Hinton, G.: Deep learning. Nature **521**(7553), 436–444 (2015)
19. Liu, B., Ferrari, V.: Active learning for human pose estimation. In: Proceedings of the IEEE International Conference on Computer Vision, pp. 4363–4372 (2017)
20. Mariusmuja: FLANN - fast library for approximate nearest neighbors, May 2020. https://github.com/mariusmuja/flann
21. Mottaghi, A., Yeung, S.: Adversarial representation active learning. arXiv preprint arXiv:1912.09720 (2019)
22. Muja, M., Lowe, D.G.: Fast approximate nearest neighbors with automatic algorithm configuration. In: VISAPP (1), vol. 2, pp. 331–340, 2 (2009)
23. Munro, R.: Human-in-the-Loop Machine Learning. Manning Publications (2020). https://books.google.fi/books?id=LCh0zQEACAAJ
24. Roy, S., Unmesh, A., Namboodiri, V.P.: Deep active learning for object detection. In: BMVC, p. 91 (2018)
25. Ruggero Ronchi, M., Perona, P.: Benchmarking and error diagnosis in multi-instance pose estimation. In: Proceedings of the IEEE International Conference on Computer Vision, pp. 369–378 (2017)
26. Sener, O., Savarese, S.: Active learning for convolutional neural networks: a core-set approach. arXiv preprint arXiv:1708.00489 (2017)
27. Settles, B.: Active learning literature survey. Technical report, University of Wisconsin-Madison Department of Computer Sciences (2009)
28. Shao, J., Wang, Q., Liu, F.: Learning to sample: an active learning framework. arXiv preprint arXiv:1909.03585 (2019)
29. Sinha, S., Ebrahimi, S., Darrell, T.: Variational adversarial active learning. In: Proceedings of the IEEE International Conference on Computer Vision, pp. 5972–5981 (2019)
30. Smith, L.N.: Cyclical learning rates for training neural networks. In: 2017 IEEE Winter Conference on Applications of Computer Vision (WACV), pp. 464–472. IEEE (2017)
31. Spotify: Annoy (approximate nearest neighbors oh yeah), May 2020. https://github.com/spotify/annoy
32. Tensorlayer: Hyperpose (Python training library, C++ inference library), June 2020. https://github.com/tensorlayer/hyperpose
33. Wagstaff, K.: Machine learning that matters. arXiv preprint arXiv:1206.4656 (2012)
34. Wei, S.E., Ramakrishna, V., Kanade, T., Sheikh, Y.: Convolutional pose machines. In: Proceedings of the IEEE conference on Computer Vision and Pattern Recognition, pp. 4724–4732 (2016)
35. Zhao, Z.Q., Zheng, P., Xu, S.T., Wu, X.: Object detection with deep learning: a review. IEEE Trans. Neural Netw. Learn. Syst. **30**(11), 3212–3232 (2019)
36. Zhu, J.J., Bento, J.: Generative adversarial active learning. arXiv preprint arXiv:1702.07956 (2017)

IterDet: Iterative Scheme for Object Detection in Crowded Environments

Danila Rukhovich[✉], Konstantin Sofiiuk, Danil Galeev, Olga Barinova, and Anton Konushin

Samsung AI Center, Moscow, Russia
{d.rukhovich,k.sofiiuk,d.galeev,o.barinova,
a.konushin}@samsung.com

Abstract. Deep learning-based detectors tend to produce duplicate detections of the same objects. After that, the detections are filtered via a non-maximum suppression algorithm (NMS) so that there remains only one bounding box per object. This simple greedy scheme is sufficient for isolated objects. However, it often fails in crowded environments since boxes for different objects should be preserved and duplicate detections should be suppressed at the same time. In this work, we propose to obtain predictions following *iterative scheme* called IterDet. At each iteration, a new subset of objects is detected. Detected boxes from all the previous iterations are considered at the current iteration to ensure that the same object would not be detected twice. This iterative scheme can be applied to both one-stage and two-stage deep learning-based detectors with minor modifications. Through extensive evaluation on 4 diverse datasets with two different baseline detectors, we prove our iterative scheme to achieve significant improvement over the baseline. On CrowdHuman and WiderPerson datasets, we obtain state-of-the-art results. The source code and the trained models are available at https://github. com/saic-vul/iterdet.

1 Introduction

In recent years, deep learning-based methods of object detection have significantly evolved and achieved solid improvements in terms of speed and accuracy [4,11,13, 14,21].

All deep learning-based detectors densely sample and independently evaluate candidate object locations. Accordingly, for a single object, they yield multiple similar boxes of varying confidence. This redundant set of detected boxes is then filtered via non-maximum suppression (NMS) or similar techniques to produce exactly one bounding box per object. This *greedy scheme* is sufficient if instances of the same class do not overlap in the image.

However, this is not always the case. Another possible scenario for object detection is so-called crowded environments that contain multiple overlapping objects of the same class (*e.g.* people in the street or bacteria in microscopy images). Crowded environments provide a challenging task for object detectors due to several reasons. First, it is extremely difficult to distinguish whether two candidate boxes belong to the same object or correspond to two overlapping objects. Second, weak visual cues of

© Springer Nature Switzerland AG 2021
A. Torsello et al. (Eds.): S+SSPR 2020, LNCS 12644, pp. 344–354, 2021.
https://doi.org/10.1007/978-3-030-73973-7_33

baseline recall: 78.8, AP: 76.81   *IterDet 1 iter.* recall: 75.9, AP: 74.28
                                          IterDet 2 iter. recall: **82.5**, AP: **79.59**

Fig. 1. The results of original Faster RCNN (left) and the proposed IterDet based on Faster RCNN (right) for the same image from CrowdHuman *test* set with *visible-body* annotations. The boxes found on the first and second iteration are marked in green and yellow, respectively. The metrics for baseline and IterDet after the first and the second iterations are listed below the images. (Color figure online)

heavily occluded instances can hardly provide sufficient information for accurate object detection.

In several works, this problem has been addressed with various modifications of the NMS algorithm [2, 8, 10, 12, 15, 22]. By NMS, both duplicate detections of the same object should be removed and the hard-to-detect occluded objects should be kept at the same time. Therefore, there is a natural trade-off between precision and recall that imposes severe restrictions on all these approaches.

In this work, we describe a novel *iterative scheme* (IterDet) for object detection. Rather than detecting all objects in the image simultaneously, we propose to yield detections iteratively. At each iteration, a new subset of objects is detected. Object boxes that have been already found at the previous iterations are passed to the neural network at the current iteration, so duplicates can be avoided. The proposed iterative scheme can be applied to any one-stage or two-stage object detection method with only minor modifications.

Figure 1 demonstrates the results of IterDet for Faster RCNN [14] on a test image from CrowdHuman dataset [17]. True positive boxes with scores above 0.1 are visualized, and false positives are omitted for visual clarity. At the second iteration, 9 additional objects (shown in yellow) out of 137 are detected, overtaking the baseline Faster RCNN by 5 true positives and 2.7% of average precision (AP). In the top-right corner of the images, there is an example of two strongly overlapping objects detected with IterDet yet missed by the baseline detector.

Recently, there have been introduced several neural architectures that handle image context thus being more suitable for crowded environments [3, 5, 9]. For instance, [20] proposed to use a special Hungarian loss function to train a convolutional-recurrent model that yields strictly one detection per iteration. In comparison, our approach is more computationally efficient. Moreover, instead of storing information about previously detected objects via LSTM, we explicitly pass it to the network in a form of object masks. Our approach guarantees that no previously detected bounding boxes are accidentally forgotten. Furthermore, compared to [20] it allows incorporating the history of detections into deeper layers of a neural network.

In PS-RCNN [3], objects are also detected iteratively: simple objects are supposed to be found on the first iteration while the second iteration is performed to explore more difficult cases. This iterative approach can be applied only for RCNN-based detectors. At the same time, our approach can be easily integrated into state-of-the-art object detection methods.

We perform extensive experiments with both one-stage (RetinaNet [11]) and two-stage (Faster RCNN [14]) object detectors on four challenging datasets (AdaptIS ToyV1 and ToyV2 [19], CrowdHuman [17], and WiderPerson [24]). To prove our ideas, we evaluate IterDet against baseline models and compare the obtained results with the results reported by competitors. On all datasets, IterDet outperforms baseline models and sets new state-of-the-art on CrowdHuman and WiderPerson datasets.

2 Related Work

Standard Methods for Object Detection. Deep learning-based object detectors can be classified as two-stage and one-stage detectors. Two-stage detectors are based on proposal-driven mechanism [4, 14]. They consist of two subnetworks: the first one outputs a sparse set of candidate object locations and the second one classifies these object locations into one of the foreground classes or a background.

One-stage methods are applied over a regular, dense sampling of object locations, scales, and aspect ratios [11, 13]. Being much faster on inference than their two-stage counterparts, recent one-stage methods achieve comparable accuracy on some datasets. Moreover, anchor-free one-stage methods [21] are more agile and less limited compared to their predecessors. However, two-stage methods still demonstrate state-of-the-art accuracy on challenging datasets.

Overall, all detectors have certain pros and cons and are applicable under certain conditions. To cover all possible scenarios, we design our iterative scheme so it can be combined with both one-stage and two-stage object detectors.

For deep learning-based methods, the detection problem is formulated as a classification task. Namely, class probabilities are estimated independently for each location for multiple candidate locations across an image. Differently, in our iterative scheme, the history of detections from the previous iterations is passed to the detector at the following iterations, providing the context for resolving ambiguities.

Modifications of NMS Algorithm. The standard NMS algorithm greedily selects detections with a higher score and removes the less confident neighbors. Thus, a wide suppression parameter improves the precision and the narrow suppression improves the recall. Consequently, crowded environments are the most challenging case for NMS since both wide and narrow suppression lead to errors. To address this, numerous modifications of the NMS algorithm have been proposed in the literature. Rothe *et al.* [15] formulated NMS as a clustering problem. Hosang *et al.* [8] suggested decreasing the confidence of detections that cover the already detected objects. In soft NMS [2], scores for object proposals depend on their overlap with a target object. In adaptive NMS [12], parameters of NMS are chosen according to the density of the objects estimated via an extra branch. Most recent R^2NMS [10] simultaneously predicts the full and visible boxes of an object.

Differently from all the listed methods, our proposed scheme is iterative that gives more freedom and flexibility. More specifically, we might miss the more difficult objects at the first iteration, since these objects can be detected later on. Accordingly, we do not need to assure high recall at each iteration as we can set wider suppression parameters to favor precision.

Neural Architectures for Crowded Environments. Several neural architectures for object detection in crowded environments have been described in the literature. Stewart *et al.* [20] used a Hungarian loss function to train an LSTM-based model that yields a sequence of detections. LSTM was also used in [6] for iterative proposal refinement in RPN-based detectors. However, performing the NMS step after all iterations negates all the benefits in case of crowded environments. Hu *et al.* [9] proposed an object relation module that processes a set of objects based on their visual appearance and geometry. Ge *et al.* [3] introduced a modification of two-stage detectors called PS-RCNN. First, it detects non-occluded objects with RCNN and then suppresses the detected instances with object-shaped masks. At the second step, another RCNN detects occluded objects.

Compared to the aforementioned methods, our iterative scheme does not imply changing neural architecture, therefore it is much easier to implement.

3 Proposed Method

The proposed iterative scheme is shown in Fig. 2. First, we introduce notation and describe the inference process. Then, we describe the modified training procedure.

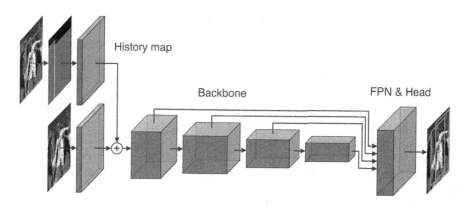

Fig. 2. Proposed iterative scheme. The unchanged meta-architecture of an arbitrary detector is marked with blue. The single convolution layer for the history map is marked green. Out of the 4 overlapping objects in the image, 2 are in the history, where they were either randomly sampled at the training step, or detected during previous iterations of the inference. The remaining 2 are predicted by the detector. (Color figure online)

Inference Process. A typical object detector D is an algorithm that maps image $I \in \mathbb{R}^{w \times h \times 3}$ to a set of bounding boxes $B = \{(x_k, y_k, w_k, h_k)\}_{k=1}^n$. Each box is represented by the coordinates of its top left corner (x, y), width w and height h. For a given set of boxes B, we define a history image $H \in \mathbb{Z}^{w \times h}$ of the same size as an input image. In $H \in \mathbb{Z}^{w \times h}$, each pixel contains the number of already detected boxes that cover that pixel:

$$H_{xy} = \sum_{k=1}^{|B|} \mathbb{1}_{\ x_k \leq x \leq x_k + w_k, \ y_k \leq y \leq y_k + h_k} \tag{1}$$

Figure 2 shows an example of the history, where its values are color-coded. We can make a detector D' history-aware if we pass the history H along with the image I as its inputs.

Let us now introduce the iterative scheme $IterDet(D')$, that, given an image I, produces a set of bounding boxes B in an iterative manner. At the first iteration $t = 1$ history H_1 is empty and D' maps an image I and H_1 to a set of bounding boxes B_1. Second, B_1 is mapped to history H_2. H_2 is then mapped to B_2 by D' at iteration $t = 2$. This process stops when the limit of iterations is reached or when $|B_m| = 0$ at some iteration m. The final prediction of $IterDet(D')$ is $B = \bigcup_{t=1}^m B_t$, where m denotes the total number of iterations.

To implement the described scheme, two modifications should be implied: 1) an arbitrary detector D should be altered to become a history-aware detector D' and 2) D' should be forced to predict different sets of objects B_t on each iteration t. Below, we explain these alterations in detail.

Architecture of a History-Aware Detector. State-of-the-art deep learning-based object detection pipelines start with passing an image to an already pre-trained backbone, *e.g.* ResNet [7], VGG [18], etc. to obtain multi-level image features. These features are then fed into trainable feature extractors, *e.g.* Region Proposal Network, Feature Pyramid Network, etc. Their outputs are further passed to the head module predicting bounding boxes. Finally, non-maximum suppression is applied. We aim to introduce minimal changes to the original network architectures and incorporate history in the deepest layers of the network.

The proposed architecture of the history-aware detector is simple yet efficient. The history is processed via one convolution layer which output sums up with the output of the first convolution layer of the backbone. This scheme can be applied to any backbone without hyperparameter tuning. For ResNet-like backbone, the image is passed through a convolution layer with 64 filters of size 7 and stride 2, Batch Normalization layer, and ReLU activation layer. We follow the design choices of ResNet and use a convolution layer with 64 filters of size 3 and stride 2.

Training Procedure. During training, we randomly split the set of ground truth bounding boxes $\hat{B}$ into two subsets B_{old} and B_{new} such that $B_{old} \cup B_{new} = \hat{B}$ and $B_{old} \cap B_{new} = \emptyset$. We map B_{old} to a history H and force D' to predict the bounding boxes B_{new} that are missing in history. Thus, we optimize the losses of D' by back-propagation of the error between the predicted boxes B and target boxes B_{new}. We do not describe losses since we do not modify this part of baseline detectors. On the one

Table 1. Average number of objects and pair-wise overlap between two instances on the four datasets used in our experiments.

	Toy V1	Toy V2	CrowdHuman	WiderPerson
Object/image	14.88	31.25	22.64	29.51
Pair/image				
IoU > 0.3	3.67	7.12	9.02	9.21
IoU > 0.4	1.95	3.22	4.89	4.78
IoU > 0.5	0.95	1.25	2.40	2.15
IoU > 0.6	0.38	0.45	1.01	0.81

hand, this method of training forces the model to exploit the history and predict only new objects at each iteration of inference. On the other hand, it provides an additional source of augmentations by sampling different combinations of B_{old} and B_{new}.

Several iterative methods predict only one object per iteration [1,20]. Our iterative scheme is also able to predict one object per iteration *e.g.* by selecting the most confident detection. However, in practice, such an approach would be inefficient since inference time is proportional to the number of objects in the image. Our experiments in Sect. 4 demonstrate that performing two iterations is enough to achieve the best accuracy. With increasing the number of iterations, the recall improves but the precision degrades, worsening mMR and AP metrics.

4 Experiments

4.1 Datasets and Implementation Details

We evaluate our proposed iterative scheme on four datasets containing images of various crowded environments: AdaptIS ToyV1 and ToyV2 [19], CrowdHuman [17] and WiderPerson [24].

AdaptIS. AdaptIS Toy V1 and Toy V2 are two synthetic datasets originally used for instance segmentation task [19]. Each image contains about 30 objects on average, with many of those severely overlapping. The datasets statistics are summarized in Table 1. For Toy V1, training and validation splits contain 10000 and 2000 images of size 96 × 96 pixels respectively. Toy V2 is split into training, validation, and test subsets with 25000, 1000, and 1000 images of size 128 × 128 pixels respectively. For both Toy datasets, we chose AP as the main metric, and also provide recall values for consistency. We do not report the mMR metric: if the average number of false positives per image is less than 1 it turns zero, thus being not representative.

CrowdHuman. The recently introduced CrowdHuman dataset has the largest number of persons per image and the largest number of pairs of intersecting bounding boxes among all datasets for human detection, according to [17]. It contains 15000, 4370, and 5000 images for training, validation, and testing, respectively. There are 23 people presenting on an average image, each annotated with 3 boxes: *full-body, visible-body*

Table 2. Experimental results on AdaptIS Toy V1 and Toy V2 dataset.

Method	Detector	Toy V1		Toy V2	
		Recall	AP	Recall	AP
Baseline	RetinaNet	95.46	94.46	96.27	95.62
IterDet, 1 iter.		95.21	95.31	96.27	94.17
IterDet, 2 iter.		**99.56**	**97.71**	**99.35**	**97.27**
Baseline	Faster RCNN	94.05	93.96	94.88	94.81
IterDet, 1 iter.		94.34	94.27	94.97	94.89
IterDet, 2 iter.		**99.60**	**99.25**	**99.29**	**99.00**

Table 3. Experimental results on CrowdHuman dataset with *full-body* annotations.

Method	Detector	Recall	AP	mMR
Baseline [17]	RetinaNet	**93.80**	80.83	63.33
IterDet, 1 iter.		79.68	76.78	**53.03**
IterDet, 2 iter.		91.49	**84.77**	56.21
Baseline [17]	Faster RCNN	90.24	84.95	50.49
Soft NMS [2, 12]		91.73	83.92	51.97
Adaptive NMS [12]		91.27	84.71	49.73
Repulsion Loss [3, 23]		90.74	85.71	–
PS-RCNN [3]		93.77	86.05	–
IterDet, 1 iter.		88.94	84.43	**49.12**
IterDet, 2 iter.		**95.80**	**88.08**	49.44

and *head* box. The most challenging and most frequently used in other works is full-body annotation, where the boxes not only overlap more strongly but also go beyond the edges of the image. We also conduct experiments on visible-body annotation, training models on the training part of the data, and benchmarking on the validation subset.

[17] also reports metrics for one-stage RetinaNet detector and the two-stage Faster RCNN detector, both using ResNet-50 as a backbone. The mMR metric is proposed as the major metric to evaluate detection quality. This metric is calculated as the logarithm of missing rate averaged over 9 points ranging from 10^{-2} to 10^{0} false positives per image. Besides, recall and average precision (AP) are reported.

WiderPerson. WiderPerson [24] is another human detection dataset collected from various sources. There are 8000, 1000, and 4382 images in train, validation, and test subsets. It contains annotations for 5 classes: pedestrians, riders, partially visible persons, crowd, and ignored regions. Following [3], we merge the last four types into one category for both training and testing.

Table 4. Experimental results on CrowdHuman dataset with *visible-body* annotations.

Method	Detector	Recall	AP	mMR
Baseline [17]	RetinaNet	**90.96**	77.19	65.47
Feature NMS [16]		–	68.65	75.35
IterDet, 1 iter.		86.91	81.24	**58.78**
IterDet, 2 iter.		89.63	**82.32**	59.19
Baseline [17]	Faster RCNN	91.51	**85.60**	55.94
IterDet, 1 iter.		87.59	83.28	**55.54**
IterDet, 2 iter.		**91.63**	85.33	55.61

Table 5. Experimental results on WiderPerson dataset.

Method	Detector	Recall	AP	mMR
Baseline [24]	RetinaNet	–	–	48.32
IterDet, 1 iter.		90.38	87.17	**43.23**
IterDet, 2 iter.		**95.35**	**90.23**	43.88
Baseline [24]	Faster RCNN	–	–	46.06
Baseline [3]		93.60	88.89	–
PS-RCNN [3]		94.71	89.96	–
IterDet, 1 iter.		92.67	89.49	**40.35**
IterDet, 2 iter.		**97.15**	**91.95**	40.78

4.2 Results and Discussion

Results on AdaptIS Datasets. Table 2 summarizes IterDet and baseline metrics on AdaptIS Toy V1 and Toy V2 datasets. For both datasets and detectors, IterDet substantially increases AP. For Faster RCNN, this increase expands 4% bringing the final AP up to 99%.

Results on CrowdHuman. Results on full-body and visible-body annotations of the CrowdHuman dataset are presented in Tables 3 and 4 respectively. We compare the proposed IterDet scheme to the methods that do not use additional data or annotations during training. According to Table 3, we achieve a significant improvement in terms of all metrics for the most challenging *full-body* annotation. More specifically, IterDet improves recall by more than 5.5%, AP by 3.1% and mMR by 1.0% w.r.t. baseline. These results remain solid even when compared to the previous state-of-the-art approaches such as Adaptive NMS and PS-RCNN. In terms of mMR, IterDet outperforms all existing methods in all four scenarios: single- and two-stage detectors, visible- and full-body annotations. For the RetinaNet detector, the quality gap exceeds 6% for both types of annotations. Notice, that such an improvement of mMR value is achieved even after the first iteration. We attribute this to the regularization provided by history-aware training. Despite a slight degradation of mMR with an increasing number of iterations, the growth of AP always remains significant. For RetinaNet, we outperform the competitors by 3.9% AP for both types of annotations.

Results on WiderPerson. The results on WiderPerson dataset are summarized in Table 5. We refer to [24] for results obtained on *hard* subset of annotations which contains all the boxes larger than 20 pixels in height. Following the protocol from [3], we do not limit height during testing which is an even more challenging task. For both detectors, we achieve significantly better results in terms of recall, AP, and mMR.

We do not conduct experiments with a larger number of iterations due to the following reasons. First, IterDet already achieves state-of-the-art performance on Crowd-Human and WiderPersons datasets after only 2 iterations. Second, the inference time of the iterative scheme is proportional to the number of iterations, and for 3 iterations the inference would be 3 times slower which is not acceptable in practice.

Figure 3 shows the results of IterDet based on Faster RCNN on the four datasets. In all examples, there are strongly overlapping objects with IoU > 0.5 which are missed by the baseline detector but found by IterDet with 2 iterations.

Fig. 3. IterDet results on ToyV1, ToyV2 (first row), CrowdHuman (with visible- and full-body annotations, second row), and WiderPerson (third row). The boxes found on the first and second iterations are marked in green and yellow respectively. The scores thresholded for visualization are above 0.1. (Color figure online)

5   Conclusion

We present an iterative scheme (IterDet) of object detection designed for crowded environments. It can be applied to both two-stage and one-stage object detectors. On challenging AdaptIS ToyV1 and ToyV2 datasets with multiple overlapping objects Iter-Det achieves almost perfect accuracy. Through extensive evaluation on CrowdHuman and WiderPerson benchmarks, we show that the proposed iterative scheme outperforms existing methods when applied to either two-stage Faster RCNN or one-stage RetinaNet detector.

References

1. Barinova, O., Lempitsky, V., Kholi, P.: On detection of multiple object instances using Hough transforms. IEEE Trans. Pattern Anal. Mach. Intell. **34**(9), 1773–1784 (2012)
2. Bodla, N., Singh, B., Chellappa, R., Davis, L.S.: Soft-NMS-improving object detection with one line of code. In: Proceedings of the IEEE International Conference on Computer Vision, pp. 5561–5569 (2017)
3. Ge, Z., Jie, Z., Huang, X., Xu, R., Yoshie, O.: PS-RCNN: detecting secondary human instances in a crowd via primary object suppression. arXiv preprint arXiv:2003.07080 (2020)
4. Girshick, R.: Fast R-CNN. In: Proceedings of the IEEE International Conference on Computer Vision, pp. 1440–1448 (2015)
5. Goldman, E., Herzig, R., Eisenschtat, A., Goldberger, J., Hassner, T.: Precise detection in densely packed scenes. In: Proceedings of the IEEE Conference on Computer Vision and Pattern Recognition, pp. 5227–5236 (2019)
6. Gong, J., Zhao, Z., Li, N.: Improving multi-stage object detection via iterative proposal refinement. In: BMVC, p. 223 (2019)
7. He, K., Zhang, X., Ren, S., Sun, J.: Deep residual learning for image recognition (2015)
8. Hosang, J., Benenson, R., Schiele, B.: Learning non-maximum suppression. In: Proceedings of the IEEE Conference on Computer Vision and Pattern Recognition, pp. 4507–4515 (2017)
9. Hu, H., Gu, J., Zhang, Z., Dai, J., Wei, Y.: Relation networks for object detection. In: Proceedings of the IEEE Conference on Computer Vision and Pattern Recognition, pp. 3588–3597 (2018)
10. Huang, X., Ge, Z., Jie, Z., Yoshie, O.: NMS by representative region: towards crowded pedestrian detection by proposal pairing. arXiv preprint arXiv:2003.12729 (2020)
11. Lin, T.Y., Goyal, P., Girshick, R., He, K., Dollár, P.: Focal loss for dense object detection. In: Proceedings of the IEEE International Conference on Computer Vision, pp. 2980–2988 (2017)
12. Liu, S., Huang, D., Wang, Y.: Adaptive NMS: refining pedestrian detection in a crowd. In: Proceedings of the IEEE Conference on Computer Vision and Pattern Recognition, pp. 6459–6468 (2019)
13. Liu, W., et al.: SSD: single shot MultiBox detector. In: Leibe, B., Matas, J., Sebe, N., Welling, M. (eds.) ECCV 2016. LNCS, vol. 9905, pp. 21–37. Springer, Cham (2016). https://doi.org/10.1007/978-3-319-46448-0_2
14. Ren, S., He, K., Girshick, R., Sun, J.: Faster R-CNN: towards real-time object detection with region proposal networks. In: Advances in Neural Information Processing Systems, pp. 91–99 (2015)
15. Rothe, R., Guillaumin, M., Van Gool, L.: Non-maximum Suppression for object detection by passing messages between windows. In: Cremers, D., Reid, I., Saito, H., Yang, M.-H. (eds.) ACCV 2014. LNCS, vol. 9003, pp. 290–306. Springer, Cham (2015). https://doi.org/10.1007/978-3-319-16865-4_19
16. Salscheider, N.O.: FeatureNMS: non-maximum suppression by learning feature embeddings. arXiv preprint arXiv:2002.07662 (2020)
17. Shao, S., et al.: CrowdHuman: a benchmark for detecting human in a crowd. arXiv preprint arXiv:1805.00123 (2018)

18. Simonyan, K., Zisserman, A.: Very deep convolutional networks for large-scale image recognition. arXiv preprint arXiv:1409.1556 (2014)
19. Sofiiuk, K., Barinova, O., Konushin, A.: AdaptIS: adaptive instance selection network. In: Proceedings of the IEEE International Conference on Computer Vision, pp. 7355–7363 (2019)
20. Stewart, R., Andriluka, M., Ng, A.Y.: End-to-end people detection in crowded scenes. In: Proceedings of the IEEE Conference on Computer Vision and Pattern Recognition, pp. 2325–2333 (2016)
21. Tian, Z., Shen, C., Chen, H., He, T.: FCOS: fully convolutional one-stage object detection. In: Proceedings of the IEEE International Conference on Computer Vision, pp. 9627–9636 (2019)
22. Tychsen-Smith, L., Petersson, L.: Improving object localization with fitness NMS and bounded IoU loss. In: Proceedings of the IEEE Conference on Computer Vision and Pattern Recognition, pp. 6877–6885 (2018)
23. Xinlong, W., Tete, X., Yuning, J., Shuai, S., Jian, S., Chunhua, S.: Repulsion loss: detecting pedestrians in a crowd. In: Proceedings of the IEEE Conference on Computer Vision and Pattern Recognition, pp. 7774–7783 (2018)
24. Zhang, S., Xie, Y., Wan, J., Xia, H., Li, S.Z., Guo, G.: WiderPerson: a diverse dataset for dense pedestrian detection in the wild. IEEE Trans. Multimed. **22**, 380–393 (2019)

Multi-modal 3D Human Pose Estimation for Human-Robot Collaborative Applications

Konstantinos Peppas$^{(\boxtimes)}$ (ID), Konstantinos Tsiolis (ID), Ioannis Mariolis (ID),
Angeliki Topalidou-Kyniazopoulou (ID), and Dimitrios Tzovaras (ID)

Information Technologies Institute,
Centre for Research and Technology Hellas - CERTH,
6th km Charilaou-Thermi Road, Thessaloniki, Greece
kppeppas@iti.gr

Abstract. We propose a multi-modal 3D human pose estimation approach which combines a 2D human pose estimation network utilizing RGB data with a 3D human pose estimation network utilizing the 2D pose estimation results and depth information, in order to predict 3D human poses. We improve upon the state-of-the-art by proposing the use of a more accurate 2D human pose estimation network, as well as by introducing squeeze-excite blocks into the architecture of the 3D pose estimation network. More importantly, we focused on the challenging application of 3D human pose estimation during collaborative tasks. In that direction, we selected appropriate sub-sets that address collaborative tasks from a large-scale multi-view RGB-D dataset and generated a novel one-view RGB-D dataset for training and testing respectively. We achieved above state-of-the-art performance among RGB-D approaches when tested on a novel benchmark RGB-D dataset on collaborative assembly that we have created and made publicly available.

Keywords: Multi-modal learning · 3D human pose estimation ·
Collaborative tasks · Deep learning · CNN

1 Introduction

Human pose estimation is a challenging problem, because there are many degrees of freedom to be estimated, such as the body shape, the view point, the clothes and occlusions. This is the reason that Deep Learning approaches have outperformed the traditional ones in recent years [16], since it is difficult to handcraft features to account for all these variables. Deep Learning approaches rely on large-scale datasets to implicitly learn these constraints. In recent years, large-scale in-the-wild RGB datasets for 2D human pose estimation have been made available [1,12], which enabled Deep Learning approaches to achieve impressive performance [3,6]. However, 3D human pose estimation is a more demanding

© Springer Nature Switzerland AG 2021
A. Torsello et al. (Eds.): S+SSPR 2020, LNCS 12644, pp. 355–364, 2021.
https://doi.org/10.1007/978-3-030-73973-7_34

task, the difficulty of capturing such a ground-truth dataset in-the-wild means that many of them are synthetically created [5,17]. The standard benchmark dataset [9] is a real one, but it is constrained in indoor environments. Still, there are many 3D human pose estimation implementations which infer 3D poses from RGB images [13,14]. However, since they do not have any input depth information, their predictions live in a scale and translation normalized frame or use assumptions to resolve the depth ambiguity. A way to resolve this is to use multi-modal RGB-D 3D pose estimation data.

Robotic applications require the inference of human poses in absolute units, which necessitates the use of multi-modal RGB-D data in order for it to be scale and translation invariant. An implementation which utilizes RGB-D data and achieves state-of-the-art performance has been developed by Zimmermann et al. [19], however their multi-modal training dataset has not been made available. It predicts 2D human poses using a robust 2D human pose estimator, fuses the 2D joint data with the input depth information and lifts these joints to 3D with a fully convolutional network. The main contributions of the proposed approach, which builds upon [19], are the following:

- it proposes a novel multi-modal 3D human pose estimator that achieves above state-of-the-art performance
- it proposes training with task-specific datasets for boosting performance of 3D human pose estimation in collaborative tasks
- creation of a novel multi-modal RGB-D dataset of a collaborative assembly task and making it publicly available.

2   Related Work

There are numerous methods that get as input just an RGB image and estimate the 3D pose of the depicted subject. The simplest way to do this is to train a network to directly regress 3D joint locations from an image [11]. Instead of directly predicting 3D pose from image, the proposed method in [4] reasons an intermediate 2D pose estimation exploiting the accuracy that deep neural networks have achieved in 2D pose estimation as well as the public availability of large 3D mocap datasets, utilised for lifting predicted 2D poses to 3D through simple memorization. Zhou et al. [18] improved this two-stage pipeline by proposing a weakly-supervised transfer learning method that uses mixed 2D and 3D labels in a unified deep neural network that presents two-stage cascaded structure. They achieved to exploit rich cues for 3D pose recovery included in the original in-the-wild 2D image information that cannot be detected in common 3D datasets, captured by mocap systems in controlled lab environments. Modern 3D human pose estimation methods rely on deep networks, which require vast amounts of data to be trained. Rhodin et al. [14] proposed a method that overcomes this problem by learning a geometry-aware body representation from multi-view images without annotations. An encoder-decoder is trained to predict an image from one viewpoint given an image from another viewpoint, outperforming most state-of-the-art approaches which require a much deeper network and therefore more training data.

Little research has gone into approaches that incorporate both depth and color modalities. Firstly, a pose estimation system was proposed in [2] which benefits from using both color and depth data. Recently, a deep learning based approach that estimates 3D human pose by combining color and depth into two steps was proposed by Zimmermann et al. [19]. Initially, given a color image as input, 2D human pose is predicted accurately by exploiting the discriminative power of keypoint detectors trained on large scale datasets of color images. Subsequently, additional information is retrieved from the depth map for calculating the real world 3D coordinates.

3 Architecture

The proposed approach takes an RGB image and a registered depth map as input and predicts 3D human poses in absolute units. It improves upon the Zimmermann et al. [19] implementation. The first step of the multi-modal prediction pipeline is to predict 2D human poses of K keypoints from an RGB image using a 2D human pose estimation network. This result is combined with the depth input to create a voxel grid of size $S \times S \times S \times (K+1)$, which is the input to the 3D human pose estimation network, where S is the size of each dimension. The final prediction utilizes the predictions and confidences from both the 2D and 3D networks to decide the output keypoints as detailed below.

3.1 2D Human Pose Estimation Network

AlphaPose [6] is a state-of-the-art multi-person 2D pose estimator on RGB images. It achieves 82.1 and 72.3 mAP on MPII [1] and COCO [12] datasets respectively, while OpenPose [3] used in Zimmermann et al. [19] reports 79.7 and 61.8 mAP. It is a top-down approach which first predicts bounding boxes of humans and then runs single-person pose estimation on them. For an image $I \epsilon \mathbb{R}^{H \times W \times 3}$, we get a prediction matrix $P \epsilon \mathbb{R}^{K \times 2}$ which holds the location of each keypoint and a confidence matrix $C_{2D} \epsilon \mathbb{R}^{K \times 1}$ of each keypoint prediction.

3.2 2D Human Pose Estimation and Depth Data Fusion

The neck, right hip or left hip joint can be a root joint r, the one whose prediction confidence is above a fixed threshold $C_r > th$ is picked, where $th = 0.5$. A voxel occupancy grid $V \epsilon \mathbb{R}^{S \times S \times S}$ is then created whose center is the backprojection b_r of the root keypoint p_r to the closest depth value in the registered depth map d_r. The size of each voxel is set to v centimeters. The registered depth map $D \epsilon \mathbb{R}^{H \times W}$ is transformed into a point cloud and the values of V are set to 1 if at least one point is found inside a voxel, or 0 otherwise. Furthermore, a voxel grid $\dot{V}_i \epsilon \mathbb{R}^{S \times S \times S}, i \epsilon K$ with the same center and size is created for each predicted keypoint. For each keypoint i, the prediction P_i is backprojected to the depth value d_r giving $b_i \epsilon \mathbb{R}^3$. The index $l_i \epsilon I^3$ is calculated by finding the voxel in which b_i is located and $l'_i \epsilon I^2$ is calculated by keeping the x and y component of

l_i. We create a smooth 2D scoremap $S_{2D}^i \in \mathbb{R}^{S \times S}$ by placing a gaussian function with $max = 1$ and $sigma = 3 * v$ on each l_i'. V_i is finally created by tiling the scoremaps S_{2d}^i along the z-axis. The input to the 3D human pose estimation network is the concatenation ($\otimes$) of all the voxel grids.

$$V_{in} = V \otimes (V_1 \otimes V_2 \otimes ... \otimes V_K), V_{in} \in \mathbb{R}^{S \times S \times S \times (1+K)} \tag{1}$$

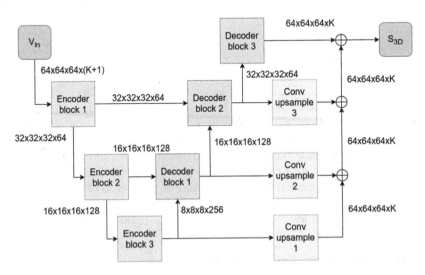

Fig. 1. The blocks of the network with the input and output dimensions noted for $S = 64$. Every encoder block consists of two 3D convolutional layers and a 3D max pooling operation with dense connections. The decoder blocks consist of a transposed 3D convolutional layer to upsample their input, concatenate it with the skip connection input (except for decoder block 3) and pass it through a 3D convolutional layer. The conv upsample layers sequentially apply a 3D convolutional layer and a transposed 3D convolutional layer to the input providing the intermediate scoremaps S_{3D}^i. S_{3D} is the element-wise addition of the intermediate scoremaps and the output of the final decoder block.

3.3   3D Human Pose Estimation Network

The 3D human pose estimation network VoxelPoseNet-SE is an improvement upon the VoxelPoseNet architecture proposed in [19] by incorporating squeeze and excitation blocks [7] into this architecture. VoxelPoseNet is a Fully Convolutional Neural Network with 3D convolutions, which is an encoder decoder architecture inspired by the U-net [15] that uses dense blocks [8] in the encoder. The encoder and decoder consist of 3 blocks each with skip connections between the ones with compatible input dimensions. Additionally, scorevolumes are predicted from each block and added element-wise to produce the final 3D scoremaps S_{3D}

(Fig. 1). The original block is augmented by changing every convolutional operation to a squeeze and excitation block (Fig. 2). VoxelPoseNet-SE takes as input V_{in} and outputs 3D scoremaps $S_{3D} \in \mathbb{R}^{S \times S \times S \times K}$, the probability of each keypoint being in the voxels of the voxel grid. The indexes of the maxima of the scoremaps $l_{3D}^i = argmax(S_{3D}[:,:,:,i]) \in I^3, \forall i \in K$ are translated to real-world coordinates $p_{3D}^i \in \mathbb{R}^3$. The values of the maxima $C_{3D}^i = max(S_{3D}[:,:,:,i]) \in \mathbb{R}, \forall i \in K$ correspond to the confidence of the 3D network for each prediction. The final 3d keypoint estimation is calculated as follows:

$$p_{fin}^i = \begin{cases} p_{3D}^i & \text{if } C_{3D}^i \geq C_{2D}^i \\ b_{2D}^i & \text{if } C_{3D}^i < C_{2D}^i \end{cases}, \tag{2}$$

where b_{2D} is the backprojection of the estimation of the 2D human pose estimation network to the depths estimated by the 3D human pose estimation network p_{3D} [2].

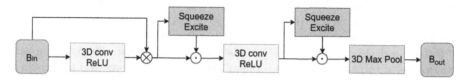

Fig. 2. The output of the 3D convolutional layers (after the ReLU activation) are passed through a squeeze and an excite operation, producing per-channel modulation weights. These are applied to the identity of this output, allowing dynamic channel-wise feature recalibration. We have used a reduction ratio of 4 in the squeeze-excite block. The proposed changes are marked in light red color. (Color figure online)

3.4 Training Details

All the trainings run with size $S = 64$ and keypoints $K = 18$. The 18 keypoints are the body landmarks of the COCO dataset [12]. The input image size $H \times W$ is 1080×1920, since the CMU dataset (see Sect. 4) is used for training, but any image size can be used at inference time. The size of the voxel is randomly sampled in the range $[0,0275 - 0,04125]$ to encourage the network to learn the 3D representation for multiple scales and avoid overfitting. The Adam optimizer is used with learning rate $lr = 0.0001$ with a decay rate of 0.1 every epoch. The training runs for $e = 5$ epochs with batch size $bs = 2$. The loss consists of the sum of l_2 losses between the 3 intermediate and final predicted scorevolumes S_{3D}^i of the network (Fig. 1) and the ground-truth scorevolumes S_{gt}.

$$L = \sum_i (S_{3D}^i - S_{gt})^2 \tag{3}$$

$S_{gt}^i \in \mathbb{R}^{64 \times 64 \times 64}$ is calculated by transforming the 3D ground-truth positions to the voxel grid's coordinates and then placing a 3D gaussian function on the index

$l_c \in I^3$ of each keypoint with $max = 1$ and $sigma = 3 * v$. S_{gt} is, finally, created by tiling all S^i_{gt} producing $S_{gt} \in \mathbb{R}^{64 \times 64 \times 64 \times 18}$.

4 Datasets

4.1 CMU Panoptic Dataset

The CMU panoptic is a massively multi-view capture consisting of 480 VGA cameras, 31 HD cameras and 10 Kinect v2 RGB+D sensors, distributed over the surface of a geodesic sphere maintained by the Carnegie Mellon University [10]. Specifically, this dataset includes videos in which more than 60 subjects participate, performing various scenarios in groups or alone. Also, the dataset is provided with 3D human pose ground truth data. Concerning our experiments, colored frames as well as depth frames, retrieved from the Kinect v2 RGB+D sensors, were utilized having a resolution of 1920×1080 and 512×424 respectively at 30 fps. Given that we are focusing on human-robot collaborative applications, sequences that depict two subjects working cooperatively were preferred for training and testing purposes. Moreover, we sampled 1 frame every 10 frames and this resulted in 875 frames per camera view. The training set consisted of a total of 7875 frames taken from 9 camera views and the testing set consisted of 875 frames from the remaining camera view.

4.2 CERTH Dataset[1]

Besides the CMU Panoptic dataset, we created a new dataset, which comprises 2 actors performing a collaborative task and only 1 viewpoint. Specifically, it shows 2 actors assembling an LCD-TV and it has to be mentioned that there was no such sequence in the CMU Panoptic dataset. An Orbbec Astra (0.6 m–8 m) was used in order to record both RGB (1280×720) and Depth (640×480) data at 30 fps. The final dataset includes 700 frames and was utilized only as a testing dataset, while 1 frame every 10 frames was sampled and was manually annotated for evaluation purposes. Given that CERTH dataset is single-view, we managed to manually annotate 70% of the total number of joints, whereas the rest of them are deemed as occluded.

5 Results

All the tests were run with size $S = 64$ and keypoints $K = 14$, where the eyes and ears keypoints were excluded because they are of lesser importance for the task in hand. The size of the voxel was $v = 0.034375$.

---

[1] https://doi.org/10.5281/zenodo.4475685

5.1   Evaluation Metrics

In order to evaluate our method, the most common metric mentioned in the literature is the mean per joint position error, known as MPJPE. For a frame f and a skeleton S, MPJPE is computed as

$$E_{MPJPE}(f, S) = \frac{1}{N_S} \sum_{i=0}^{N_S} \|m_{f,S}^{(f)}(i) - m_{gt,S}^{(f)}(i)\|_2, \qquad (4)$$

where N_S is the number of joints in skeleton S. For a set of frames the error is the average over the MPJPEs of all frames. We also report the percentage of bodies that were predicted out of the total ground-truth bodies. A body is missed if any of its root joints can not be found during the 2D human pose estimation, which means that all 3 joints' (neck, left hip, right hip) predictions had a confidence lower than $th = 0.5$.

5.2   CMU Test Dataset

The first benchmark is the held-out sequence from the CMU panoptic dataset. Although it is set in the same environment and with the same people as the training dataset, it is taken from a different view point. The tested models are the baseline implementation [19], referred to from now on as "VoxelPoseNet-Op", the same trained on the CMU dataset, referred to as "VoxelPoseNet-Op-CMU", an implementation with AlphaPose and VoxelPoseNet trained on the CMU dataset, "VoxelPoseNet-Ap-CMU", and our final implementation with VoxelPoseNet-SE, reffered to as "VoxelPoseNet-SE-Ap-CMU".

Table 1. MPJPE on the CMU test dataset

	MPJPE (m)	Bodies found %
VoxelPoseNet-Op	0.197	**97.0**
VoxelPoseNet-Op-CMU	0.0912	**97.0**
VoxelPoseNet-Ap-CMU	0.0792	96.1
VoxelPoseNet-SE-Ap-CMU	**0.0782**	96.1

It is expected that training on the CMU dataset greatly improves the accuracy of the model, which explains the drop to more than half of the baseline error (Table 1). Using Alphapose as a 2D human pose estimation network further reduces the error by 13%, while predicting only 1% less bodies. Finally, incorporating the squeeze-excite blocks further improves the error by 1% in the CMU test dataset. A qualitative comparison can be seen in Fig. 3.

Fig. 3. Qualitative comparison between VoxelPoseNet-Op and VoxelPoseNet-SE-Ap-CMU on CMU dataset

5.3   CERTH Dataset

The second benchmark is the novel CERTH collaborative dataset. It is a different task set in an unseen environment, with unseen people and is taken with a different camera capturing smaller image resolution. Thus, it serves as the perfect dataset to test the models' generalization capability.

Table 2. MPJPE on the CERTH dataset

	MPJPE (m)	Bodies found %
VoxelPoseNet-Op	0.219	81.7
VoxelPoseNet-Op-CMU	0.0884	81.7
VoxelPoseNet-Ap-CMU	0.0781	**88.0**
VoxelPoseNet-SE-Ap-CMU	**0.0709**	**88.0**

Fig. 4. Qualitative comparison between VoxelPoseNet-Op and VoxelPoseNet-SE-Ap-CMU on CERTH dataset

Training the model on the CMU dataset greatly reduces the error on the CERTH dataset (59.6%) with respect to the baseline implementation (Table 2), even though less than half the number of training samples were used (7875 vs 18000). This is a strong indication that using the CMU dataset is a significant contribution to the collaborative tasks, while helping the model generalize better to other collaborative tasks. Using Alphapose as a 2D human pose estimation network further decreases the error by 11.7% and at the same time predicts 6.3% more bodies. AlphaPose detects the poses with higher confidence than OpenPose in this challenging dataset, where the resolution is lower and the viewpoint is different, which points to AlphaPose being more robust to diverse input. This was not as evident in the CMU test dataset where the resolution was higher and there were multiple view points. Finally, incorporating the squeeze-excite blocks further improves the error by 9.2% compared to VoxelPoseNet-Ap-CMU. It is important to note that the respective improvement in the CMU test dataset is 1%, which highlights the squeeze-excite modification's contribution to the generalization capability of the network. A qualitative comparison can be seen in Fig. 4.

6 Conclusion

In conclusion, we have utilized a large-scale multi-view dataset of 2 people performing a collaborative task for 3D multi-modal human pose estimation training and testing, created a novel one-view collaborative testing dataset and made it publicly available. The former is large-scale and diverse enough to train models that generalize well to other view points and datasets, while the latter serves us a public benchmark for testing other 3D human pose estimation methods, especially ones that target collaborative tasks. Furthermore, we have improved threefold upon the state-of-the-art on RGB-D 3D human pose estimation, scoring up to 67.6% reduction in MPJPE and 6.3% increase in predicted bodies in an unseen collaborative testing dataset. This was achieved by utilizing a more accurate 2D human pose estimator, by enriching the architecture of the 3D human pose estimation network and by training on the novel extended dataset.

Acknowledgement. This work has been supported by the European Union's Horizon 2020 research and innovation programme funded project namely: "Co-production CeLL performing Human-Robot Collaborative AssEmbly (CoLLaboratE)" under the grant agreement with no: 820767.

References

1. Andriluka, M., Pishchulin, L., Gehler, P., Schiele, B.: 2D human pose estimation: new benchmark and state of the art analysis, June 2014. https://doi.org/10.1109/CVPR.2014.471
2. Buys, K., Cagniart, C., Baksheev, A., De Laet, T., De Schutter, J., Pantofaru, C.: An adaptable system for RGB-D based human body detection and pose estimation. J. Vis. Commun. Image Represent. **25**(1), 39–52 (2014)
3. Cao, Z., Hidalgo Martinez, G., Simon, T., Wei, S., Sheikh, Y.A.: OpenPose: realtime multi-person 2D pose estimation using part affinity fields. IEEE Trans. Pattern Anal. Mach. Intell. **43**(1), 172–186 (2019)

4. Chen, C.H., Ramanan, D.: 3D human pose estimation = 2D pose estimation + matching. In: Proceedings of the IEEE Conference on Computer Vision and Pattern Recognition, pp. 7035–7043 (2017)

5. Fabbri, M., Lanzi, F., Calderara, S., Palazzi, A., Vezzani, R., Cucchiara, R.: Learning to detect and track visible and occluded body joints in a virtual world. In: Ferrari, V., Hebert, M., Sminchisescu, C., Weiss, Y. (eds.) ECCV 2018. LNCS, vol. 11208, pp. 450–466. Springer, Cham (2018). https://doi.org/10.1007/978-3-030-01225-0_27

6. Fang, H.S., Xie, S., Tai, Y.W., Lu, C.: RMPE: regional multi-person pose estimation. In: ICCV (2017)

7. Hu, J., Shen, L., Sun, G., Albanie, S.: Squeeze-and-excitation networks. IEEE Trans. Pattern Anal. Mach. Intell. (2017). https://doi.org/10.1109/TPAMI.2019.2913372

8. Huang, G., Liu, Z., van der Maaten, L., Weinberger, K.: Densely connected convolutional networks (07 2017). https://doi.org/10.1109/CVPR.2017.243

9. Ionescu, C., Papava, D., Olaru, V., Sminchisescu, C.: Human3.6M: large scale datasets and predictive methods for 3D human sensing in natural environments. IEEE Trans. Pattern Anal. Mach. Intell. **36**, 1325–2339 (2014). https://doi.org/10.1109/TPAMI.2013.248

10. Joo, H., et al.: Panoptic studio: a massively multiview system for social interaction capture. IEEE Trans. Pattern Anal. Mach. Intell. **41**(1), 190–204 (2019)

11. Li, S., Chan, A.B.: 3D human pose estimation from monocular images with deep convolutional neural network. In: Cremers, D., Reid, I., Saito, H., Yang, M.-H. (eds.) ACCV 2014. LNCS, vol. 9004, pp. 332–347. Springer, Cham (2015). https://doi.org/10.1007/978-3-319-16808-1_23

12. Lin, T.-Y., et al.: Microsoft COCO: common objects in context. In: Fleet, D., Pajdla, T., Schiele, B., Tuytelaars, T. (eds.) ECCV 2014. LNCS, vol. 8693, pp. 740–755. Springer, Cham (2014). https://doi.org/10.1007/978-3-319-10602-1_48

13. Martinez, J., Hossain, R., Romero, J., Little, J.J.: A simple yet effective baseline for 3D human pose estimation. In: Proceedings of the IEEE International Conference on Computer Vision, pp. 2640–2649 (2017)

14. Rhodin, H., Salzmann, M., Fua, P.: Unsupervised geometry-aware representation for 3D human pose estimation. In: Ferrari, V., Hebert, M., Sminchisescu, C., Weiss, Y. (eds.) ECCV 2018. LNCS, vol. 11214, pp. 765–782. Springer, Cham (2018). https://doi.org/10.1007/978-3-030-01249-6_46

15. Ronneberger, O., Fischer, P., Brox, T.: U-Net: convolutional networks for biomedical image segmentation. In: Navab, N., Hornegger, J., Wells, W.M., Frangi, A.F. (eds.) MICCAI 2015. LNCS, vol. 9351, pp. 234–241. Springer, Cham (2015). https://doi.org/10.1007/978-3-319-24574-4_28

16. Toshev, A., Szegedy, C.: DeepPose: human pose estimation via deep neural networks. In: Proceedings of the IEEE Conference on Computer Vision and Pattern Recognition, pp. 1653–1660 (2014)

17. Varol, G., et al.: Learning from synthetic humans. In: 2017 IEEE Conference on Computer Vision and Pattern Recognition (CVPR), July 2017

18. Zhou, X., Huang, Q., Sun, X., Xue, X., Wei, Y.: Towards 3D human pose estimation in the wild: a weakly-supervised approach. In: Proceedings of the IEEE International Conference on Computer Vision, pp. 398–407 (2017)

19. Zimmermann, C., Welschehold, T., Dornhege, C., Burgard, W., Brox, T.: 3D human pose estimation in RGBD images for robotic task learning, pp. 1986–1992, May 2018. https://doi.org/10.1109/ICRA.2018.8462833

Image = Structure + Few Colors

Darshan Batavia[1]([✉]), Rocio Gonzalez-Diaz[2], and Walter G. Kropatsch[1]

[1] TU Wien, Pattern Recognition and Image Processing Group 193/03,
Vienna, Austria
{darshan,krw}@prip.tuwien.ac.at
[2] Applied Math I, University of Seville, Seville, Spain
rogodi@us.es

Abstract. Topology plays an important role in computer vision by capturing the structure of the objects. Nevertheless, its potential applications have not been sufficiently developed yet. In this paper, we combine the topological properties of an image with hierarchical approaches to build a topology preserving irregular image pyramid (TIIP). The TIIP algorithm uses combinatorial maps as data structure which implicitly capture the structure of the image in terms of the critical points. Thus, we can achieve a compact representation of an image, preserving the structure and topology of its critical points (maxima, the minima and the saddles). The parallel algorithmic complexity of building the pyramid is $O(\log d)$ where d is the diameter of the largest object. We achieve promising results for image reconstruction using only a few color values and the structure of the image, although preserving fine details including the texture of the image.

1 Introduction

Critical points and curves connecting them are an effective means of communicating topological information, which governs the structure of an image. An image compact representation can be achieved by using a surface topology based data structure as mentioned in [11]. For example, Nackman in [17] represented a surface in form of graphs of critical points, subdividing the surface into *slope districts*. Other approaches include: Reeb graphs [21], hierarchical decomposition of Morse-Smale complexes into piecewise linear 2-manifolds [9].

Compact representation has become necessary with the increase of digital data, resulting in down-sampling for compactness and up-sampling for surface reconstruction/approximation. Gaussian and Laplacian pyramids [6] are the most basic regular image pyramid methods. Irregular sampling and image pyramids are an excellent tool for topological representation. Cerman *et al.*, in [7], developed a topology-based image segmentation algorithm. In [18], Maia *et al.* use hierarchical watershed for image segmentation. The most common method for topology simplification using graph representation is by repeated application of the fundamental operation *edge contraction* [12]. Simplification of data is broadly divided into two major types: the topological simplification and the geometric simplification. Persistence measure is the most famous technique used

© Springer Nature Switzerland AG 2021
A. Torsello et al. (Eds.): S+SSPR 2020, LNCS 12644, pp. 365–375, 2021.
https://doi.org/10.1007/978-3-030-73973-7_35

in 2D to decide the priority for simplification [4,10]. For example, in [28], an efficient image down-sampling and up-sampling technique based on interpolation is developed. Other approaches include [8] and [13]. Besides down-sampling, up-sampling is equally important for applications like image super-resolution, image enhancement and denoising. In [22], the anchored neighborhood regression for learning-based super-resolution is used. Other related researches related to image super-resolution using simple functions are [25] and [26].

Motivation and Contribution: In this paper, we propose a *Topology-preserving Irregular Image Pyramid (TIIP)* algorithm and a hierarchical method to build an irregular image pyramid that preserves the critical points and their connections. For a surface, the topology of its contours changes at the function values of its critical points. For example, the surface contours will collapse to a point at a non-degenerated extremum and multiple contours will intersect at a saddle point. Therefore, we preserve the critical points and their function values to preserve the topology of an image. The algorithm operates on combinatorial maps [5] which implicitly encode the structure of the image on the higher level of the pyramid with a compact representation. The use of combinatorial maps supports parallel processing [23] with time complexity of $O(\log(d))$, where d is the diameter of the longest object in the image. The approach is reinforced by a concrete theory of cellular decomposition into cells called slope regions[1] [2,15]. The TIIP algorithm is explained in two distinct algorithms: (a) the bottom-up construction of the pyramid (*REDUCE*) and (b) the top-down expansion of the higher level of the pyramid (*EXPAND*), terminologies as introduced in [6]. The TIIP algorithm achieved a perceptually superior reconstruction especially in the focused region and preserved the texture information maintaining structure and features similar to the original image. Thus an image can be efficiently reconstructed by its structure and a few colors.

This paper is organized as follows: Sect. 2 introduces the basic definitions and terminologies required together with the proposed TIIP algorithm. In Sect. 3, we show results for image reconstruction and its comparison with other algorithms. In Sect. 4, we end with some conclusions and future work.

2 Reduce and Expand Operations in the TIIP Algorithm

We introduce now a method to efficiently build an irregular image pyramid which preserves structural and topological information. We first explain the bottom-up REDUCE operation (TIIP Algorithm 1) for building the topology preserving image pyramid. Later, the top-down EXPAND operation is explained, depending on the application: (1) image segmentation and (2) image recovery.

A discrete 2D image P where the gray value of a pixel p is denoted by $g(p)$, can be represented as a 4-neighborhood graph $G_0(V, E)$. The labels of a segmentation are often stored in form of a label image where each region has a distinct

---

[1] Slope regions are the surfaces in which every pair of points can be connected by a monotonic curve.

label. The adjacencies of the regions are described by the *region adjacency graph (RAG)* $G = (V, E)$. Taking every pixel of the image as a (smallest) region, the neighborhood graph can also be interpreted as a RAG. Every pixel p in the image P corresponds to a vertex $v \in V$ with gray value $g(v) := g(p)$. Vertex v is connected to its adjacent vertices[2] by edges of G, being the degree of v, denoted by degree(v), the number of edges incident to v. An edge $e \in E$ is a *boundary edge* if e is in the border of the unbounded face. The endpoints of a boundary edge are *boundary vertices*. The *orientation of an edge* $e = (v, w) \in E$ is directed from vertex $v \in V$ to vertex $w \in V$ iff $g(v) > g(w)$. When $g(v) = g(w)$, the edge e is not oriented. The *weight* of edge e is its contrast $g(v) - g(w)$. A connected subgraph of G having the same gray value for all the vertices is referred to as a *plateau region* where every pair of vertices $v, w \in V$ of the subgraph satisfies $g(v) = g(w)$. A *path* $\pi(v_1, v_2, \ldots, v_r) = (V_\pi, E_\pi)$ is a non empty subgraph of G, where $V_\pi = \{v_1, v_2, \ldots, v_r\} \subseteq V$ and $E_\pi = \{(v_1, v_2), (v_2, v_3), \ldots, (v_{r-1}, v_r)\} \subseteq E$. The path π is *monotonic* if all the oriented edges of E_π have the same orientation, e.g. from v_1 to v_r or from v_r to v_1. The path π is a *level curve* if $g(v_i) = g(v_{i+1})$ for all i. A level curve can be a part of monotonic paths. A face is a *slope region* if the edges in its border can be divided in two disjoint sets forming two monotonic paths with same orientation.

The orientation of edges can be used to categorize a vertex $v \in V$ into critical (maximum, minimum, saddle), or non-critical (slope). A vertex $v \in V$ is a *local maximum (2-max)* if all the edges incident to v are oriented outwards. Analogously, a boundary vertex $v \in V$ is a local boundary maximum *(1-max)* if the two boundary edges incident to v are oriented outwards. A vertex $v \in V$ is a *local minimum (2-min)* if all the edges incident to v are oriented inwards. Analogously, a boundary vertex $v \in V$ is a local boundary minimum *(1-min)* if the two boundary edges incident to v are oriented inwards. A vertex $v \in V$ is a *degenerated critical vertex* if all the edges incident to v are non-oriented. A non-boundary vertex $v \in V$ is a slope vertex if there are exactly two changes in the orientation of edges incident to v, when traversed circularly (clockwise or counter-clockwise direction). In this case, a slope vertex is a *singular slope vertex* if all the oriented edges incident to v have the same orientation except one. Otherwise, it is a *regular slope vertex*. Observe that degenerated critical vertices always belong to plateau regions. A boundary vertex that is neither a local boundary maximum nor a local boundary minimum is considered a singular slope vertex. A vertex $v \in V$ is a *saddle vertex* if it is not a local maximum, nor a local minimum, neither a plateau nor a slope vertex.

REDUCE Operation. It basically comprises edge contraction [14] and edge removal operations on the graph G_0, forming a graph pyramid. Significance of steps in Algorithm 1 (the enumerations are correlated to the algorithm step numbers).

---

[2] $\mathcal{N}(v)$ denotes the set of vertices adjacent to v.

Algorithm 1. Topology preserving Irregular Image Pyramid

1: **Input:** A digital image P.
2: **Initialize:** Generate the 4-connected neighborhood graph G_0.
3: Insertion of hidden saddle vertices.
4: Contraction of level curves connecting boundary vertices.
5: Contraction of plateau regions.
6: Categorising the vertices into critical and non-critical.
7: Replicating regular slope vertices.
8: **while** #non-critical vertices > 0 **do**
9:    **if** #regular slope vertices > 0 **then**
10:       replicate regular slope vertices.
11:    **end if**
12:    I. **Select contraction kernel** giving priority to edges with lowest weight.
      II. **Decide surviving vertices** for edges connecting:
      (a) non-critical - non-critical vertices.
      (b) critical - non-critical vertices.
13:    **Edge contraction**
14:    **Reduction function:** Recomputing the edge weights.
15:    **Simplification:** Edge removal such that the resulting region is a slope region.
16: **end while**
17: Contraction of edges connecting critical vertices (if required).
18: **end**

---

3. Insertion of Hidden Saddle Vertices:

If $\pi = (V_\pi, E_\pi)$ being $V_\pi = \{a, b, c, d\}$ and $E_\pi = \{(a,b), (b,c), (c,d), (d,a)\}$, is a closed path of G_0, satisfying that $g(a) < g(b)$, $g(a) < g(d)$, $g(c) < g(b)$ and $g(c) < g(d)$, then a new vertex r called *hidden saddle vertex* together with edges $(r,a), (r,b), (r,c), (r,d)$ are added to the RAG (see Fig. 1) with a new gray value $\max(g(a), g(c)) < g(r) < \min(g(b), g(d))$ (see [7,16]).

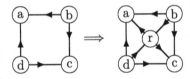

Fig. 1. Inserting a hidden saddle.

4. Contraction of Level Curves Connecting Boundary Vertices:

The shaded region in Fig. 2.a exhibits a plateau region. It connects two disconnected parts of the image boundary. In such cases, contracting the plateau region will result into partitioning the image into

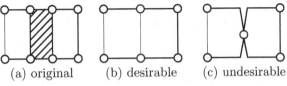

(a) original   (b) desirable   (c) undesirable

Fig. 2. A plateau region and two contractions of it.

two regions connected through a single cut vertex as shown in Fig. 2.c. This result is undesirable because the connections of all the vertices through the plateau region is now concentrated in a single vertex and then the disconnected parts of the image boundary are now connected. To avoid such undesirable occurrences, we first contract the level curves on the boundary and then proceed by contracting the plateau region not allowing any two remaining vertices on the boundary to be contracted. The resulting level, curve showed in Fig. 2.b, preserves the connection through the plateau region.

5. Contraction of Plateau Regions: In this step (which is similar to the superpixel hierarchy), we cluster the vertices with the same attribute (gray value) and represent them by a single vertex. Details of the contraction of plateau regions spread across the image are explained in [3, Section 2.1]. In the case of binary images and segmented images, this step usually contains the maximum number of edge contractions and will result in a connected component labelling.

6. Categorising the Vertices into Critical and Non-critical: To preserve the topological properties, it is important to categorize the vertices into critical (local maximum, local minimum, saddle) and non-critical (slope) vertices. This step is executed after the contraction of plateau regions so that we avoid the misclassification of the degenerated critical vertices. We use the orientation of the edges incident to a vertex to categorize it as mentioned in Sect. 2.

7. Replicating the Regular Slope Vertices: The replication operation of a regular slope vertex v consists of an edge de-contraction [23], i.e., replacing v by an edge e with endpoints x and y having the same gray value than v (the edges incident to v being now incident to one of the two endpoints of e) followed by the removal of e. After the replication operation of v, the two vertices x and y are also slope vertices with lower degrees. The process is repeated until all regular slope vertices become singular slopes vertices.

12. I. Select the Contraction Kernel: With combinatorial maps as the data structure, we can contract multiple edges in parallel, as a result of which we can have larger contraction kernels consequently lowering the height of the pyramid. The constraints for computation of the contraction kernels is application dependent. For the contraction process, this paper gives priority to the edges with the lower edge weight. This constraint can be included in the computation of the contraction kernel. Contraction of edges with lower weights preserves the high frequency components of the data preserving sharp edges and contours.

12. II. Decide Surviving Vertices: Below are the few aspects which we take into account before the edge contraction operation:

(a) To guarantee that critical vertices are preserved and that the vertices do not change from critical to non-critical or vice-versa, an edge is selected for contraction if it is incident to a singular slope vertex v and have a different orientation than the rest of the edges incident to v.

(b) Before contraction of an edge connecting a critical and a non-critical vertex, we mark the critical vertex as the surviving vertex and the non-critical vertex as the non-surviving vertex.

(c) Before contraction of an edge connecting two non-critical vertices, decision of the surviving and the non-surviving vertex is rather application dependent. For example, in the case of Connected Component Labelling (CCL) of a binary image, it is not significant to categorize the surviving and non-surviving vertices. In contrast, for visualization and analysis, the vertex closer to the centroid (visual center for concave polygon) would be the surviving vertex. As a result the vertex corresponding to the centroid will represent the segment.

13. Edge Contraction: In this step, we contract the edges selected in step 12 of Algorithm 1. Given an edge e connecting two vertices s, n of a RAG $G_k = (V_k, E_k)$, the contraction of e will result in merging the survivor vertex s and the non-survivor vertex n (see [14]). After the contraction of e, all the edges, except e, previously incident to vertex n will now be incident to vertex s in graph G_{k+1}.

14. Reduction Function: It computes the weights of the edges previously connected to the non-surviving vertices. While contracting an edge $e = (s, n)$, where s is a surviving vertex and n is a non-surviving vertex, if s is critical, we preserve the gray value of s; if s is non-critical, we compute the gray value of s by $g(s)$: $\max\{g(n) \leq g(s) | n \in \mathcal{N}(s)\} \leq g(s) < \min\{g(n) > g(s) | n \in \mathcal{N}(s)\}$.

15. Simplification: The simplification operation removes the redundant edges (merging the two respective faces sharing the same edge) which leads to an increase in the degree of the faces in the RAG. The empty self-loops[3] are always removed. In [3], a more generalized version of of this step is shown. Edge removal simplifies the graph for visualization and also eliminates the redundant information. We do not remove boundary edges to preserve the image boundary.

17. Contraction of Edges Connecting Critical Vertices: Unlike edge contraction mentioned in steps 4, 5 and 13 of Algorithm 1, in this step we contract the edges connecting the critical vertices. It is usually observed in a natural image that approximately 30% of the total vertices at the base level of the image pyramid are critical. For further reduction in information, we need to contract edges connecting critical vertices, and this may affect the topology of the data. The selection criteria for the contraction kernel depends on the application. For example: if the given image is noiseless and smooth and the application is segmentation, then the contraction of edges with lower weight is preferred. In contrary, if the input is a natural image with salt and pepper noise, then the edges with higher weight are preferred. Topological persistence can be used for the selection process [10].

Graph of Critical Vertices Only: All the faces in a RAG are slope regions. Besides, faces after edge contraction, edge removal and vertex replication continue to be slope regions. At the end of the process, only critical vertices survive as the following result states. See the naive example showed in Fig. 3.

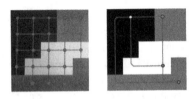

Fig. 3. Base and top level of a pyramid.

---

[3] An empty self-loop is a self-loop that does not encapsulate a subgraph inside.

Theorem 1. *Algorithm 1 (without step 17) computes a graph pyramid with the property that all the vertices on the top are critical and the number of critical vertices is constant along the pyramid after plateaus are contracted.*

Proof. Let us prove that critical vertices are preserved in the all pyramid. Basically, there are three steps applied recursively to compute the graph pyramid, namely, regular slope vertices replication, edge contraction (step 13) and simplification. Let us see that these operations do not modify the critical vertices. With edge contraction, the number of slope vertices is reduced, and hence, critical vertices are not modified. With simplification, which is an edge removal between two vertices which are not both critical, the degree of the two endpoints of the removed edge decreases, with the possibility that such vertices change their category to singular slope vertices and they can be removed in the next step in the contraction process. With the replication process, which is applied on regular slope vertices only and it does not modify critical vertices, a regular slope vertex is divided in two slope vertices. It increases the number of slope vertices but reduces their degrees (recall that the incident edges connected to the regular slope vertex before the replication process are later distributed between the two new slope vertices). Therefore a new singular slope can show up and be removed in the next contraction step. Observe that given a regular slope vertex v, if n edges incident to v are oriented inwards and m edges incident to v are oriented outwards with $n < m$, then the replicating process can be applied $n-1$ times to obtain n singular slope vertices that are contracted in the next step. Let F_k be the number of edges of G_k with at least one endpoint being non-critical. Observe that to obtain G_{k+1}, the replication process does not increase F_k whereas the contraction process reduces F_k to one unit for each singular slope vertex. Therefore, $F_{k+1} < F_k$ and the process finishes when there are no edges in the graph with at least one endpoint being non-critical, then the graph has only critical vertices and they are the original critical vertices because the applied operations did not modify them during all the process. □

EXPAND Operation: In contrast to the REDUCE operation, in the EXPAND operation, we project the information from the higher level to the lower level of the pyramid. After eliminating the inserted hidden saddles (introduced in step 3 of Algorithm 1), the RAG can be reconverted into a reconstructed image. For the purpose of image reconstruction, we aim to reconstruct a perceptually superior image with a high structural similarity as compared to the original image. For the EXPAND operation, we define edge de-contraction and insertion [7] which are the inverse operations for edge contraction and removal used in the REDUCE operation. The top level of the pyramid conceives all the critical vertices (as proved in Theorem 1) and the monotonic paths connecting critical vertices. Knowing the fact that the monotonic paths are bounded by the critical vertices, the monotonic path connecting two critical vertices can be interpolated between the intensity range of the two critical vertices for promising reconstructed images. To further reduce the error, the weights of the contracted edges can be stored externally for a perfect reconstruction of an image as explained in Burt's Laplacian pyramid [6]. Certainly there is a trade-off between accuracy and required memory.

3   Experiments and Results

In this section we analyse the performance of our TIIP algorithm for image reconstruction and compare it with the state-of-art CNN models for image super-resolution. Since super resolution algorithms are tested by first down-scaling the image through the regular pyramids and then up-scaling, it was closest to the application of image reconstruction mention in this paper. Hence the TIIP algorithm is compared with the super resolution algorithms. The methods for comparison include RCAN [29], DRLN+ [1], A+ANR [22], and EnhanceNet [20]. For evaluations of a (regular grid based) down-scaling factor of 3× was adopted and the parameters used in [1, 22] are untouched. For TIIP algorithm, the irregular pyramid was constructed until the surviving vertices were approximately 33% of the total number of pixels in the original image. In this paper we have chosen minimal contrast as the criteria for selection of the contraction kernels. The chosen reduction function preserves the contrast between the endpoints of the surviving edges. This choice preserves the critical vertices and the topology of the image. The image was reconstructed by implementing the EXPAND operation on the reduced graph and the pixel intensity (RGB) information of the surviving vertices only.

The algorithms were tested on all the 100 images of BSD100 - Berkeley segmentation data set [19] which is widely used publicly available data set for various image processing and computer vision tasks. For quality assessments, we used Structural Similarity Index Measure (SSIM) [24] and Feature Similarity Index Measure (FSIM) [27] which are perceptual metric based on the visible structures in the image. In contrast to these assessment, to measure the global degradation of image, we calculated Peak Signal to Noise Ratio (PSNR). Table 1 is showed to communicate the advantages of preserving the structural properties through irregular pyramids over regular pyramids[4].

Table 1. Quality assessment of reconstructed images on BSD100 (3×).

Quality measure\ method	A+ANR [22]	EnchanceNet [20]	RCAN [29]	DRLN+ [1]	TIIP
SSIM	0.75	0.73	0.811	0.812	0.92
FSIM	0.86	0.84	–	–	0.93
PSNR	26.64	27.50	29.3	29.4	36.26

The example *Leopard* and *Horses* in Fig. 4 clearly shows that TIIP successfully preserves the high frequency texture information, such as the thin lines

---

[4] Quality measures of RCAN and DRLN+ algorithms are not given because the code did not compile. The FSIM measures are not mentioned in [1, 29].

Fig. 4. For each image, original (from BSD 100) and reconstructed images from A+ANR [22], EnhanceNet [20] and TIIP algorithms with their zoom visuals.

due to hairs and the fencing. While, in the example of *Penguin* there appears pastelized especially in the smooth areas where the focus is lower. The results are also reflected in the Table 1 which shows that TIIP algorithm successfully preserves the structure and the features of the image.

Observations from the Experiments: 1. Irregular pyramids are more suitable to preserve and analyse the structural information of the image compared to regular pyramids. In contrary, low resolution images cannot be displayed properly since a non-grid graph is used to represent the image. 2. The TIIP algorithm performs better on images (or patch of image) with high texture, because they contain a higher number of critical points which are preserved by TIIP. In contrary, there is a considerable amount of perceptual inconsistencies in smooth regions of images which are out of focus. The inconsistent regions are usually comprised of the background and homogeneous regions of an image.

4   Conclusions and Future Work

The most important steps, controlling the hierarchy in the proposed algorithm, are the selection of the contraction kernels and the choice of the reduction function. In this paper, we have chosen minimal contrast for selecting the contraction kernels. The chosen reduction function preserves the contrast between the endpoints of the surviving edges. This choice allows us to preserve the critical vertices and the topology of the image. As a result, the algorithm generates superior results by reconstructing the fine details and the high frequency texture information of the image, which is typically lost after smoothing. Thus an image is equivalent to the combination of its structure and few colors. The hierarchy provides a (multi-resolution) structural overview of the image.

Future work includes speeding up the algorithm by implementing it on a GPU. We also plan to derive application-specific constraints and combine them with the TIIP algorithm to achieve better results for image segmentation, connected component labelling, etc. By merging machine learning with the TIIP algorithm, we anticipate learning the edge weights corresponding to the edges and derive a promising contraction and removal kernel, thereby achieving even better application-specific results.

References

1. Anwar, S., Barnes, N.: Densely residual laplacian super-resolution. IEEE Trans. Pattern Anal. Mach. Intell. (2020)
2. Batavia, D., Hladůvka, J., Kropatsch, W.G.: Partitioning 2D images into prototypes of slope region. In: Vento, M., Percannella, G. (eds.) CAIP 2019. LNCS, vol. 11678, pp. 363–374. Springer, Cham (2019). https://doi.org/10.1007/978-3-030-29888-3_29
3. Batavia, D., Kropatsch, W.G., Casablanca, R.M., Gonzalez-Diaz, R.: Congratulations! Dual graphs are now orientated!. In: Conte, D., Ramel, J.-Y., Foggia, P. (eds.) GbRPR 2019. LNCS, vol. 11510, pp. 131–140. Springer, Cham (2019). https://doi.org/10.1007/978-3-030-20081-7_13
4. Bremer, P.T., Hamann, B., Edelsbrunner, H., Pascucci, V.: A topological hierarchy for functions on triangulated surfaces. IEEE TVCG 10(4), 385–396 (2004)
5. Brun, L., Kropatsch, W.: Irregular pyramids with combinatorial maps. In: Ferri, F.J., Iñesta, J.M., Amin, A., Pudil, P. (eds.) SSPR /SPR 2000. LNCS, vol. 1876, pp. 256–265. Springer, Heidelberg (2000). https://doi.org/10.1007/3-540-44522-6_27
6. Burt, P.J., Adelson, E.H.: The Laplacian pyramid as a compact image code. IEEE TC 31(4), 532–540 (1983)
7. Cerman, M., Janusch, I., Gonzalez-Diaz, R., Kropatsch, W.G.: Topology-based image segmentation using LBP pyramids. MVA 27(8), 1161–1174 (2016)
8. Dai, Q., Chopp, H., Pouyet, E., Cossairt, O., et al.: Adaptive image sampling using deep learning and its application on x-ray fluorescence image reconstruction. IEEE TM 22(10), 2564–2578 (2019)
9. Edelsbrunner, H., Harer, J., Zomorodian, A.: Hierarchical morse-smale complexes for piecewise linear 2-manifolds. DCG 30(1), 87–107 (2003)
10. Edelsbrunner, H., Letscher, D., Zomorodian, A.: Topological persistence and simplification. In: Proceeding of International Conference on IEEE FOCS, pp. 454–463 (2000)
11. Helman, J.L., Hesselink, L.: Visualizing vector field topology in fluid flows. IEEE CGA 11(3), 36–46 (1991)
12. Hoppe, H.: Progressive meshes. In: Proceeding of CGIT, pp. 99–108. ACM (1996)

13. Iliadis, M., Spinoulas, L., Katsaggelos, A.K.: Deep fully-connected networks for video compressive sensing. Digital Signal Process. **72**, 9–18 (2018)

14. Kropatsch, W.G.: From equivalent weighting functions to equivalent contraction kernels. In: International Conference on DIP, vol. 3346, pp. 310–320 (1998)

15. Kropatsch, W.G., Casablanca, R.M., Batavia, D., Gonzalez-Diaz, R.: On the space between critical points. In: Couprie, M., Cousty, J., Kenmochi, Y., Mustafa, N. (eds.) DGCI 2019. LNCS, vol. 11414, pp. 115–126. Springer, Cham (2019). https://doi.org/10.1007/978-3-030-14085-4_10

16. Latecki, L.J.: 3D well-composed pictures. CVGIP **59**(3), 164–172 (1997)

17. Lee, R.N.: Two-dimensional critical point configuration graphs. IEEE TPAMI **4**, 442–450 (1984)

18. Maia, D.S., Cousty, J., Najman, L., Perret, B.: Recognizing hierarchical watersheds. In: Couprie, M., Cousty, J., Kenmochi, Y., Mustafa, N. (eds.) DGCI 2019. LNCS, vol. 11414, pp. 300–313. Springer, Cham (2019). https://doi.org/10.1007/978-3-030-14085-4_24

19. Martin, D., Fowlkes, C., Tal, D., Malik, J.: A database of human segmented natural images and its application to evaluating segmentation algorithms and measuring ecological statistics. In: Proceeding of ICCV, vol. 2, pp. 416–423. IEEE (2001)

20. Sajjadi, M.S., Scholkopf, B., Hirsch, M.: EnhanceNet: single image super-resolution through automated texture synthesis. In: Proceeding of IEEE ICCV, pp. 4491–4500 (2017)

21. Shinagawa, Y., Kunii, T.L., Kergosien, Y.L.: Surface coding based on morse theory. IEEE Comput. Graph. Appl. **11**(5), 66–78 (1991)

22. Timofte, R., De Smet, V., Van Gool, L.: A+: adjusted anchored neighborhood regression for fast super-resolution. In: Cremers, D., Reid, I., Saito, H., Yang, M.-H. (eds.) ACCV 2014. LNCS, vol. 9006, pp. 111–126. Springer, Cham (2015). https://doi.org/10.1007/978-3-319-16817-3_8

23. Torres, F., Kropatsch, W.G.: Canonical encoding of the combinatorial pyramid. In: Proceeding of CVWW, pp. 118–125 (2014). ISBN: 978-80-260-5641-6

24. Wang, Z., Bovik, A.C., Sheikh, H.R., Simoncelli, E.P.: Image quality assessment: from error visibility to structural similarity. IEEE TIP **13**(4), 600–612 (2004)

25. Yang, C.Y., Yang, M.H.: Fast direct super-resolution by simple functions. In: Proceeding of IEEE ICCV, pp. 561–568 (2013)

26. Zeyde, R., Elad, M., Protter, M.: On single image scale-up using sparse-representations. In: Boissonnat, J.D., et al. (eds.) Curves and Surfaces 2010. LNCS, vol. 6920, pp. 711–730. Springer, Heidelberg (2012). https://doi.org/10.1007/978-3-642-27413-8_47

27. Zhang, L., Zhang, L., Mou, X., Zhang, D.: FSIM: a feature similarity index for image quality assessment. IEEE TIP **20**(8), 2378–2386 (2011)

28. Zhang, Y., Zhao, D., Zhang, J., Xiong, R., Gao, W.: Interpolation-dependent image downsampling. IEEE TIP **20**(11), 3291–3296 (2011)

29. Zhang, Y., Li, K., Li, K., Wang, L., Zhong, B., Fu, Y.: Image super-resolution using very deep residual channel attention networks. In: Ferrari, V., Hebert, M., Sminchisescu, C., Weiss, Y. (eds.) ECCV 2018. LNCS, vol. 11211, pp. 294–310. Springer, Cham (2018). https://doi.org/10.1007/978-3-030-01234-2_18

Author Index

Printed in the United States
by Baker & Taylor Publisher Services